Lecture Notes in Computer Science 9223

Commenced Publication in 1973
Founding and Former Series Editors:
Gerhard Goos, Juris Hartmanis, and Jan van Leeuwen

More information about this series at http://www.springer.com/series/7407

Frank Drewes (Ed.)

Implementation and Application of Automata

20th International Conference, CIAA 2015
Umeå, Sweden, August 18–21, 2015
Proceedings

 Springer

Editor
Frank Drewes
Umeå University
Umeå
Sweden

ISSN 0302-9743   ISSN 1611-3349   (electronic)
Lecture Notes in Computer Science
ISBN 978-3-319-22359-9   ISBN 978-3-319-22360-5   (eBook)
DOI 10.1007/978-3-319-22360-5

Library of Congress Control Number: 2015945122

LNCS Sublibrary: SL1 – Theoretical Computer Science and General Issues

Printed on acid-free paper

Springer International Publishing AG Switzerland is part of Springer Science+Business Media
(www.springer.com)

Preface

This volume contains the papers presented at the 20th International Conference on Implementation and Application of Automata (CIAA 2015), which was organized by the Department of Computing Science at Umeå University, Sweden, and took place at Umeå Folkets hus during August 18–21, 2015.

The CIAA conference series is the major international venue for the dissemination of new results in the implementation, application, and theory of automata. The previous 19 conferences were held in various locations all around the globe: Blois (2011), Giessen (2014), Halifax (2013), Kingston (2004), London Ontario (WIA 1997, WIA 1996, and 2000), Nice (2005), Porto (2012), Potsdam (WIA 1999), Prague (2007), Pretoria (2001), Rouen (WIA 1998), San Francisco (2008), Santa Barbara (2003), Sydney (2009), Taipei (2006), Tours (2002), and Winnipeg (2010).

The topics of this volume include cover automata, counter automata, decision algorithms on automata, descriptional complexity, expressive power of automata, homing sequences, jumping finite automata, multidimensional languages, parsing and pattern matching, quantum automata, realtime pushdown automata, random generation of automata, regular expressions, security issues, sensors in automata, transducers, transformation of automata, and weighted automata.

In total, 49 papers were submitted by authors in 20 different countries: Brazil, Canada, Czech Republic, Finland, France, Germany, Hungary, India, Israel, Italy, Japan, South Korea, Norway, Poland, Portugal, Russia, South Africa, Sweden, the UK, and the USA. Each of these papers was reviewed by at least three reviewers and thoroughly discussed by the Program Committee, which resulted in the selection of 22 papers for presentation at the conference and publication in this volume. Four invited talks were given by Benedikt Bollig, Christof Löding, Andreas Maletti, and Bruce Watson. In addition to these contributions, the volume contains two short papers about tool demonstrations that were given at the conference.

I am very thankful to all invited speakers, authors of submitted papers, system demonstrators, Program Committee members, and external reviewers for their valuable contributions and help. Without them, CIAA 2015 could not have been realized. The entire process from the original submissions to collecting the final versions of papers was greatly simplified by the use of the EasyChair conference management system.

I would furthermore like to thank the editorial staff at Springer, and in particular Alfred Hofmann and Anna Kramer, for their guidance and help during the process of publishing this volume, and Camilla Andersson at the conference site Umeå Folkets hus for her help with all the practical preparations.

CIAA 2015 was financially supported by (a) the Department of Computing Science at Umeå University, (b) the conference fund of Umeå Municipality, the County Council of Västerbotten and Umeå University, (c) the Faculty of Science and Technology at Umeå University, and (d) the Swedish Research Council, who provided generous funding for invited speakers.

Last but by no means least, I wish to thank the local Organizing Committee consisting of the members of the research group Foundations of Language Processing, namely, Suna Bensch, Henrik and Johanna Björklund, Loek Cleophas, Petter Ericson, Yonas Woldemariam, and Niklas Zechner for their help.

We are now looking forward to CIAA 2016 at Yonsei University, Seoul, in South Korea.

August 2015   Frank Drewes

Organization

CIAA 2015 was organized by the Department of Computing Science at Umeå University, Sweden, and took place at Umeå Folkets hus.

Invited Speakers

Benedikt Bollig   Université Paris-Saclay, France
Christof Löding   RWTH Aachen, Germany
Andreas Maletti   University of Stuttgart, Germany
Bruce Watson   University of Stellenbosch, South Africa

Program Committee

Parosh Aziz Abdulla   Uppsala University, Sweden
Marie-Pierre Beal   Université Paris-Est Marne-la-Vallée, France
Cezar Câmpeanu   University of Prince Edward Island, Canada
Pascal Caron   Université de Rouen, France
Jean-Marc Champarnaud   Université de Rouen, France
David Chiang   University of Notre Dame, USA
Stefano Crespi-Reghizzi   Politecnico di Milano, Italy
Jürgen Dassow   Otto von Guericke University Magdeburg, Germany
Frank Drewes   Umeå University, Sweden
Rudolf Freund   TU Wien, Austria
Yo-Sub Han   Yonsei University, South Korea
Markus Holzer   Justus Liebig University Giessen, Germany
Oscar Ibarra   University of California, Santa Barbara, USA
Helmut Jürgensen   The University of Western Ontario, Canada
Martin Kutrib   Justus Liebig University Giessen, Germany
Andreas Maletti   University of Stuttgart, Germany
Sebastian Maneth   University of Edinburgh, UK
Wim Martens   Bayreuth University, Germany
Denis Maurel   Université François Rabelais Tours, France
Carlo Mereghetti   Università degli Studi di Milano, Italy
Brink van der Merwe   University of Stellenbosch, South Africa
Cyril Nicaud   Université Paris-Est, France
Alexander Okhotin   University of Turku, Finnland
Daniel Reidenbach   Loughborough University, UK
Rogerio Reis   Universidade do Porto, Portugal
Kai T. Salomaa   Queen's University, Canada
Klaus Sutner   Carnegie Mellon University, USA
Sophie Tison   Université de Lille 1, France
György Vaszil   University of Debrecen, Hungary

Heiko Vogler   TU Dresden, Germany
Hsu-Chun Yen   National Taiwan University, Taiwan

Steering Committee

Jean-Marc Champarnaud   Université de Rouen, France
Markus Holzer   Justus Liebig University Giessen, Germany
Oscar Ibarra   University of California, Santa Barbara, USA
Denis Maurel   Université François Rabelais Tours, France
Kai T. Salomaa   Queen's University, Canada
Hsu-Chun Yen   National Taiwan University, Taiwan

External Reviewers

Cyriac Aiswarya
Mohamed Faouzi Atig
Johanna Björklund
Henrik Björklund
Benedikt Bollig
Sabine Broda
Cezar Campeanu
Arnaud Carayol
Yi-Jun Chang
Christian Choffrut
Julien Clément
Maxime Crochemore
Wojciech Czerwiński
Julien David
Toni Dietze
Francesco Dolce
Mike Domaratzki
Marianne Flouret

Travis Gagie
Lukas Holik
Sebastian Jakobi
Artur Jeż
Sang-Ki Ko
Marco Kuhlmann
Ondrej Lengal
Lvzhou Li
Sylvain Lombardy
Eva Maia
Andreas Malcher
Luca Manzoni
Tomas Masopust
Ian McQuillan
Katja Meckel
Ludovic Mignot
Nelma Moreira
Timothy Ng

Kim Nguyen
Florent Nicart
Damien Nouvel
Faissal Ouardi
Balasubramanian
  Ravikumar
Klaus Reinhardt
Martin Schuster
Shinnosuke Seki
Jari Stenman
Till Tantau
Bianca Truthe
Vojtěch Vorel
Johannes Waldmann
Matthias Wendlandt
Lynette Van Zijl

Organizing Committee

Suna Bensch
Henrik Björklund
Johanna Björklund

Frank Drewes
Loek Cleophas
Petter Ericson

Yonas Woldemariam
Niklas Zechner

Sponsoring Institutions

Department of Computing Science, Umeå University
Faculty of Science and Technology, Umeå University
Umeå Municipality, the County Council of Västerbotten and Umeå University
The Swedish Research Council

Invited Papers

Automata and Logics for Concurrent Systems: Five Models in Five Pages

Benedikt Bollig

LSV, ENS Cachan, CNRS & Inria
bollig@lsv.ens-cachan.fr

Abstract. We survey various automata models of concurrent systems and their connection with monadic second-order logic: finite automata, class memory automata, nested-word automata, asynchronous automata, and message-passing automata.

Resource Automatic Structures for Verification of Boundedness Properties
Extended Abstract

Christof Löding

RWTH Aachen, Germany
loeding@cs.rwth-aachen.de

Automatic structures are (possibly infinite) structures that can be represented by means of finite automata [1, 10]. The elements of the domain of the structure are encoded as words and form a regular language. The relations of the structure are recognized by synchronous automata with several input tapes (the number of the tapes corresponding to the arity of the relation). A typical example of such a structure is $(\mathbb{N}, +, <)$, the natural numbers with addition and order. The natural numbers are encoded by words corresponding to their binary representation (or any other base). The order and the addition (as ternary relation) can then be accepted by synchronous automata with the corresponding number of input tapes.

Another class of examples are configuration graphs of pushdown automata with reachability relation. The vertices of a pushdown graph are naturally encoded as words (a control state followed by a stack content). The set of reachable configurations (from the initial configuration) forms a regular language [4], and more generally, the reachability relation is automatic, that is, there is a finite two-tape automaton that accepts those pairs of configurations such that the second one is reachable from the first one [7].[1]

Automatic structures are interesting in verification because their first-order theory (FO) is decidable: the atomic formulas are already given by automata, and the closure properties of finite automata can be used for an inductive translation of composed formulas.

In [13] we have introduced the notion of resource automatic structures. In this model of resource structures, a relation is not a set of tuples but a function that assigns to each tuple a natural number or ∞, where the value ∞ corresponds to the classical case of not being in the relation. A value n for a tuple can be seen as a cost for being in the relation.

As an illustration, we extend the above example of pushdown graphs with reachability relation: Assume that the transitions of the pushdown system are annotated with operations on resources. For each type of resource, a transition can either consume one unit of this resource, completely replenish the resource at once, or not use the resource at all. Then we can associate the cost of a finite path through the pushdown graph to be the maximal number of units consumed from a resource without being replenished in

---

[1] The result in [7] is for the more general case of ground term rewriting systems, which include pushdown automata as special case.

between. This corresponds to the size of the reservoir required for the resource to execute the path. We naturally obtain a resource relation that assigns to each pair of configurations the cost of a cheapest path between these two configurations (and ∞ if there is no path between the configurations).

In [13] it is shown that this resource reachability relation for pushdown automata can be defined by an automaton model called B-automata. The transitions of these automata are annotated by actions on counters that either increment the counter, reset the counter to value 0, or leave the counter unchanged. In this way, a cost is assigned to each input word that is accepted by the automaton as follows. The cost of a run is the maximal value that one of the counters assumes during the run. The cost of the input word is the minimal cost of an accepting run for this input word (and ∞ if the word is not accepted).

The class of resource automatic structures [13] is defined to be the class of resource structures that can be encoded by B-automata, thus pushdown graphs with the resource reachability relation are resource automatic structures.

As logic over these structures, we consider FO+RR, first-order logic with resource relations, which is standard FO logic without negation. Similar to the resource relations, a formula of FO+RR has a value (instead of being true or false). Intuitively, this value corresponds to the cost for making this formula true: the value of the atomic formulas is given by the resource relations, disjunction and conjunction are translated to min and max, and existential and universal quantifiers are translated to inf and sup.

The intention of this logic is to be able to formalize and solve boundedness properties for resource structures. Taking again the example of pushdown graphs with resource reachability, a typical question would be the bounded reachability problem: given two regular sets A, B of configurations, does there exist a bound K such that from each configuration in A there is a path to a configuration in B with cost at most K. The corresponding formula is

$$\forall x \in A \exists y \in B \ : \ x \to^* y$$

where $\to^*$ is the resource reachability relation. According to the semantics of FO+RR, the value of the formula is not ∞ if, and only if, the above bounded reachability property holds.

Similar to the translation of classical FO formulas over automatic structures into finite automata, FO+RR formulas can be translated into B-automata preserving boundedness [13] (using the closure of B-automata under the operations max, min, inf, sup). Thus, for deciding whether the value of a formula is finite, it suffices to check the boundedness property on B-automata.

Boundedness properties for finite automata have been studied in the context of the star-height problem for regular expressions [8, 11]. Given a regular language and a number h, the question of whether there exists a regular expression of star-height at most h for this language, can be reduced to a boundedness question of B-automata. The boundedness (or limitedness) problem for B-automata is the question whether there is a bound on the cost of the accepted words. It is shown to be decidable in [11], where the automata are called distance-desert automata. The name of B-automata originates from a model introduced in [3] for describing boundedness properties of infinite words. Based on the decidability results for B-automata, one obtains the decidability of the boundedness problem for FO+RR formulas over resource automatic structures.

In [14] the class of resource automatic structures is studied in more detail. It is shown that there is a complete resource automatic structure (each other resource automatic structure can be obtained from this complete structure by interpretations in FO+RR logic). Furthermore, connections between FO+RR over resource automatic structures and cost monadic second-order logic (cost MSO) [5] and cost FO [12] over words are established that generalize the standard setting over words without costs.

The model of B-automata and the corresponding decidability results can be extended to finite trees [6], which leads to the class of resource tree automatic structures. In recent work [9], it is shown that an extension of FO+RR with an operator for testing boundedness of formulas, can be used to capture weak cost MSO [5] and weak MSO+U [2], obtaining alternative proofs for the decidability of these logics.

References

1. Blumensath, A., Grädel, E.: Automatic structures. In: Proceedings of the 15th IEEE Symposium on Logic in Computer Science, LICS 2000, pp. 51–62. IEEE Computer Society Press (2000)
2. Bojańczyk, M.: Weak MSO with the unbounding quantifier. Theory Comput. Syst. **48**(3), 554–576 (2011). http://dx.doi.org/10.1007/s00224-010-9279-2
3. Bojańczyk, M., Colcombet, T.: Bounds in w-regularity. In: 21th IEEE Symposium on Logic in Computer Science (LICS 2006), Seattle, WA, USA, Proceedings, 12–15 August 2006, pp. 285–296. IEEE Computer Society (2006)
4. Büchi, J.R.: Regular canonical system. Archiv für Mathematische Grundlagenforschung **6**, 91–111 (1964)
5. Colcombet, T.: Regular cost functions, part I: logic and algebra over words. Log. Methods Comput. Sci. **9**(3) (2013). http://dx.doi.org/10.2168/LMCS-9(3:3)2013
6. Colcombet, T., Löding, C.: Regular cost functions over finite trees. In: Twenty-Fifth Annual IEEE Symposium on Logic in Computer Science, LICS 2010, pp. 70–79. IEEE Computer Society (2010)
7. Dauchet, M., Tison, S.: The theory of ground rewrite systems is decidable. In: Proceedings of the Fifth Annual IEEE Symposium on Logic in Computer Science, LICS 1990, pp. 242–248. IEEE Computer Society Press (1990)
8. Hashiguchi, K.: Algorithms for determining relative star height and star height. Inf. Comput. **78**(2), 124–169 (1988)
9. Kaiser, Ł., Lang, M., Löding, C., Leßenich, S.: A unified approach to boundedness properties in MSO (2015, submitted)
10. Khoussainov, B., Nerode, A.: Automatic presentations of structures. In: Leivant, D. (ed.) LCC 1994. LNCS, vol. 960, pp. 367–392. Springer, Heidelberg (1995)
11. Kirsten, D.: Distance desert automata and the star height problem. RAIRO – Theor. Inf. Appl. **3**(39), 455–509 (2005)
12. Kuperberg, D.: Study of classes of regular cost functions. Ph.D. thesis, LIAFA Paris, December 2012
13. Lang, M., Löding, C.: Modeling and verification of infinite systems with resources. Log. Methods Comput. Sci. **9**(4) (2013)
14. Lang, M., Löding, C., Manuel, A.: Definability and transformations for cost logics and automatic structures. In: Csuhaj-Varjú, E., Dietzfelbinger, M., Ésik, Z. (eds.) MFCS 2014, Part I. LNCS, vol. 8634, pp. 390–401. Springer, Heidelberg (2014). http://dx.doi.org/10.1007/978-3-662-44522-8

Finite-State Technology in Natural Language Processing
Extended Abstract

Andreas Maletti

Institute for Natural Language Processing, Universität Stuttgart
Pfaffenwaldring 5b, 70569 Stuttgart, Germany
maletti@ims.uni-stuttgart.de

Finite-state technology is at the core of many standard approaches in natural language processing [11, 15]. However, the terminology and the notations differ significantly between theoretical computer science (TCS) [8] and natural language processing (NLP) [13]. In this lecture, inspired by [11, 13], we plan to illustrate the close ties between formal language theory as discussed in TCS and its use in mainstream applications of NLP. In addition, we will try to match the different terminologies in three example tasks. Overall, this lecture shall serve as an introduction to (i) these tasks and (ii) the use of finite-state technology in NLP and shall encourage closer collaboration between TCS and NLP.

We will start with the task of part-of-speech tagging [11, Chapter 5], in which given a natural language sentence the task is to derive the word category (the part-of-speech, e.g. noun, verb, adjective, etc.) for each occurring word in the sentence. The part-of-speech information is essential for several downstream applications like co-reference resolution [11, Chapter 21] (i.e., detecting which entities in a text refer to the same entities), automatic keyword detection [11, Chapter 22] (i.e., finding relevant terms for a document), and sentiment analysis [18] (i.e., the process of determining whether a text speaks favorably or negatively about a subject). Along the historical development of systems for this task [9] we will discuss the main performance breakthrough (in the mid 80s) that led to the systems that are currently state-of-the-art for this task. This breakthrough was achieved with the help of statistical finite-state systems commonly called *hidden Markov models* [11, Chapter 6], which roughly equate to probabilistic finite-state transducers [17]. We will outline the connection and also demonstrate how various well-known algorithms like the forward and backward algorithms relate to TCS concepts.

Second, we will discuss the task of parsing [11, Chapter 13], in which a sentence is given and its syntactic structure is to be determined. The syntactic structure is beneficial in several applications including syntax-based machine translation [14] or natural language understanding [11, Chapter 18]. In parsing, a major performance breakthrough was obtained in 2005 by adding finite-state information to probabilistic context-free grammars [16]. The currently state-of-the-art models (for English) are

Supported by the German Research Foundation (DFG) grant MA/ 4959 / 1-1.

probabilistic context-free grammars with latent variables, which are known as prob-abilistic finite-state tree automata [10] in TCS. We will review the standard process [7] (expectation maximization), which determines the hidden finite-state information in the hope that similar processes might be helpful also in the TCS community. In addition, we will recall a spectral learning approach [6], which builds on the minimization of nondeterministic field-weighted tree automata [3]. Similarly, advanced evaluation mechanisms like coarse-to-fine parsing [19] that have been developed in NLP should be considered in TCS.

Finally, we will cover an end-user application in NLP. The goal of machine translation [14] is the provision of high-quality and automatic translations of input sentences from one language into another language. The main formalisms used in NLP in this area are *probabilistic synchronous grammars* [5], which originate from the seminal syntax-based translation schemes of [1]. These grammars correspond to certain subclasses of probabilistic finite-state transducers [17] or probabilistic tree transducers [10]. So far, only local versions (grammars without latent variables) are used in state-of-the-art systems, so the effective inclusion of finite-state information remains an open problem in this task. However, the requirements of syntax-based machine translation already spurred a lot of research in TCS because the models traditionally studied had significant shortcomings [12]. In the other direction, advanced models like multi bottom-up tree transducers [2] have made reasonable impact in syntax-based machine translation [4].

References

1. Aho, A.V., Ullman, J.D.: The Theory of Parsing, Translation, and Compiling. Prentice Hall (1972)
2. Arnold, A., Dauchet, M.: Morphismes et bimorphismes d'arbres. Theor. Comput. Sci. **20**, 33–93 (1982)
3. Bozapalidis, S.: Effective construction of the syntactic algebra of a recognizable series on trees. Acta Informatica **28**(4), 351–363 (1991)
4. Braune, F., Seemann, N., Quernheim, D., Maletti, A.: Shallow local multi-bottom-up tree transducers in statistical machine translation. In: Proceedings of 51st ACL, pp. 811–821. Association for Computer Linguistics (2013)
5. Chiang, D.: An introduction to synchronous grammars. In: Proceedings of 44th ACL. Association for Computational Linguistics (2006), part of a tutorial given with Kevin Knight
6. Cohen, S.B., Stratos, K., Collins, M., Foster, D.P., Ungar, L.: Spectral learning of latent-variable PCFGs. In: Proceedings of 50th ACL, pp. 223–231. Association for Computational Linguistics (2012)
7. Dempster, A.P., Laird, N.M., Rubin, D.B.: Maximum likelihood from incomplete data via the EM algorithm. J. Roy. Stat. Soc. Series B (Methodol.) **39**(1), 1–38 (1977)
8. Droste, M., Kuich, W., Vogler, H. (eds.): Handbook of Weighted Automata. Springer (2009)
9. Francis, W.N., Kučera, H., Mackie, A.W.: Frequency Analysis of English Usage: Lexicon and Grammar. Houghton Mifflin (1982)
10. Fülöp, Z., Vogler, H.: Weighted tree automata and tree transducers. In: Droste et al. [8], chap. 9, pp. 313–403
11. Jurafsky, D., Martin, J.H.: Speech and Language Processing, 2nd edn. Prentice Hall (2008)

12. Knight, K.: Capturing practical natural language transformations. Mach. Transl. **21**(2), 121–133 (2007)

13. Knight, K., May, J.: Applications of weighted automata in natural language processing. In: Droste et al. [8], chap. 14, pp. 571–596

14. Koehn, P.: Statistical Machine Translation. Cambridge University Press (2010)

15. Manning, C., Schütze, H.: Foundations of Statistical Natural Language Processing. MIT Press (1999)

16. Matsuzaki, T., Miyao, Y., Tsujii, J.: Probabilistic CFG with latent annotations. In: Proceedings of 43rd ACL, pp. 75–82. Association for Computational Linguistics (2005)

17. Mohri, M.: Weighted automata algorithms. In: Droste et al. [8], chap. 6, pp. 213–254

18. Pang, B., Lee, L.: Opinion mining and sentiment analysis. Found. Trends Inf. Retr. **2**(1–2), 1–135 (2008)

19. Petrov, S.: Coarse-to-Fine Natural Language Processing. Ph.D. thesis, University of California at Bekeley, Berkeley, CA, USA (2009)

Hardware Implementations
of Finite Automata and Regular Expressions
Extended Abstract

Bruce W. Watson

FASTAR Group, Department of Information Science
Stellenbosch University, South Africa
bruce@fastar.org

Abstract. This extended abstract sketches some of the most recent advances in hardware implementations (and surrounding issues) of finite automata and regular expressions. The traditional application areas for automata and regular expressions are compilers, text editors, text programming languages (for example Sed, AWK, but more recently Python, and Perl), and text processing in general purpose languages (such as Java, C++ and C#). In all these cases, while the regular expression implementation should be efficient, it rarely forms the performance bottleneck in resulting programs and applications. Even more exotic application areas such as computational biology are not particularly taxing on the regular expression implementation — provided some care is taken while crafting the regular expressions [5].

Contents

Tool Demonstration Papers

Invited Papers

Automata and Logics for Concurrent Systems: Five Models in Five Pages

Benedikt Bollig$^{(\boxtimes)}$

LSV, ENS Cachan, CNRS & Inria, Cachan, France
bollig@lsv.ens-cachan.fr

Abstract. We survey various automata models of concurrent systems and their connection with monadic second-order logic: finite automata, class memory automata, nested-word automata, asynchronous automata, and message-passing automata.

1 Introduction

A variety of automata models have emerged over the years to provide a basis for the study of various types of concurrent systems. These models capture several communication paradigms such as shared memory or message passing, and they can deal with finite-state or infinite-state processes. In this paper, we consider several of these models in a unifying framework.

The study that we conduct here is driven by questions that arise in the area of verification. Concurrent systems are often safety-critical and come with a requirements specification to be fulfilled. In the automata-theoretic approach, which we adopt here, a system is modeled as an automaton $\mathcal{A}$, and the specification is given as a formula φ in a high-level language such as temporal logic or monadic second-order logic. In order for both, the system model and the specification, to be comparable, they should have a common domain. We associate with an automaton its language $L(\mathcal{A})$, representing the set of possible behaviors. Similarly, the specification determines a set $L(\varphi)$ of models, namely the set of behaviors that satisfy it. Then, correctness of the system can be expressed as the inclusion problem $L(\mathcal{A}) \subseteq L(\varphi)$, which is commonly referred to as the *model-checking problem*. In the different approach of *realizability*, only the specification φ is given, and we aim at an automaton $\mathcal{A}$ such that $L(\mathcal{A}) = L(\varphi)$. In that case, $\mathcal{A}$ may serve as a system model that can be considered correct by construction. The models that we cover here owe much of their success to the fact that model checking and realizability have positive solutions. In particular, they all enjoy logical characterizations in terms of (an expressive fragment of) monadic second-order (MSO) logic and come, possibly under restrictions, with a decidable emptiness problem.

But what is actually a behavior? A behavior is the collection of events that we observe during an execution. There are essentially two approaches: In the

---

Not including introduction and references.

© Springer International Publishing Switzerland 2015
F. Drewes (Ed.): CIAA 2015, LNCS 9223, pp. 3–12, 2015.
DOI: 10.1007/978-3-319-22360-5_1

interleaving approach, one imposes an order on a priori independent events. In the graph-based approach, a behavior reveals causal dependencies between them and explicitly records, in terms of edges/binary relations, any access to a data structure such as the current state, a channel, or a stack. We adopt here the graph-based approach. As argued convincingly in [1], it is more natural and expressive when one wants to reason about system properties beyond reachability.

Signature. One crucial parameter of a system model and its behaviors is the underlying *signature*, which consists of a nonempty finite set Σ of *actions* and a nonempty finite set R of binary relation symbols. Intuitively, a letter $a \in \Sigma$ describes an action executed by a system such as "send request to the server", or "call procedure P". Moreover, binary relations $\lhd_r$, with r ranging over R, reflect the functionality of a system. To model message passing, for example, a behavior comes with a binary relation that links a send event with the corresponding receive event. In a system involving recursive processes, there will be a binary relation connecting a procedure call with the corresponding return. Once a signature is fixed, we obtain both a canonical automata model and a canonical monadic second-order logic (see below).

Behavior. To give a meaning to the signature σ, we first have to define what a behavior is, representing one possible execution of a system. A σ-*behavior* is a tuple $B = (E, \lambda, (\lhd_r)_{r \in R})$ where E is a nonempty finite set of *events*, $\lambda : E \to \Sigma$ is a labeling function, and $\lhd_r \subseteq E \times E$ is a binary relation, for every $r \in R$. We interpret each $\lhd_r$ as edge relation and call its elements r-*edges*. As a behavior describes the progress of an execution, it is natural to require that $\bigcup_{r \in R} \lhd_r$ is acyclic (and, in particular, irreflexive). Moreover, we assume that access to a data structure is well-defined: for all $r \in R$, each event has at most one outgoing r-edge and at most one incoming r-edge. Example behaviors can be found in Figs. 1, 2, 3 and 4. Later, depending on the automata model that we consider, we will restrict σ-behaviors further (e.g., to model FIFO queues or pushdown stacks).

We conclude this paragraph with a definition that will be useful later on. For $Q \subseteq R$, we say that behavior B is *disconnected* wrt. Q if, informally speaking, each event belongs to at most one Q-edge: for all $r, r' \in Q$, $(e, f) \in \lhd_r$, and $(e', f') \in \lhd_{r'}$, either $r = r' \wedge e = e' \wedge f = f'$ or $\{e, f\} \cap \{e', f'\} = \emptyset$.

Automata. A σ-automaton $\mathcal{A}$ has a finite set of *states* S. A *run* of $\mathcal{A}$ on a σ-behavior $B = (E, \lambda, (\lhd_r)_{r \in R})$ is just a mapping $\rho : E \to S$. Intuitively, for $e \in E$, $\rho(e)$ is the state taken *after* executing e. Of course, this assignment has to conform with a finite set of *transitions*. Actually, when executing e, the automaton $\mathcal{A}$ has access to some states taken previously in the run, and this access is determined by the relations associated with R. In fact, a transition is a triple $(pred, a, s)$ where $a \in \Sigma$, $s \in S$, and $pred : R \rightharpoonup S$ is a *partial* mapping that allows the automaton to access the state taken at a $\lhd_r$-predecessor of e. Formally, we require that, for every $e \in E$, there is a transition $(pred, a, s)$ such that $\lambda(e) = a$, $\rho(e) = s$, and, for all $r \in R$, the following hold:

- if e does not have a $\lhd_r$-predecessor, then $pred(r)$ is undefined, and
- if e has a $\lhd_r$-predecessor f (which is then unique), then $\rho(f) = pred(r)$.

Note that initial states can be implicitly modeled by functions $pred$ that are partially or entirely undefined. To take into account several automata models in a unifying framework, we assume a quite general *acceptance condition*, which is a set of tuples $(O, (T_r)_{r \in R})$ where $O \subseteq \Sigma$ and $T_r \subseteq S$ for all $r \in R$. Run ρ is *accepting* if the acceptance condition contains some tuple $(O, (T_r)_{r \in R})$ such that $O = \{\lambda(e) \mid e \in E\}$ and, for all $r \in R$ and events e without $\lhd_r$-successor, we have $\rho(e) \in T_r$. In some of the settings discussed below, the *occurrence set* O can be used to guarantee that a process makes at least one move. Note that, in the simple framework of finite automata, an acceptance condition will be no more expressive than just assuming a single set of final states and requiring that a run ends in a final state. To summarize, a σ-automaton consists of a finite set of states, a finite set of transitions, and an acceptance condition. By $L(\mathcal{A})$, we denote the set of σ-behaviors that allow for an accepting run of $\mathcal{A}$.

Logic. Given a signature σ, we assume the canonical MSO logic for σ, which we denote by $\mathrm{MSO}(\sigma)$. There are infinite supplies of first-order variables $x, y, \ldots$ and second-order variables $X, Y, \ldots$ They allow us to quantify over events and sets of events, respectively, using formulas $\exists x \varphi$ and $\exists X \varphi$ (where φ is again an $\mathrm{MSO}(\sigma)$ formula). Furthermore, we can use the boolean operators negation and disjunction (and, therefore, conjunction and universal quantification). The formula $\lambda(x) = a$ expresses that event x executes $a \in \Sigma$. Finally, we have access to the binary relations in terms of formulas $x \lhd_r y$, with $r \in R$, and we include $x = y$ with the obvious meaning. The set of σ-behaviors that satisfy a given $\mathrm{MSO}(\sigma)$ sentence φ (without free variables) is defined as usual and denoted by $L(\varphi)$. We refer the reader to [21] for an introduction to MSO logic.

Later, we may have to consider fragments of MSO logic to match the expressive power of an automata model. The existential fragment of $\mathrm{MSO}(\sigma)$, denoted by $\mathrm{EMSO}(\sigma)$, contains the formulas of the form $\exists X_1 \ldots \exists X_n \varphi$ where φ does not use any second-order quantifier. When, in addition, we use only two first-order variables (which, however, can occur several times in a formula), we deal with the fragment $\mathrm{EMSO}_2(\sigma)$ of $\mathrm{EMSO}(\sigma)$. In those fragments, it is sometimes impossible, or not obvious, to encode the reflexive transitive closure of a binary predicate. We may add such predicates explicitly and write, for example, $\mathrm{EMSO}_2(\sigma + \lhd_{r_1}^* + \lhd_{r_2}^*)$ when we allow access to $\lhd_{r_1}^*$ and $\lhd_{r_2}^*$.

Realizability and Emptiness Checking. Each of the following sections will describe a particular system model. For each setting, we proceed as follows. We will first fix a signature σ and define a class $\mathcal{B}$ of σ-behaviors. Recall that both an automata model and MSO logic are already determined by σ.

For a set $\mathcal{F} \subseteq \mathrm{MSO}(\sigma)$, we will then state two kinds of results: When we write σ-*automata and $\mathcal{F}$ are expressively equivalent over* $\mathcal{B}$, we mean that, for all $L \subseteq \mathcal{B}$, the following statements are equivalent:

- There is a σ-automaton $\mathcal{A}$ such that $L(\mathcal{A}) \cap \mathcal{B} = L$.
- There is sentence $\varphi \in \mathcal{F}$ such that $L(\varphi) \cap \mathcal{B} = L$.

In all the cases that we consider here, the transformations from formulas to automata, and back, are effective. When we write *emptiness of* σ-*automata over* $\mathcal{B}$ *is* decidable, we mean that one can decide whether, for some given σ-automaton $\mathcal{A}$, we have $L(\mathcal{A}) \cap \mathcal{B} = \emptyset$. We do not go into complexity considerations here, as this would require a more detailed treatment of $\mathcal{B}$.

Note that an effective translation of logic formulas into automata that have a decidable emptiness problem usually allows us to solve, positively, the model-checking problem (provided the logic is closed under negation and automata are closed under intersection).

2   Finite Automata

To illustrate the general framework, we first recall the basic setting of finite automata running on finite words. Apart from the finite alphabet Σ, the corresponding signature σ_w will just contain one single relation symbol succ to represent the direct successor relation in a word, i.e., $R = \{\text{succ}\}$. Thus, a *word (structure)* is a σ_w-behavior $(\{1, \ldots, n\}, \lambda, \lhd_{\text{succ}})$, with $n \geq 1$, such that $\lhd_{\text{succ}} = \{(e, e+1) \mid e \in \{1, \ldots, n-1\}\}$. With this definition, a σ_w-automaton running on words is just a *finite automaton*: in terms of $\lhd_{\text{succ}}$, it can only access the *current* state, i.e., the state assigned to the previous position. In our framework, the famous classical connection between finite automata and MSO logic reads as follows:

Theorem 1 (Büchi-Elgot-Trakhtenbrot [9,11,22]). *Finite automata (i.e., σ_w-automata), $\text{EMSO}(\sigma_w)$, and $\text{MSO}(\sigma_w)$ are expressively equivalent over words.*

3   Class Memory Automata

In this section, we consider systems that consist of an unbounded number of processes. An execution of such a system is naturally described as a *data word*. Usually, a data word is defined as a word over an infinite alphabet, where the latter is used to represent an unbounded number of process identifiers. In our framework, however, it will be more convenient to equip a word over a *finite* alphabet with an equivalence relation, where an equivalence class captures those positions that are executed by one and the same process. The signature σ_{dw} contains, apart from an arbitrary finite alphabet Σ, the relation symbols succ (with the same meaning as in words) and class (connecting *consecutive* positions in an equivalence class). Thus, $R = \{\text{succ}, \text{class}\}$. The idea is that a σ_{dw}-automaton can access a sort of global state (in terms of $\lhd_{\text{succ}}$) and the current local state of the executing process (in terms of $\lhd_{\text{class}}$). Note that the acceptance condition of σ_{dw} actually allows us to fix a set of global final states and a set of local final states.

Accordingly, a *data word* is a σ_{dw}-behavior $B = (\{1, \ldots, n\}, \lambda, \lhd_{\text{succ}}, \lhd_{\text{class}})$, with $n \geq 1$, such that $\lhd_{\text{succ}} = \{(e, e+1) \mid e \in \{1, \ldots, n-1\}\}$. Note that $\lhd_{\text{class}}^* \cup (\lhd_{\text{class}}^{-1})^*$ is the equivalence relation on $\{1, \ldots, n\}$ induced by $\lhd_{\text{class}}$. A data

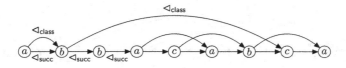

Fig. 1. A data word

word over $\Sigma = \{a, b, c\}$ is depicted in Fig. 1. The straight arrows denote succ-edges, and the curved arrows denote class-edges. Running on data words, σ_{dw}-automata actually correspond to *class memory automata* [4].

Theorem 2 (Bojanczyk et al. [5]). *Class memory automata (i.e., σ_{dw}-automata) and* $\mathrm{EMSO}_2(\sigma_{\mathrm{dw}} + \lhd^*_{\mathsf{succ}} + \lhd^*_{\mathsf{class}})$ *are expressively equivalent over data words. Moreover, emptiness of σ_{dw}-automata is decidable over data words.*

Theorem 2 was actually shown for the model of *data automata*, which are expressively equivalent to class memory automata [4]. Note that, over data words, the logic $\mathrm{EMSO}_2(\sigma_{\mathrm{dw}} + \lhd^*_{\mathsf{succ}} + \lhd^*_{\mathsf{class}})$ (and, therefore, σ_{dw}-automata) is not closed under negation/complementation [5]. However, for model checking, we can still use its first-order fragment.

4   Nested-Word Automata

We will now consider a setting with a fixed finite set of *recursive* processes, which are usually modeled as pushdown automata or, equivalently, *nested-word automata*. Nested-word automata have access to binary nesting (or call-return) relations, which link a function call with the corresponding return position. Since we have several processes, this gives rise to the notion of *multiply nested words*, which come with one nesting relation per process.

Formally, we assume a finite set $Proc = \{1, \ldots, m\}$ of *processes* with $m \geq 1$. The signature σ_{nw} consists of any finite alphabet Σ as well as the relation symbols $R = \{\mathsf{succ}, \mathsf{cr}_1, \ldots, \mathsf{cr}_m\}$. Then, a *(multiply) nested word* is a σ_{nw}-behavior $B = (\{1, \ldots, n\}, \lambda, \lhd_{\mathsf{succ}}, \lhd_{\mathsf{cr}_1}, \ldots, \lhd_{\mathsf{cr}_m})$, with $n \geq 1$, such that

- $\lhd_{\mathsf{succ}} = \{(e, e+1) \mid e \in \{1, \ldots, n-1\}\}$ (as usual),
- for each $p \in Proc$, the relation $\lhd_{\mathsf{cr}_p}$ is well-nested: if $e \lhd_{\mathsf{cr}_p} f$, $e' \lhd_{\mathsf{cr}_p} f'$, and $e < e' < f$, then $f' < f$ (where $<$ is the canonical strict total order on $\{1, \ldots, n\}$),
- B is disconnected wrt. $\{\mathsf{cr}_1, \ldots, \mathsf{cr}_m\}$.

A nested word with $m = 2$ and $\Sigma = \{a, b, c\}$, is depicted in Fig. 2.

Theorem 3 (Alur-Madhusudan [2]). *Suppose $m = 1$, i.e., there is only one nesting relation. In that case, nested-word automata (i.e., σ_{nw}-automata) and* $\mathrm{MSO}(\sigma_{\mathrm{nw}})$ *are expressively equivalent over nested words. Moreover, emptiness of σ_{nw}-automata is decidable over nested words.*

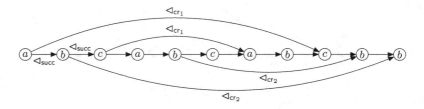

Fig. 2. A nested word

Theorem 4. *Suppose $m \geq 2$. Then, $\mathrm{MSO}(\sigma_{\mathrm{nw}})$ is strictly more expressive than σ_{nw}-automata over nested words [6]. Moreover, emptiness of σ_{nw}-automata is undecidable over nested words (as one can easily simulate a Minsky machine).*

Theorem 5. *Suppose $m = 2$. Then, σ_{nw}-automata and $\mathrm{EMSO}(\sigma_{\mathrm{nw}})$ are expressively equivalent over nested words [6].*

To recover a robust automata model in the presence of multiple stacks/ nesting relations, a fruitful approach has been to restrict (i.e., under-approximate) the set of possible behaviors. We will present only one restriction here. However, note that a variety of other restrictions have been considered in the literature, which essentially lead to the same positive results [3,10,15–18].

Namely, we impose a bound on the number $k \geq 1$ of *phases* that a nested word may traverse. In a phase, only one dedicated process is allowed to perform a return/pop. Let $B = (\{1, \ldots, n\}, \lambda, \lhd_{\mathsf{succ}}, \lhd_{\mathsf{cr}_1}, \ldots, \lhd_{\mathsf{cr}_m})$ be a nested word. An *interval* of B is a set of events of the form $\{e, e + 1, \ldots, f\}$ where $e \leq f$. An interval I is called a *phase* if, for all $e, e' \in \{1, \ldots, n\}$, $f, f' \in I$, and $p, p' \in Proc$ such that $e \lhd_{\mathsf{cr}_p} f$ and $e' \lhd_{\mathsf{cr}_{p'}} f'$, we have $p = p'$. Finally, B is *k-phase-bounded* if there are phases $I_1, \ldots, I_k$ of B such that $I_1 \cup \ldots \cup I_k = \{1, \ldots, n\}$. The nested word from Fig. 2 is 2-phase-bounded, witnessed by the phases $\{1, \ldots, 9\}$ and $\{10, 11\}$.

Theorem 6 (La Torre-Madhusudan-Parlato [14]). *Let $k \geq 1$. Then, σ_{nw}-automata and $\mathrm{MSO}(\sigma_{\mathrm{nw}})$ are expressively equivalent over k-phase-bounded nested words. Moreover, emptiness of σ_{nw}-automata is decidable over k-phase-bounded nested words.*

5   Asynchronous Automata

In this section, we deal with *asynchronous automata* [23], whose behaviors are *Mazurkiewicz traces*. As opposed to the previous models, asynchronous automata have a rather distributed flavor, since we will no longer assume that events of a behavior are totally ordered in terms of some relation $\lhd_{\mathsf{succ}}$. We fix a finite set of processes $Proc = \{1, \ldots, m\}$, $m \geq 1$, and a nonempty finite set A. Let us define the signature σ_{t}. The alphabet Σ consists of all pairs (a, P) where $a \in A$ and $P \subseteq Proc$ is a nonempty set of processes. The idea is that P contains those

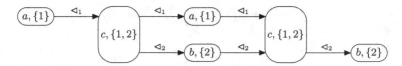

Fig. 3. A trace

processes that are involved in the execution of an event. This may indeed model common access to a shared resource. Moreover, we set $R = Proc$. For $p \in R$, the relation $\lhd_p$ will connect two *consecutive* events that are executed by process p. Note that we may have $e \lhd_p f$ and $e \lhd_{p'} f$ for distinct processes p and p'.

Consider a σ_t-behavior $B = (E, \lambda, \lhd_1, \ldots, \lhd_m)$. For $p \in R$, let $E_p := \{e \in E \mid \lambda(e) = (a, P)$ for some $(a, P) \in \Sigma$ such that $p \in P\}$. Then, B is a *(Mazurkiewicz) trace* if, for all $p \in R$, $\lhd_p$ is the direct-successor relation of a total order on E_p. An example of a trace, where $m = 2$ and $A = \{a, b, c\}$, is depicted in Fig. 3.

When running on traces, σ_t-automata are actually *asynchronous automata* or *asynchronous cellular automata* [23].

Theorem 7 (Thomas [19]). *Asynchronous automata (i.e., σ_t-automata) are expressively equivalent to* $\mathrm{MSO}(\sigma_t)$ *over traces.*

Note that emptiness of σ_t-automata over traces is easily shown decidable, since asynchronous automata are essentially finite-state systems.

6   Message-Passing Automata

The last model that we consider adopts the message-passing paradigm and goes back to Brand and Zafiropulo [8]. It is the natural counterpart of asynchronous automata in a message-passing environment. Again, we fix a finite set of processes $Proc = \{1, \ldots, m\}$ (but now assuming $m \geq 2$). The set $Proc$ determines the set $Ch = \{(p, q) \in Proc \times Proc \mid p \neq q\}$ of *channels*, which we assume to be reliable, FIFO, and (a priori) unbounded. The set of labels is $\Sigma = \{p!q \mid (p, q) \in Ch\} \cup \{q?p \mid (p, q) \in Ch\}$. Here, $p!q$ will label an event of process p that sends a message to q. The complementary receive event is then labeled by $q?p$. The behaviors that we consider have two kinds of edges. For $p \in Proc$, the relation $\lhd_p$ connects consecutive events executed by p. Moreover, for $(p, q) \in Ch$, the relation $\lhd_{(p,q)}$ links a send event with a receive event, which models the exchange of a message sent from p to q through the channel (p, q). Thus, our signature σ_{msc} assumes the set of relation symbols $R = Proc \cup Ch$.

Let $B = (E, \lambda, (\lhd_r)_{r \in R})$ be a σ_{msc}-behavior. For $p \in Proc$, we set $E_p := \{e \in E \mid$ there is $q \in Proc \setminus \{p\}$ such that $\lambda(e) \in \{p!q, p?q\}\}$. We call B a *message sequence chart* if

– for every $p \in Proc$, $\lhd_p$ is the direct-successor relation of a total order on E_p,
– for all $(p, q) \in Ch$ and $(e, f) \in \lhd_{(p,q)}$, we have $\lambda(e) = p!q$ and $\lambda(f) = q?p$,

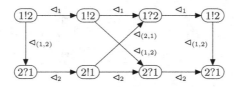

Fig. 4. A message sequence chart

- for all $(p,q) \in Ch$ and $(e,f), (e',f') \in \lhd_{(p,q)}$, we have $e \lhd_p^* e'$ iff $f \lhd_q^* f'$ (which models FIFO),
- for every $e \in E$, there are $(p,q) \in Ch$ and $f \in E$ such that $e \lhd_{(p,q)} f$ or $f \lhd_{(p,q)} e$ (i.e., every event is either a send or a receive event), and
- B is disconnected wrt. Ch.

Fig. 4 depicts a message sequence chart where $m = 2$.

A σ_{msc}-automaton running over message sequence charts is essentially a *message-passing automaton*, aka *communicating automaton* or *communicating finite-state machine* [8].

Theorem 8. *Message-passing automata (i.e., σ_{msc}-automata) are expressively equivalent to* $\mathrm{EMSO}(\sigma_{\mathrm{msc}})$ *over message sequence charts, but strictly less expressive than* $\mathrm{MSO}(\sigma_{\mathrm{msc}})$ *[7]. Emptiness of σ_{msc}-automata over message sequence charts is undecidable [8].*

It is an open problem whether message-passing automata are expressively equivalent to $\mathrm{EMSO}(\sigma_{\mathrm{msc}} + \lhd_1^* + \ldots + \lhd_m^*)$ over message sequence charts.

Again, restricting the set of message sequence charts further, one obtains positive results wrt. MSO logic and model-/emptiness-checking. The restrictions we consider here rely on the notion of a *linearization* of a message sequence chart $B = (E, \lambda, (\lhd_r)_{r \in R})$, which is any total order $\preceq$ on E such that $(\bigcup_{r \in R} \lhd_r)^* \subseteq \preceq$. For $k \geq 1$, linearization $\preceq$ is called *k-bounded*, if, for all $(p,q) \in Ch$,

$$\max_{g \in E} |\{(e,f) \in \lhd_{(p,q)} \mid e \preceq g \prec f\}| \leq k.$$

Intuitively, along the linearization, there are, at any time and in any channel, no more than k pending messages. We call B *universally k-bounded* if every of its linearizations is k-bounded. Accordingly, we call B *existentially k-bounded* if at least one of its linearizations is k-bounded. The message sequence chart from Fig. 4 is universally 2-bounded (but not universally 1-bounded), and it is existentially 1-bounded.

Theorem 9 (Henriksen-Mukund-Narayan Kumar-Sohoni-Thiagarajan [13]). *Let $k \geq 1$. Message-passing automata (i.e., σ_{msc}-automata) are expressively equivalent to* $\mathrm{MSO}(\sigma_{\mathrm{msc}})$ *over universally k-bounded message sequence charts. Moreover, emptiness of σ_{msc}-automata over universally k-bounded message sequence charts is decidable.*

Theorem 10 (Genest-Kuske-Muscholl [12]). *Let $k \geq 1$. Message-passing automata (i.e., σ_{msc}-automata) are expressively equivalent to $MSO(\sigma_{msc})$ over existentially k-bounded message sequence charts. Moreover, emptiness of σ_{msc}-automata over existentially k-bounded message sequence charts is decidable.*

7 Conclusion

Note that many more automata models enjoy logical characterizations in terms of (fragments of) MSO logic, such as tree automata or certain automata running on graphs [20]. In this paper, however, we focused on automata that may serve as models of concurrent systems and whose semantics is a set of possible executions. The apparent similarities between these models suggest that more general meta results may be possible. It would be interesting to identify signatures σ and general classes of σ-behaviors for which σ-automata and $MSO(\sigma)$ (or suitable fragments) are expressively equivalent.

References

1. Aiswarya, C., Gastin, P.: Reasoning about distributed systems: WYSIWYG. In: FSTTCS 2014, Leibniz International Proceedings in Informatics, vol. 29, pp. 11–30. Leibniz-Zentrum für Informatik (2014)
2. Alur, R., Madhusudan, P.: Adding nesting structure to words. J. ACM **56**(3), 1–43 (2009)
3. Atig, M.F., Bollig, B., Habermehl, P.: Emptiness of multi-pushdown automata Is 2ETIME-complete. In: Ito, M., Toyama, M. (eds.) DLT 2008. LNCS, vol. 5257, pp. 121–133. Springer, Heidelberg (2008)
4. Björklund, H., Schwentick, T.: On notions of regularity for data languages. Theor. Comput. Sci. **411**(4–5), 702–715 (2010)
5. Bojanczyk, M., David, C., Muscholl, A., Schwentick, T., Segoufin, L.: Two-variable logic on data words. ACM Trans. Comput. Log. **12**(4), 27 (2011)
6. Bollig, B.: On the expressive power of 2-stack visibly pushdown automata. Logical Methods in Computer Science 4(4:16), 1–35 (2008)
7. Bollig, B., Leucker, M.: Message-passing automata are expressively equivalent to EMSO logic. Theor. Comput. Sci. **358**(2–3), 150–172 (2006)
8. Brand, D., Zafiropulo, P.: On communicating finite-state machines. J. ACM **30**(2), 323–342 (1983)
9. Büchi, J.: Weak second order logic and finite automata. Z. Math. Logik, Grundlag. Math. **5**, 62–66 (1960)
10. Cyriac, A., Gastin, P., Naryanan Kumar, K.: MSO decidability of multi-pushdown systems via split-width. In: Koutny, M., Ulidowski, I. (eds.) CONCUR 2012. LNCS, vol. 7454, pp. 547–561. Springer, Heidelberg (2012)
11. Elgot, C.C.: Decision problems of finite automata design and related arithmetics. Trans. Am. Math. Soci. **98**, 21–52 (1961)
12. Genest, B., Kuske, D., Muscholl, A.: A Kleene theorem and model checking algorithms for existentially bounded communicating automata. Infor. Comput. **204**(6), 920–956 (2006)

13. Henriksen, J.G., Mukund, M., Narayan, K., Sohoni, M., Thiagarajan, P.S.: A theory of regular MSC languages. Infor. Comput. **202**(1), 1–38 (2005)
14. La Torre, S., Madhusudan, P., Parlato, G.: A robust class of context-sensitive languages. In: LICS 2007, pp. 161–170. IEEE Computer Society Press (2007)
15. La Torre, S., Madhusudan, P., Parlato, G.: The language theory of bounded context-switching. In: López-Ortiz, A. (ed.) LATIN 2010. LNCS, vol. 6034, pp. 96–107. Springer, Heidelberg (2010)
16. La Torre, S., Napoli, M., Parlato, G.: Scope-bounded pushdown languages. In: Shur, A.M., Volkov, M.V. (eds.) DLT 2014. LNCS, vol. 8633, pp. 116–128. Springer, Heidelberg (2014)
17. La Torre, S., Napoli, M., Parlato, G.: A unifying approach for multistack pushdown automata. In: Csuhaj-Varjú, E., Dietzfelbinger, M., Ésik, Z. (eds.) MFCS 2014, Part I. LNCS, vol. 8634, pp. 377–389. Springer, Heidelberg (2014)
18. Madhusudan, P., Parlato, G.: The tree width of auxiliary storage. In: POPL 2011, pp. 283–294. ACM (2011)
19. Thomas, W.: On logical definability of trace languages. In: Proceedings of Algebraic and Syntactic Methods in Computer Science (ASMICS), Report TUM-I9002, Technical University of Munich, pp. 172–182 (1990)
20. Thomas, W.: Elements of an automata theory over partial orders. In: POMIV 1996, vol. 29, DIMACS. AMS (1996)
21. Thomas, W.: Languages, automata and logic. In: Salomaa, A., Rozenberg, G. (eds.) Handbook of Formal Languages, vol. 3, pp. 389–455. Springer, Heidelberg (1997)
22. Trakhtenbrot, B.A.: Finite automata and monadic second order logic. Siberian Math. J. **3**, 103–131 (1962); In Russian; English translation in Amer. Math. Soc. Transl. **59**, 23–55 (1966)
23. Zielonka, W.: Notes on finite asynchronous automata. R.A.I.R.O. Informatique Théorique et Applications **21**, 99–135 (1987)

Hardware Implementations of Finite Automata and Regular Expressions
Extended Abstract

Bruce W. Watson[(⊠)]

FASTAR Group, Department of Information Science, Stellenbosch University,
Stellenbosch, South Africa
bruce@fastar.org

1 Introduction

This extended abstract sketches some of the most recent advances in hardware implementations (and surrounding issues) of finite automata and regular expressions. The traditional application areas for automata and regular expressions are compilers, text editors, text programming languages (for example Sed, AWK, but more recently Python, and Perl), and text processing in general purpose languages (such as Java, C++ and C#). In all these cases, while the regular expression implementation should be efficient, it rarely forms the performance bottleneck in resulting programs and applications. Even more exotic application areas such as computational biology are not particularly taxing on the regular expression implementation — provided some care is taken while crafting the regular expressions [5].

One application domain stands out in its requirement of very high performance — regular expression processing of network traffic. Such processing is required in a variety of contexts: network security (intrusion detection and prevention), protocol detection, policy enforcement, load balancing/traffic differentiation, and quality of service. Given that it usually involves regular expression pattern matching over the network packet 'payload', it is often known as *deep packet inspection* (DPI). Currently, all network equipment vendors (and several software vendors) provide DPI products using regular expressions. Despite its age, [11] still gives the best introduction to the algorithmic and implementation intricacies of networks.

Current network speeds at a typical switch are 40 Gbits/s. Full regular expression processing must therefore be done at 4 Gbytes/s after accounting for overheads — one byte per clock cycle on a fast 4 GHz processor. The latency requirements vary per application (e.g. telephony and banking require low latency, while video and music streaming can allow for higher latency provided the variability is low) — meaning that significantly delaying a packet for processing is typically unacceptable. Network packet sizes vary dramatically from hundreds of bytes to tens of kilobytes. Packets from various network flows (e.g. some from a web-browsing session, ftp, mail, and a web application) are interspersed and may arrive out of order, implying that any regular expression

© Springer International Publishing Switzerland 2015
F. Drewes (Ed.): CIAA 2015, LNCS 9223, pp. 13–17, 2015.
DOI: 10.1007/978-3-319-22360-5_2

processor must 'context switch' appropriate at the beginning and end of a new packet. Lastly, the number of regular expressions (relatively small Perl-like regular expressions) being matched is usually in the range from a few hundred to a few thousand, making it infeasible to deal with them individually. Occasionally, the regular expression set changes, giving the additional challenge of updating the processor, either in batch mode or incrementally when only few of the regular expressions have been edited.

Unfortunately, from a performance perspective, network speed and volume has been outpacing Moore's law for computational performance.

2   Typical Solutions

As mentioned earlier, Varghese [11] remains an excellent introduction to the algorithmics and implementation aspects of high-performance networking, including regular expression processing. Recently, [14] gives an overview of the latest developments in DPI for networking in virtualized (and cloud-based) environments. Essentially, all solutions share a common set of abstractions grounded in formal languages, and then vary based on implementation.

2.1   Abstractions

While occasional attempts have been made to implement regular expressions directly in hardware[1], most require the 'compilation' of the regular expression(s) to some form of finite automaton, with the predictable tradeoffs:

- Nondeterministic automata — requiring space linear in the size of the regular expressions. DPI does not allow for backtracking simulation of the automaton, meaning that all paths are pursued in parallel (processing a byte can take up to time linear in the size of the automaton) and the 'current state set' is a significant data-structure overhead which must be stored/restored during context (also potentially taking time linear in the size of the automaton).
- Deterministic automata — requiring space potentially exponential in the size of the regular expression. Processing a byte of network traffic requires a small number of clock cycles (largely independent in the size of the automaton, though memory caching can affect this slightly), as does a context switch.

2.2   Implementations

The above-mentioned abstractions underlie most of the software implementations of DPI[2]. While there is some variation in the CPU speed, cache memory,

---

[1] Most such attempts decompose the regular expressions in a set of much smaller ones which are then mapped to *content-addressable memory* (CAM) implementations. None of these implementations have yet proven competitive in practice.

[2] See [13] for one of many treatments of automata and regular expression implementations in software.

etc., eventually all such implementations are outpaced by the network traffic, leading DPI implementers to consider *acceleration* options.

The first option is to use the *graphics processing unit* (GPU) [8]. Numerous such DPI accelerations can be found in the literature (indeed, it appears to be a favourite student project), all showing impressive performance improvements in large packets arriving in-order. The architecture of the GPU (SIMD, meaning that numerous smaller processing elements execute the same instructions in lockstep) and the interface to the CPU (network traffic being transferred over this interface) impair the performance in realistic networks, which involve widely varying packet sizes and frequent context switches. This largely limits GPU accelerations to open-source and software only DPI.

Instead of a general purpose CPU, most network equipment vendors use domain-specific *network processing units* (NPUs)[3]. Most NPUs have been designed for the breadth of packet processing tasks (routing, packet verification, etc.), with relatively little memory and silicon real-estate devoted to DPI, and such DPI implementations tend to suffer from the same performance limitations as on CPUs[4].

Any remaining acceleration is only achievable with custom hardware, which broadly falls into two categories: *reconfigurable hardware*[5] and *application specific integrated circuits* (ASICs). Several vendors provide for FPGA solutions, and the relatively low cost of implementation makes it also an attractive student project [7, Chapter 34]. The regular expression set is usually compiled on a CPU (see [12] for a variety of such compilation algorithms) to an automaton or to circuit structures encoding the automaton, which are then downloaded to the FPGA. The chosen circuit structures are usually optimized for high-speed processing (fewest clock cycles per byte of network traffic), or least silicon real-estate — though the cost of updating the regular expressions is usually high due to the compilation on the CPU and the CPU-FPGA bandwidth for reconfiguring the FPGA.

ASIC solutions typically use a circuit structure resembling a generic automaton (with additional circuitry to simulate it), allowing for rapid updating of the automaton as the regular expressions are changed. As such, the ASIC solution has only a few advantages over FPGAs: higher density and performance, lower volume costs and lower power consumption, but much higher development costs.

2.3   Gaps in Current Solutions

Clearly, all current solutions involve trading off byte-processing time against silicon real-estate, and the ease of updating the regular expression set.

---

[3] See the websites of prominent vendors such as Cisco, Netronome (which took over Intel's NPU product line) and IBM.

[4] A notable exception is Netronome's NPU which includes SIMD processing — in turn having the same performance characteristics as DPI on GPUs.

[5] In the form of *field programmable gate arrays* (FPGAs).

3   New Implementations

Homogeneous automata[6] are (not necessarily deterministic) ones in which any given state has in-transitions on the same alphabet symbol (byte). This allows for an efficient encoding of the transition relation — without node labels, as an adjacency matrix — and with a mapping from each state to 'its symbol'. The bit-matrix and -vector operations (see [13] for implementation details, then in software) map extremely efficiently to digital circuits and will be discussed in detail in this talk. In particular, the bit vectors are linear in the total regular expression size and allow for single clock cycle bit-vector operations to pursue all nondeterministic automaton paths simultaneously. Furthermore, context switching can be done rapidly using burst transfers of the bit-vector to/from memory.

Interestingly, *dual homogeneous* automata[7] enjoy a similarly compact encoding. The resulting mapping is subtly different from that of homogeneous automata, with occasional circuit real-estate and power savings.

The compilation algorithm mapping a regular expression to a homogeneous automaton is virtually identical to that mapping to a dual homogeneous one. Our most recent work (included in this talk) encodes the compilation algorithm in the circuitry with a minimal overhead. For the first time, this enables an embedded DPI device to be fed regular expressions for compilation directly in silicon — a significant win over first compiling on a CPU and then downloading the automaton (which is typically much larger than the regular expression).

4   Ongoing and Future Work

Brzozowski's algorithm for constructing a deterministic automaton are both elegant and efficient in practice [3]. Recent work led by Strauss and Kourie [10] has given a parallel version of Brzozowski's algorithm as *communicating sequential processes* (CSP). Coincidentally, Brzozowski's career has included lines of research into mapping CSP-like programs to delay-insensitive (unclocked) circuits — see [4], though numerous others have also worked on such mappings and circuitry. This talk also covers the use of such mappings to directly compile Brzozowski's construction algorithm to a delay-insensitive circuit.

References

1. Aho, A.V., Sethi, R., Ullman, J.D.: Compilers: Principles, Techniques and Tools. Addison-Wesley, Reading (1988)
2. Berry, G., Sethi, R.: From regular expressions to deterministic automata. Theoretical Comput. Sci. **48**, 117–126 (1986)

---

[6] These automata (and variants thereof) were discovered by [1,2,6,9] and are detailed in most treatments of automata construction algorithms.

[7] Where any given state has *out-transitions* on the same alphabet symbol, see [12] where they are referred to as *reduced finite automata*.

3. Brzozowski, J.A.: Regular expression techniques for sequential circuits. Ph.D. thesis, Princeton University, Princeton, New Jersey, June 1962
4. Brzozowski, J.A., Seger, C.J.: Asynchronous Circuits. Springer (1995)
5. Friedl, J.: Mastering Regular Expressions, 3rd edn. O'Reilly Media Inc., Sebastopol (2006)
6. Glushkov, V.: The abstract theory of automata. Russ. Math. Surveys **16**, 1–53 (1961)
7. Hauck, S., DeHon, A. (eds.): Reconfigurable Computing: The Theory and Practice of FPGA-Based Computation. Morgan Kaufmann, San Francisco (2007)
8. Kirk, D.B., Hwu, W.M.W.: Programming Massively Parallel Processors: A Hands-On Approach. Morgan Kaufmann, San Francisco (2010)
9. McNaughton, R., Yamada, H.: Regular expressions and state graphs for automata. IEEE Trans. Electron. Comput. **9**(1), 39–47 (1960)
10. Strauss, T., Kourie, D.G., Watson, B.W.: A concurrent specification of Brzozowski's DFA construction algorithm. Int. J. Found. Comput. Sci. **19**(1), 125–135 (2008)
11. Varghese, G.: Network Algorithmics: An Interdisciplinary Approach to Designing Fast Networked Devices. Morgan Kaufmann, San Francisco (2004)
12. Watson, B.W.: A taxonomy of finite automata construction algorithms. Technical Report 43, Faculty of Computing Science, Eindhoven University of Technology, the Netherlands (1993)
13. Watson, B.W.: The design of the FIRE Engine: A C++ toolkit for FInite automata and Regular Expressions. Technical Report 22, Faculty of Computing Science, Eindhoven University of Technology, the Netherlands (1994)
14. Watson, B.W.: Elastic deep packet inspection. In: Brangetto, P., Maybaum, M., Stinissen, J. (eds.) 6th International Conference on Cyber Conflict, pp. 241–253. IEEE, Tallinn (2014)

Regular Papers

Complexity of Inferring Local Transition Functions of Discrete Dynamical Systems

Abhijin Adiga[1]([✉]), Chris J. Kuhlman[1], Madhav V. Marathe[1], S.S. Ravi[2],
Daniel J. Rosenkrantz[2], and Richard E. Stearns[2]

[1] Virginia Bioinformatics Institute, Virginia Tech, Blacksburg, VA, USA
{abhijin,ckuhlman,mmarathe}@vbi.vt.edu
[2] Computer Science Department, University at Albany – SUNY, Albany, NY, USA
sravi@albany.edu, {drosenkrantz,thestearns2}@gmail.com

Abstract. We consider the problem of inferring the local transition functions of discrete dynamical systems from observed behavior. Our focus is on synchronous systems whose local transition functions are threshold functions. We assume that the topology of the system is known and that the goal is to infer a threshold value for each node so that the system produces the observed behavior. We show that some of these inference problems are efficiently solvable while others are **NP**-complete, even when the underlying graph of the dynamical system is a simple path. We also identify a fixed parameter tractable problem in this context.

1 Introduction

1.1 Motivation

Methods that use observations of systems to calibrate or infer models and model properties arise in many contexts. These properties help explain system behaviors, and parameterized models may be transferable to other contexts [29]. A case in point is a protest in Spain in 2011, in which people demonstrated against economic austerity measures [13]. In that work on information spread via Twitter, threshold behavior was used to model tweeting about the protest. If a person v received t_v tweets (from unique users that she followed) about the event, and then tweeted for the first time before v received the $(t_v + 1)$'th tweet, that person's threshold for participation was chosen as t_v. The inference of thresholds enabled the full specification of a model for the underlying socio-technical phenomenon. Several other studies along the same lines have also been carried out (e.g., [25,30]). The threshold model for social networks, introduced in [15], is used in many applications (see e.g. [9]). Discrete dynamical systems [4,22], which generalize cellular automata, represent a rigorous and convenient abstract model to study socio-technical phenomena.

The problem of inferring the components of a system from observed behavior has also received a lot of attention in the theoretical computer science literature. For example, many researchers have studied the problem of learning automata from sets of accepted strings (see e.g. [23]). Likewise, the problem of learning

© Springer International Publishing Switzerland 2015
F. Drewes (Ed.): CIAA 2015, LNCS 9223, pp. 21–34, 2015.
DOI: 10.1007/978-3-319-22360-5_3

CNF and DNF formulas have also been studied extensively in the literature (see e.g. [18]). Additional information on these topics will be provided in Sect. 1.3.

1.2   Problems Considered and Summary of Results

Here, we consider an inference problem that arises in the context of discrete dynamical systems. In particular, we focus on one such model, namely *synchronous* discrete dynamical systems (SyDSs). We provide an informal description of a SyDS here; a formal description is given later. A SyDS consists of an undirected graph whose vertices represent entities (agents) and edges represent local interactions among entities. Each vertex has a state value chosen from a finite domain (e.g. {0,1}). In addition, each vertex v also has a local transition function whose inputs are the current state of v and those of its neighbors; the output of this function is the next state of v. The vector consisting of the state values of all the nodes at each time instant is referred to as the **configuration** of the system at that instant. In each time step, all nodes of a SyDS compute and update their states *synchronously*. Starting from a (given) initial configuration, the time evolution of a SyDS consists of a sequence of successive configurations. Models similar to SyDSs have been used in several applications, including the propagation of diseases and social phenomena (e.g. [9,21]).

Several researchers have studied the complexity of various **analysis problems** for discrete dynamical systems (e.g. [3,5,16,20,28]). Informally, the goal of an analysis problem is to predict the behavior of a system given its static description. An example of an analysis problem is that of **reachability**: given a SyDS $\mathcal{S}$ and two configurations $\mathcal{C}_1$ and $\mathcal{C}_2$, will $\mathcal{S}$ reach $\mathcal{C}_2$ starting from $\mathcal{C}_1$? Such analysis questions are studied by considering the **phase space** of the SyDS, which is a directed graph with one vertex for each possible configuration and a directed edge (x,y) from a vertex x to vertex y if the SyDS can transition from the configuration corresponding to x to the one corresponding to y in one time step. In such a case, y is the **successor** of x and x is a **predecessor** of y. Each self loop in the phase space of a SyDS represents a **stable** configuration or a **fixed point** of the system[1] (i.e., a configuration in which the system will stay forever). Any configuration $\mathcal{C}$ whose successor is different from $\mathcal{C}$ is called an **unstable configuration**. A **trajectory** in phase space is a directed path with one or more edges. Also, any vertex in the phase space with no incoming edges represents a **Garden–of–Eden** (GE) configuration which can arise only as the initial configuration of the system.

Here, our focus is on a problem which may be considered as an *inverse* of the analysis problem. In such a problem, the goal is to infer some aspect of the structure of a partially specified system given a description of its observed behavior. In particular, we focus on inferring the local transition functions of a SyDS, given the underlying graph and appropriate behavior patterns. We consider several

---

[1] We will use the term "stable configuration" instead of "fixed point" throughout this paper since we will be using the word "fixed" in the context of fixed parameter tractability.

different behavior patterns (e.g. collection of stable configurations, collection of trajectories, collection of unstable configurations) and study the complexity of identifying the local transition functions that can explain the observed behavior. In particular, we assume that each local transition function f_i is a r_i-threshold function for some non-negative integer r_i (see Sect. 2 for the definition of threshold functions).

We present results for threshold inference problems for two categories of observed behavior, namely homogeneous collections (e.g. a set of stable configurations, a set of unstable configurations) and heterogeneous collections (e.g. a collection of stable configurations and a set of unstable configurations). Our results establish the complexity of the corresponding decision and optimization problems (see Sects. 3 and 4). For heterogeneous behavior specifications, we also establish a fixed parameter tractability result (see Sect. 4.2). Thus, our results provide insights regarding the efficient solvability and computational intractability of threshold inference problems for SyDSs. For space reasons, proofs of some results are omitted in this version; they can be found in [2].

1.3   Related Work

Several inference problems have been explored in the context of disease, information and meme spread. One direction of work considers estimating model parameters given the traces of a diffusion process, network and a class of models (e.g. Bailon et al. [13]). Abraho et al. [1], Gomez et al. [12], Soundarajan and Hopcroft [27] consider the problem of inferring the network structure given the model. Recently, there has been a lot of work on source detection, where the goal is to find the source of infection given limited information about the network, diffusion model and the set of infected nodes [26]. Most of the these problems turn out to be hard even for simple models (such as progressive systems [19]) and networks.

Learning finite automata and Boolean functions are two rich areas which consider problems with a similar flavor. In the case of learning finite automata, the problem is to infer a finite (stochastic) automaton given a set of strings labeled as either in the language or not [18,23]. Similarly, in concept learning (or learning Boolean functions), the task is to infer a Boolean function given information about its values on some points, together with the knowledge that it belongs to a particular class of functions [18].

As mentioned earlier, many researchers have addressed the computational aspects of testing phase space properties of discrete dynamical systems. Goles and Martínez [11] provide bounds on the lengths of transients (i.e., trajectories that end in stable configurations) and cycles in threshold dynamical systems. The complexity of determining whether a given configuration y of a deterministic SyDS has a predecessor has been studied in [5]. Problems similar to predecessor existence have also been considered in the context of cellular automata [8,16]. Researchers have also studied various questions for dynamical systems under the sequential update model, where the vertex functions are applied according to a specified order [6,20,22].

2 Definitions and Problem Formulation

2.1 Formal Definition of the SyDS Model

Let $\mathbb{B}$ denote the Boolean domain $\{0,1\}$. A **Synchronous Dynamical System** (SyDS) $\mathcal{S}$ over $\mathbb{B}$ is specified as a pair $\mathcal{S} = (G, \mathcal{F})$, where

(a) $G(V, E)$, an undirected graph with $|V| = n$, represents the underlying graph of the SyDS, and

(b) $\mathcal{F} = \{f_1, f_2, \dots, f_n\}$ is a collection of functions in the system, with f_i denoting the **local transition function** associated with node v_i, $1 \le i \le n$.

Each node of G has a state value from $\mathbb{B}$. Each function f_i specifies the local interaction between node v_i and its neighbors in G. The inputs to function f_i are the state of v_i and those of the neighbors of v_i in G; function f_i maps each combination of inputs to a value in $\mathbb{B}$. This value becomes the next state of node v_i. It is assumed that each local function can be computed efficiently. In a SyDS, all nodes compute and update their next state *synchronously*. Other update disciplines (e.g. sequential updates) for discrete dynamical systems have also been considered in the literature (e.g. [3, 22]). At any time t, the **configuration** $\mathcal{C}$ of a SyDS is the n-vector $(s_1^t, s_2^t, \dots, s_n^t)$, where $s_i^t \in \mathbb{B}$ is the state of node v_i at time t $(1 \le i \le n)$.

The local function f_v associated with node v of a SyDS $\mathcal{S}$ is a t_v-**threshold** function for some integer $t_v \ge 0$ if the following condition holds: the value of f_v is 1 if the number of 1's in the input to f_v is *at least* t_v; otherwise, the value of the function is 0. Thus, the state of a node may change from 0 to 1 or from 1 to 0. Throughout this paper, we assume that the local transition function for each node is a threshold function.

We let d_v denote the degree of node v, and t_v denote the threshold of node v. The number of inputs to the function f_v varies from 1 to $d_v + 1$. We assume without loss of generality that $0 \le t_v \le d_v + 2$. (The threshold values 0 and $d_v + 2$ allow us to realize functions that always output 1 and 0 respectively.)

Example: Consider the graph shown in Fig. 1. Suppose the local transition functions at each of the nodes v_1, v_4, v_5, v_6 is the **1-threshold function**, the function at v_3 is the **2-threshold function** and that at v_2 is the **3-threshold function**. Assume that initially, v_1, v_4 and v_6 are in state 1 and all other nodes are in state 0. During the first time step, the state of node v_5 changes to 1 since its neighbor v_6 is in state 1; the states of other nodes do not change. The configurations at subsequent time steps are shown in the figure. The configuration $(1, 1, 1, 1, 1, 1)$ at time step 3 is a stable configuration for this system.   □

2.2 Additional Terminology and Notation

A **trajectory** of a SyDS $\mathcal{S}$ is a sequence of configurations $\langle C_1, C_2, \dots, C_r \rangle$, with $r \ge 2$, such that C_{i+1} is the successor of C_i, for $1 \le i \le r - 1$. When the last two configurations in a trajectory are identical, the trajectory ends in a stable

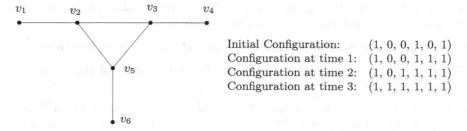

Initial Configuration: (1, 0, 0, 1, 0, 1)
Configuration at time 1:   (1, 0, 0, 1, 1, 1)
Configuration at time 2:   (1, 0, 1, 1, 1, 1)
Configuration at time 3:   (1, 1, 1, 1, 1, 1)

Note: Each configuration has the form $(s_1, s_2, s_3, s_4, s_5, s_6)$, where s_i is the state of node v_i, $1 \le i \le 6$. The configuration at time 3 is a stable configuration.

Fig. 1. An example of a SyDS

configuration. A stable configuration can be thought of as a special trajectory consisting of two identical configurations. Given a configuration $\mathcal{C}$ and a node v, we let $\mathcal{C}(v)$ denote the value of v in $\mathcal{C}$, and $\mathcal{C}^v$ denote the number of 1's in the input to f_v in $\mathcal{C}$.

A problem is **fixed parameter tractable** (FPT) with respect to a parameter k if there is an algorithm for the problem with a running time of $O(h(k)N^{O(1)})$, where N is the size of the problem instance and the function $h(k)$ depends only on k (see e.g. [24]). In particular, the function h does not depend on N. Many combinatorial problems are known to be fixed parameter tractable [24]. In Sect. 4 we show that one of the inference problems for SyDSs is fixed parameter tractable.

Given a bipartite graph $G(V_1, V_2, E)$, a **matching** in G is a subset M of edges such that no two edges of M share an end point. A matching of largest cardinality is called a **maximum matching**. It is well known that for any bipartite graph with n nodes and m edges, a maximum matching can be found in $O(m\sqrt{n})$ time [7]. We will use this result in Sect. 4.2.

2.3   Problem Formulations

In all of the threshold inference problems considered in this paper, it is assumed that the underlying graph of the SyDS is given and that each local function is a threshold function with an unknown threshold value. We start with the definitions of threshold inference problems with *homogeneous* behavior specifications. In this case, the input representing behavior is a set Q along with a type tag; all elements of Q are of the type specified by the tag. For example, a set Q with type tag "stable configuration" indicates that each element of Q is a stable configuration. Similar interpretations can be given for the type tags "unstable configuration", "trajectory" and "GE configuration". We say that a SyDS $\mathcal{S}$ **exhibits the behavior specified by** Q if $\mathcal{S}$ satisfies the property specified by the type tag of Q for each element of Q. For example, if the type tag of Q is "stable configuration", then $\mathcal{S}$ exhibits the behavior specified by Q if each element of Q is a stable configuration of $\mathcal{S}$. The general problem formulation for the homogeneous case can now be stated as follows.

Inferring Thresholds from Homogeneous Behavior Specifications

Given: A partially specified SyDS $\mathcal{S}$ over $\{0, 1\}$ consisting of the underlying graph G and a set Q with a type tag.

Question: Is there an assignment of threshold values to the nodes of $\mathcal{S}$ such that the resulting fully specified SyDS $\mathcal{S}$ exhibits the behavior corresponding to the type tag of Q?

A number of specific problems can be derived from the above general definition. For example, when the type tag of Q is "stable configuration", we refer to the resulting problem as **Inferring Thresholds from Stable Configurations** (ITSC). Likewise, when the type tag of Q is "trajectory" or "unstable configuration" or "GE configuration", we refer to the corresponding problems as Inferring Thresholds from Trajectories (ITT), Unstable Configurations (ITUC) and GE Configurations (ITGE) respectively.

When the answer to an instance of a decision problem such as ITSC is "no", it is natural to consider a maximization version where the goal is to find a threshold assignment that makes a largest subset of Q to be stable configurations. We use the prefix "Max" in naming these problems. Thus, we denote the maximization versions of ITSC, ITT, ITUC and ITGE by Max-ITSC, Max-ITT, Max-ITUC and Max-ITGE respectively.

When the behavior specification consists of two or more sets, each with a different type tag, we obtain inference problems for *heterogeneous* behavior specifications. Many such inference problems can be formulated by considering combinations of type tags. Here, we focus on the problem where the observed behavior is specified by two sets Q_1 and Q_2 with type tags "stable configuration" and "unstable configuration" respectively. We say that a SyDS $\mathcal{S}$ exhibits the behavior specified by Q_1 and Q_2 if $\mathcal{S}$ exhibits the behavior specified by Q_1 as well as Q_2. We refer to the corresponding problem as **Inferring Thresholds from Stable and Unstable Configurations** (ITSUC).

2.4 Preliminary Results

Here, we present some preliminary results which will be used throughout this paper. Proofs of these results appear in [2].

Lemma 1. *Every SyDS $\mathcal{S}$ where each local transition function is a threshold function has at least one stable configuration. Furthermore, a stable configuration can be found in polynomial time.*

Lemma 2. *Given the underlying graph G of a SyDS $\mathcal{S}$ over $\{0, 1\}$ and a configuration D of $\mathcal{S}$, there is a linear time algorithm for constructing an assignment of threshold values to the nodes of $\mathcal{S}$ such that D is the successor of every configuration of $\mathcal{S}$.*

3   Threshold Inference from Homogeneous Behavior Specifications

3.1   Inferring Thresholds from Stable Configurations

In this section, we present an efficient algorithm for the problem of inferring thresholds from stable configurations (ITSC). We also show that Max-ITSC is **NP**-complete and that it cannot be approximated to within any nontrivial ratio unless **P = NP**.

Theorem 1. *The ITSC problem can be solved efficiently. When there is a solution, an assignment of threshold values to the nodes can also be obtained efficiently.*

Proof: Let $Q = \{\mathcal{C}_1, \mathcal{C}_2, \ldots, \mathcal{C}_r\}$ be the given set with type tag "stable configuration". We can consider the threshold assignment separately for each node. For any node v, let Q_v^0 (Q_v^1) be the subset of Q such that for each configuration in Q_v^0 (Q_v^1) the value of v is 0 (1). We use f_v to denote the threshold function at node v.

1. If Q_v^0 is nonempty, we consider each configuration $\mathcal{C}_i \in Q_v^0$. Since $\mathcal{C}_i$ is a stable configuration, the threshold t_v must satisfy the condition $t_v > \mathcal{C}_i^v$, i.e., $t_v \geq \mathcal{C}_i^v + 1$. Let $t_v^{low} = 1 + \max_{\mathcal{C}_i \in Q_v^0} \mathcal{C}_i^v$. If Q_v^0 is empty, we let $t_v^{low} = 0$.

2. If Q_v^1 is nonempty, we consider each configuration $\mathcal{C}_j \in Q_v^1$. Since $\mathcal{C}_j$ is a stable configuration, the threshold t_v must satisfy the condition $t_v \leq \mathcal{C}_j^v$. Let $t_v^{high} = \min_{\mathcal{C}_j \in Q_v^1} \mathcal{C}_j^v$. If Q_v^1 is empty, we let $t_v^{high} = d_v + 2$.

3. The two steps above provide a collection of constraints that can be satisfied provided $t_v^{low} \leq t_v^{high}$. When there is a solution, any value for t_v can be chosen such that $t_v^{low} \leq t_v \leq t_v^{high}$.

Clearly, the above computations can be done in polynomial time.   ∎

Also note that for a given node v, if Q_v^0 is nonempty, then $t_v^{low} \geq 1$; and if Q_v^1 is nonempty, then $t_v^{high} \leq d_v + 1$. Thus, if there is a solution, then there is a solution where none of the local functions are constant functions, i.e., there is a solution where for each node v, $1 \leq t_v \leq d_v + 1$.

Recall that in the Max-ITSC problem, the goal is to choose threshold values so that a maximum number of elements in the set Q are stable points of $\mathcal{S}$. We now show a hardness result for this problem which holds even when the underlying graph of a SyDS is a simple path.

The idea behind the proof is the following. Let us say that two configurations $\mathcal{C}_i$ and $\mathcal{C}_j$ **conflict** if there is a node v such that $\mathcal{C}_i(v) = 0, \mathcal{C}_j(v) = 1$, and $\mathcal{C}_i^v \geq \mathcal{C}_j^v$. Note that if $\mathcal{C}_i$ and $\mathcal{C}_j$ conflict, then there is no assignment of thresholds under which both $\mathcal{C}_i$ and $\mathcal{C}_j$ are stable. On the other hand, if a set of configurations is conflict-free (i.e., no pair of configurations in the set conflict), then thresholds can be assigned so that all the configurations in the set are stable. Given an ITSC problem instance, define the **conflict graph** for the instance to be the undirected graph with a node for each configuration in Q and an edge between each pair of nodes whose corresponding configurations conflict. Each maximum independent set [10] in the conflict graph gives a maximum cardinality subset Q' of Q of configurations that can be made stable.

Theorem 2. *Max-ITSC is **NP**-complete. Further, for any $\epsilon > 0$, there is no polynomial time $O(n^{1-\epsilon})$ approximation algorithm for Max-ITSC, unless $\boldsymbol{P} = \boldsymbol{NP}$. (Here n is the number of configurations in the given set Q.) These results hold even when the underlying graph of the SyDS is a simple path.*

Proof: It can be seen that Max-ITSC is in **NP**. To establish **NP**-hardness, we use a reduction from the Maximum Independent Set (MIS) problem: given an undirected graph $H(V_H, E_H)$ and an integer $K \leq |V_H|$, does H have an independent set of size at least K? MIS is known to be **NP**-complete [10].

The key to the **NP**-hardness reduction is to construct an ITSC problem instance from the MIS in such a way that the conflict graph (defined above) for the constructed Max-ITSC problem instance is identical to the graph in the given MIS problem instance. We do this as follows. There is a configuration for each node of H, with configuration $\mathcal{C}_i$ corresponding to node $v_i \in V_H$. We let $Q = \{\mathcal{C}_1, \mathcal{C}_2, \ldots, \mathcal{C}_n\}$, where $n = |V_H|$. Let $E_H = \{e_1, e_2, \ldots, e_m\}$. For each edge $e_k \in E_H$, the underlying graph G for SyDS $\mathcal{S}$ has two nodes denoted by w_k^1 and w_k^2 along with the edge $\{w_k^1, w_k^2\}$. Suppose the edge $e_k \in E_H$ joins nodes v_i and v_j of H. We set $\mathcal{C}_i(w_k^1) = 0$, $\mathcal{C}_i(w_k^2) = 1$, $\mathcal{C}_j(w_k^1) = 1$, and $\mathcal{C}_j(w_k^2) = 0$. All other configurations in Q have value 0 for both w_k^1 and w_k^2. Nodes w_k^1 and w_k^2 induce a conflict between $\mathcal{C}_i$ and $\mathcal{C}_j$, but do not induce a conflict between any other pair of configurations. Thus, any independent set of V_H corresponds to a conflict-free subset of configurations of $\mathcal{S}$ and vice versa.

To ensure that the underlying graph of SyDS $\mathcal{S}$ is a simple path, we add $m - 1$ new nodes denoted by $z_1, z_2, \ldots, z_{m-1}$. For each j, $1 \leq j \leq m-1$, node z_j is adjacent to w_j^2 and w_{j+1}^1. In each of the n configurations constructed above, the values of the nodes $z_1, z_2, \ldots, z_{m-1}$ are all 0.

It can be verified that H has an independent set of size $\geq K$ if and only if there is a conflict-free subset of Q with at least K configurations. This proves the **NP**-hardness of Max-ITSC.

It is well known that for any $\epsilon > 0$, there is no polynomial time $O(n^{1-\epsilon})$-approximation algorithm for the MIS problem, unless $\boldsymbol{P} = \boldsymbol{NP}$ [17]. Since our construction preserves approximations (i.e., for each r, any independent set of size r in H leads to a subset of r conflict-free configurations of $\mathcal{S}$), the same negative result holds for Max-ITS as well. ∎

Results similar to Theorems 1 and 2 can also be proven when the set Q consists of trajectories; see [2].

3.2 Inferring Thresholds from Unstable Configurations

Here, we consider the ITUC problem where the goal is to infer thresholds from a given set Q of unstable configurations. In this case, we show that both ITUC and Max-ITUC can be solved efficiently.

Theorem 3. *The ITUC problem can be solved efficiently. When there is a solution, an assignment of threshold values to the nodes can also be obtained efficiently.*

Proof: If Q contains every possible configuration of $\mathcal{S}$, (i.e., $|Q| = 2^n$, where n is the number of nodes in the underlying graph of $\mathcal{S}$), then, by Lemma 1, there is no solution to the problem.

Suppose that Q excludes at least one configuration of $\mathcal{S}$. Then there is always a solution to the problem, as outlined below. Let $\mathcal{C}$ be a configuration that is not in Q. From Lemma 2, we can set the thresholds so that the successor of *every* configuration is the configuration $\mathcal{C}$; that is, every configuration other than $\mathcal{C}$ is an unstable configuration.   ■

The following result for Max-ITUC is a simple consequence of the above theorem.

Corollary 1. *Max-ITUC is efficiently solvable.*

Results similar to Theorem 3 and Corollary 1 can also be proven when the set Q consists of Garden of Eden configurations. These results appear in [2].

4   Inference from Heterogeneous Collections of Behavior

4.1   The Complexity of ITSUC

We now consider the ITSUC problem, where there are two sets Q_1 and Q_2 of configurations, and the requirement is to find a threshold value for each node of $\mathcal{S}$ such that each configuration in Q_1 is stable and each configuration in Q_2 is unstable. As indicated in the following theorem, this decision problem is **NP**-complete, even when the underlying graph of $\mathcal{S}$ is a simple path.

Theorem 4. *The ITSUC problem is **NP**-complete even when the underlying graph of the given SyDS is a simple path.*

Proof Idea: A reduction from 3SAT [10] appears in [2].   ■

4.2   Fixed Parameter Tractability of ITSUC

We now show that ITSUC is fixed parameter tractable with respect to the number of unstable configurations specified in the problem instance, with no restrictions on the underlying graph of the given SyDS. Given a set $A = \{a_1, a_2, \ldots, a_r\}$, let $P(A)$ denote a **partition** of A. Let each subset in $P(A)$ be called a **block**. We use $\pi(A)$ to denote the collection of all partitions of A. For a set A with r elements, it is known that $|\pi(A)| = O((r/\log r)^r)$ [14].

Let Q_1 and Q_2 denote the set of stable and unstable configurations respectively in the given ITSUC instance. Let $q = |Q_2|$ and let n be the number of nodes in the given SyDS $\mathcal{S}$. From the proof of Theorem 1, it can be seen that the configurations in Q_1 impose constraints on the threshold value of each node of $\mathcal{S}$. Given any configuration $\mathcal{C} = (s_1, s_2, \ldots, s_n)$ in Q_2, we can try to make $\mathcal{C}$ an unstable configuration by choosing a threshold t_{v_i} for node v_i so that in the successor $\mathcal{C}'$ of $\mathcal{C}$, the state of v_i is different from s_i, $1 \leq i \leq n$. Such a choice must also satisfy the constraints imposed on t_{v_i} by the configurations of Q_1. Given an instance of ITSUC, we say that a node v of $\mathcal{S}$ is **compatible** with a configuration $\mathcal{C} \in Q_2$ if a value for t_v can be chosen so that (i) t_v satisfies all the constraints imposed by the collection Q_1 and (ii) this choice makes $\mathcal{C}$ an unstable configuration (regardless of the threshold values assigned to the other nodes). Extending this definition, we say that a node v is **compatible with a subset** R of Q_2 if v is compatible with every configuration in R. These definitions are used in our algorithm (Alg-ITSUC) shown in Fig. 2. We now establish the correctness of the algorithm and its running time.

Input: Graph $G(V, E)$ of a SyDS $\mathcal{S}$, and two sets of configurations Q_1 and Q_2 of $\mathcal{S}$.

Requirement: Output "Yes" if there is a threshold value t_v for each $v \in V$ such that in the resulting SyDS, all the configurations in Q_1 are stable and all the configurations in Q_2 are unstable. Otherwise, output "No".

Steps:
1. **for** each partition P in $\pi(Q_2)$ **do**
    (a) Let k denote the number of blocks in P and let B_1, B_2, $\ldots$ B_k denote the blocks themselves.
    (b) Construct the bipartite graph $H_P(V, V_P, E_P)$ where V_P has one node for each block in P and $E_P = \{\{x, y\} : x \in V, y \in V_P$ and node x of $\mathcal{S}$ is compatible with the block of P represented by node $y \}$.
    (c) **if** H_P has a matching with k edges **then output** "Yes" and **stop**.
  endfor
2. **Output** "No".

Fig. 2. Algorithm Alg-ITSUC to show the fixed parameter tractability of ITSUC

Lemma 3. *Algorithm Alg-ITSUC given in Fig. 2 correctly decides whether the ITSUC instance has a solution.*

Proof: We first consider the case where the algorithm returns the answer "Yes" and show that there is a solution to the ITSUC instance. From the description of the algorithm, we note that in this case, there is a partition P of Q_2 with k blocks such that there is a matching with k edges in the corresponding bipartite graph H_P. In such a matching, each node y that corresponds to block P_y of P is matched to some node x of the SyDS $\mathcal{S}$. By our construction, node x is compatible with block P_y. By the definition of compatibility, there is a threshold value t_x for node x such that t_x satisfies all the constraints imposed by the configurations in Q_1, and further, this value of t_x for x makes all the configurations in P_y unstable. Since every block of P is matched to a distinct node of $\mathcal{S}$, it follows that threshold values can be chosen for each node of $\mathcal{S}$ independently to satisfy the required conditions. In other words, there is a solution to the ITSUC instance.

We now consider the case when the algorithm returns the answer "No" and show that there is no solution to the ITSUC instance. The proof is by contradiction. Suppose there is a solution to the ITSUC instance which assigns the threshold value t_{v_i} for each node v_i of $\mathcal{S}$. Consider the following bipartite graph $W(V_1, V_2, E_W)$.

(1) V_1 and V_2 are in one-to-one correspondence with V and Q_2 respectively. Let $x_i \in V_1$ be the node corresponding to $v_i \in V$ and let $y_j \in V_2$ be the node corresponding to $\mathcal{C}_j \in Q_2$.

(2) Let $E_W = \{\{x_i, y_j\}$: the threshold value t_{v_i} assigned by the given solution to node v_i of $\mathcal{S}$ allows v_i to be compatible with $\mathcal{C}_j\}$.

Since the given threshold assignment is a solution to ITSUC, each node $y_j \in V_2$ must have at least one edge to some node in V_1. For each node $y_j \in V_2$, choose one such edge $\{x_i, y_j\}$ of E_W arbitrarily; let us call x_i the "dominator" of y_j. By this method, we assign a (possibly empty) subset, say D_i, of V_2 to each node x_i of V_1. (The subset D_i contains all the nodes that are dominated by x_i.) Since each node of V_2 was assigned only one dominator, the collection of subsets D_1, D_2, ..., D_n is pairwise disjoint. Thus, the non-empty subsets in this collection create a partition of V_2 or equivalently a partition of Q_2. Let P denote this partition of Q_2. Further, for each dominator x_i, the threshold value t_{v_i} of the corresponding node $v_i \in V$ ensures that v_i is compatible with all the configurations of Q_2 in the block assigned to v_i. That is, each block of P has a dominator in V and no two blocks have the same dominator. Since Alg-ITSUC considers all the partitions of Q_2, when it considers P, the bipartite graph H_P constructed from P has a matching with k edges, where k is the number of blocks of P. Thus, the algorithm will output the answer "Yes", contradicting our initial assumption. This completes the proof of the lemma.   ■

Lemma 4. *Algorithm Alg-ITSUC can be implemented to run in time $O(h(q) N^{O(1)})$, where q is the number of unstable configurations, N is the size of the problem instance and the function h depends only on q.*

Proof: Let Q_1 (configurations to be made stable) contain r configurations. We first discuss some preprocessing steps. For any node v and any configuration $\mathcal{C}$ in Q_1, the constraint on the threshold t_v of v imposed by $\mathcal{C}$ can be found in $O(n)$ time (since each configuration has n state values). Thus, the constraints on t_v imposed by all r configurations in Q_1 can be found on $O(nr)$ time. As indicated in the proof of Theorem 1, all of these constraints can be combined into a single constraint of the form $a_v \leq t_v \leq b_v$ for appropriate integers a_v and b_v in $O(nr)$ time. Thus, obtaining such a single constraint for each of the n nodes can be done in $O(n^2 r)$ time. These preprocessing steps can be done before starting the execution of the **for** loop in Step 1 of the algorithm.

We now estimate the time needed to check whether a node v is compatible with a configuration $\mathcal{C}$ in Q_2. For a node v, the constraint on t_v needed to make $\mathcal{C}$

an unstable configuration (regardless of the threshold values of the other nodes) can be determined in $O(n)$ time. Checking whether this constraint also satisfies the constraint on t_v obtained by considering the configurations in Q_1 can be done in $O(1)$ time. Thus, checking whether a node v is compatible with a configuration in Q_2 can be done in $O(n)$ time. The results from these compatibility checks can be stored in an $n \times q$ Boolean matrix M so that during the execution of the **for** loop of the algorithm, we can determine whether a node v is compatible with a configuration $\mathcal{C}$ of Q_2 in $O(1)$ time.

We now estimate the time used for each iteration of the **for** loop of Fig. 2. For each partition P in $\pi(Q_2)$, the algorithm constructs an appropriate bipartite graph H_P and checks whether the graph has a matching whose size is equal to the number of blocks of P. If P has k blocks, the number of nodes and edges in H_P is $n + k$ and the number of edges is at most nk. For each node $x \in V$, the blocks with which x is compatible can be found in $O(q)$ time from the precomputed matrix M. Thus, constructing the graph H_P can be done in time $O(nkq) = O(nq^2)$ since $k \leq q$. Since H_P has $n + k \leq n + q$ nodes and at most $nk \leq nq$ edges, as mentioned in Sect. 2.2, a maximum matching in H_P can be found in $O(nq\sqrt{n+q})$ time. Thus, each iteration of the **for** loop can be implemented to run in time $O(nq^2 + nq\sqrt{n+q})$ time.

Since the number of iterations of the **for** loop is at most $|\pi(Q_2)|$, the running time of Step 1 is $O(|\pi(Q_2)|nq^2 + nq\sqrt{n+q})$. The overall running time of the algorithm, including the preprocessing steps, is $O(n^2(r+q) + |\pi(Q_2)|nq^2 + nq\sqrt{n+q})$. As mentioned earlier, $|\pi(Q_2)| = O((q/\log q)^q)$. Since $n + r + q \leq N$, where N is the size of the given ITSUC instance, it follows that the running time of our algorithm for ITSUC has the form $O(h(q)N^{O(1)})$, where $h(q) = (q/\log q)^q$ depends only on q. ∎

The following theorem is a direct consequence of Lemmas 3 and 4.

Theorem 5. *The ITSUC problem is fixed parameter tractable where the parameter is the number of unstable configurations.* ∎

5   Future Research Directions

We conclude by mentioning two general directions for future work. One direction is to consider inference problems for other forms of observed behavior such as a collection of snapshots of the system, where each snapshot specifies a time and the configuration of the system at that time. Another direction is to consider inference problems assuming more powerful local functions.

Acknowledgments. We thank the reviewers for carefully reading the manuscript and providing valuable suggestions. This work has been partially supported by DTRA Grant HDTRA1-11-1-0016 and DTRA CNIMS Contract HDTRA1-11-D-0016-0010, NSF NetSE Grant CNS-1011769, NSF SDCI Grant OCI-1032677 and NIH MIDAS Grant 5U01GM070694-11.

References

1. Abrahao, B., Chierichetti, F., Kleinberg, R., Panconesi, A.: Trace complexity of network inference. In: Proceedings of the 19th ACM SIGKDD, pp. 491–499. ACM (2013)
2. Adiga, A., Kuhlman, C.J., Marathe, M.V., Ravi, S.S., Rosenkrantz, D.J., Stearns, R.E.: Complexity of inferring local transition functions of discrete dynamical systems. Technical report NDSSL-TR-15-048, NDSSL, Virginia Bioinformatics Institute, Virginia Tech, Blacksburg, VA, June 2015
3. Barrett, C.L., Hunt III, H.B., Marathe, M.V., Ravi, S.S., Rosenkrantz, D.J., Stearns, R.E.: Complexity of reachability problems for finite discrete dynamical systems. J. Comput. Syst. Sci. **72**(8), 1317–1345 (2006)
4. Barrett, C.L., Hunt III, H.B., Marathe, M.V., Ravi, S.S., Rosenkrantz, D.J., Stearns, R.E.: Modeling and analyzing social network dynamics using stochastic discrete graphical dynamical systems. Theo. Comput. Sci. **412**(30), 3932–3946 (2011)
5. Barrett, C.L., Hunt III, H.B., Marathe, M.V., Ravi, S.S., Rosenkrantz, D.J., Stearns, R.E., Thakur, M.: Predecessor existence problems for finite discrete dynamical systems. Theo. Comput. Sci. **386**(1–2), 3–37 (2007)
6. Barrett, C.L., Hunt III, H.B., Marathe, M.V., Ravi, S.S., Rosenkrantz, D.J., Stearns, R.E., Tosic, P.T.: Gardens of Eden and fixed points in sequential dynamical systems. In: DM-CCG, pp. 95–110 (2001)
7. Cormen, T., Leiserson, C., Rivest, R., Stein, C.: Introduction to Algorithms. MIT Press and McGraw-Hill, Cambridge (2009)
8. Durand, B.: A random NP-complete problem for inversion of 2D cellular automata. Theor. Comput. Sci. **148**(1), 19–32 (1995)
9. Easley, D., Kleinberg, J.: Networks, crowds and markets: reasoning about a highly connected world. Cambridge University Press, New York (2010)
10. Garey, M.R., Johnson, D.S.: Computers and Intractability: A Guide to the Theory of NP-completeness. W. H. Freeman and Co., San Francisco (1979)
11. Goles, E., Martínez, S.: Neural and automata networks. Kluwer, Dordrecht (1990)
12. Gomez Rodriguez, M., Leskovec, J., Krause, A.: Inferring networks of diffusion and influence. In: Proceedings of the 16th ACM SIGKDD, pp. 1019–1028. ACM (2010)
13. Gonzalez-Bailon, S., Borge-Holthoefer, J., Rivero, A., Moreno, Y.: The Dynamics of Protest Recruitment Through an Online Network. In: Nature Scientific Reports, pp. 1–7 (2011), doi:10.1038/srep00197
14. Graham, R., Knuth, D., Patashnik, O.: Concrete Mathematics. Addison-Wesley, Reading (1994)
15. Granovetter, M.: Threshold models of collective behavior. Am. J. Sociol. **83**(6), 1420–1443 (1978)
16. Green, F.: NP-complete problems in cellular automata. Complex Syst. **1**(3), 453–474 (1987)
17. Håstad, J.: Clique is hard to approximate within $n^{1-\epsilon}$. Acta Mathematica **182**, 105–142 (1999)
18. Kearns, M.J., Vazirani, U.V.: An Introduction to Computational Learning Theory. MIT Press, Cambridge (1994)
19. Kleinberg, J.: Cascading behavior in networks: Algorithmic and economic issues. In: Nissan, N., Roughgarden, T., Tardos, E., Vazirani, V. (eds.) Algorithmic Graph Theory. ch. 24. Cambridge University Press, New York (2008)

20. Kosub, S., Homan, C.M.: Dichotomy results for fixed point counting in Boolean dynamical systems. In: Proceedings of the ICTCS, pp. 163–174 (2007)
21. Macy, M., Willer, R.: From factors to actors: Computational sociology and agent-based modeling. Ann. Rev. Sociol. **28**, 143–166 (2002)
22. Mortveit, H.S., Reidys, C.M.: An Introduction to Sequential Dynamical Systems. Springer, Heidelberg (2007)
23. Murphy, K.P.: Passively learning finite automata. Technical report, Computer Science Department, UC Davis (1995)
24. Neidermeier, R.: Invitation to Fixed Parameter Algorithms. Oxford University Press, New York (2006)
25. Romero, D.M., Meeder, B., Kleinberg, J.: Differences in the mechanics of information diffusion across topics: idioms, political hashtags, and complex contagion on twitter. In: Proceedings of the 20th WWW, pp. 695–704. ACM (2011)
26. Shah, D., Zaman, T.: Rumors in a network: Who's the culprit? IEEE Trans. Inf. Theory **57**(8), 5163–5181 (2011)
27. Soundarajan, S., Hopcroft, J.E.: Recovering social networks from contagion information. In: Kratochvíl, J., Li, A., Fiala, J., Kolman, P. (eds.) TAMC 2010. LNCS, vol. 6108, pp. 419–430. Springer, Heidelberg (2010)
28. Sutner, K.: Computational classification of cellular automata. Int. J. Gen. Syst. **41**(6), 595–607 (2012)
29. Trucano, T.G., Swiler, L.P., Igusa, T., Oberkampf, W.L., Pilch, M.: Calibration, validation and sensitivity analysis: What is what. Reliab. Eng. Syst. Saf. **91**, 1331–1357 (2006)
30. Ugander, J., Backstrom, L., Marlow, C., Kleinberg, J.: Structural diversity in social contagion. PNAS **109**(9), 5962–5966 (2012)

From Ambiguous Regular Expressions to Deterministic Parsing Automata

Angelo Borsotti[1], Luca Breveglieri[1], Stefano Crespi Reghizzi[2](✉),
and Angelo Morzenti[1]

[1] Dip. di Elettronica, Informazione e Bioingegneria (DEIB),
Politecnico di Milano, Piazza Leonardo Da Vinci N. 32,
20133 Milano, Italy
angelo.borsotti@mail.polimi.it, {luca.breveglieri,
stefano.crespireghizzi,angelo.morzenti}@polimi.it
[2] Dip. di Elettronica, Informazione e Bioingegneria (DEIB),
CNR-IEIIT, Politecnico di Milano, Piazza Leonardo Da Vinci N. 32,
20133 Milano, Italy

Abstract. This new parser generator for ambiguous regular expressions
(RE) formally extends the Berry-Sethi (BS) algorithm into a finite-state
device that specifies the syntax tree(s). We extend the local testability
property of the marked RE's from terminal strings to linearized syntax
trees. The generator supports disambiguation, i.e., selecting a preferred
tree in case of ambiguity. The selection is parametric with respect to
the *Greedy* or *POSIX* criterion. The parser is proved correct and has
linear-time complexity. The generator is available as an interactive *SW*
tool (on *GitHub* - see http://github.com/breveglieri/ebs/README).

Keywords: Regular expression · RE · Syntax tree · Berry-Sethi · Ambiguity · Parsing

1 Introduction

The popularity of regular expressions (RE) as a formalism for specifying a text
pattern comes from the efficient algorithms available for string recognition; see
e.g., [1] for a partial survey. Most such methods transform a RE into a finite
automaton, deterministic (DFA) or not (NFA), which acts as language recognizer, i.e., a yes/no algorithm. This is not enough for modern applications, which
additionally require to build the syntax tree(s) of the input and, when the RE
is ambiguous, to select one tree in accordance with a predefined criterion, most
often either the *Greedy* or the *POSIX* one.

In the course of time, the parsers for RE's have progressed from the original
naive and inefficient idea of exploiting a nondeterministic push-down parser for
the language generated by an ambiguous RE. Progress has brought more compact representations of parse trees, linear-time complexity, and clever methods

---

Work partially supported by PRIN "Automi e Linguaggi Formali", Italy.

F. Drewes (Ed.): CIAA 2015, LNCS 9223, pp. 35–48, 2015.
DOI: 10.1007/978-3-319-22360-5_4

for directly building the tree of interest while skipping all the others. Space prevents us from fully discussing former valuable proposals, and we briefly explain how our research goes beyond them.

Our distinguishing feature is that we formally extend the well known Berry-Sethi (*BS*) Algorithm [2] for constructing a *DFA* recognizer of an *RE*. With a minimal overhead and the same conceptual approach based on local languages [3], we transform the *BS* recognizer into a device that acts as a parser and builds the linearized syntax tree. Other proposals have been based, as well, on classic construction methods: a Thompson-based *NFA* recognizer [6] or the Brzozowski derivative method [11]. But their use of a recognizer-construction method is limited: just as a conceptual reference, or as a sort of subsystem on top of which they provide the methods to build the tree. A notable exception is in [9], which has inspired our work yet differs from it by using an *NFA* instead of a *DFA*. We also refer to [5] for a similar approach for extended *BNF* grammars.

A so-called "problematic" *RE* yields infinitely many syntax trees for a string, when it contains a Kleene star with a nullable argument. Our solution (similar to former ones) uniformly deals with any *RE*, and the resulting parser just skips the irrelevant ambiguous trees, by means of a simple test that dismisses repeated empty strings. Parametricity with respect to the ambiguity resolution criterion is a feature unavailable in the existing proposals, which are either *Greedy*-oriented [7] or *POSIX*-oriented [9,11]. Each disambiguation criterion assigns a priority to the partial tree it builds, as first shown for the *Greedy* case in [7]. Our algorithm and tool express the *Greedy* and *POSIX* criteria for selecting a tree node.

About efficiency, our parser works in linear time as all the recent ones do, yet it is faster and, on unambiguous strings, it has a reduced overhead. An interactive light-weight parser generator *SW* tool is available to demonstrate our method.

Paper Organization. Section 2 lists the basic definitions, shows a running example (Fig. 1) and introduces the marked *RE*'s, wherein terminals and meta-symbols are made distinct to obtain a local language. Section 3 presents the parser generator algorithm and hints at the correctness proof of the recognizer. Section 4 explains how to build the syntax tree and to deal with infinite ambiguity, and it also includes complexity analysis. Section 5 formalizes parsing ambiguity, explains how to detect it, and outlines the technique used for selecting one out of the ambiguous trees, by the *Greedy* or *POSIX* criterion. The parser generator *SW* tool is briefly described. Section 6 summarizes the main results.

2   Basic Definitions

Since the basic concepts of the theory of regular expressions (*RE*) are well-known and easily available, e.g., in [10], it suffices to list the terms and notation we use here. The powerset of a finite set S is denoted by $\wp(S)$. The *terminal alphabet* Σ contains the *terminal symbols*, denoted by a, b, ..., and the *end-of-text* by $\dashv$. The *empty* string is denoted by ε. The length of a string x is denoted by $|x| \geq 0$ and the number of occurrences of a symbol a in x is denoted by $|x|_a \geq 0$.

The *metasymbol alphabet* is $M = \{\, 0, 1, \text{`}|\text{'}, \text{`}\cdot\text{'}, \text{`}*\text{'}, \text{`}+\text{'}, \text{`}(\text{'}, \text{`})\text{'} \,\}$ and its elements are generically denoted by m. Symbols 0 and 1 denote the empty set $\emptyset$ and the string ε, respectively. Symbols "$|$", "$\cdot$", "$*$" and "$+$" denote union, concatenation (optional), and the Kleene star and cross, respectively. Assume that it holds $\Sigma \cap M = \emptyset$ and define a new alphabet $\Omega = \Sigma \cup M$. The elements of the alphabet Ω are generically denoted by s.

We will consider RE's, denoted as usual, over the alphabet Ω. Union and concatenation are associative and may form chains of any length. The operator priority is this: union (lowest), concatenation, and Kleene star / cross (highest). The argument of an iterative operator (star / cross) is always parenthesized, e.g., $(a)^*$. A RE e *generates* a language $L(e) \subseteq \Sigma^*$, and both are *nullable* if $\varepsilon \in L(e)$.

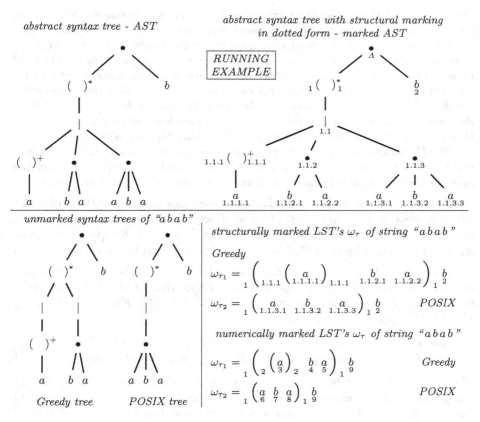

abstract syntax tree - AST

abstract syntax tree with structural marking in dotted form - marked AST

RUNNING EXAMPLE

unmarked syntax trees of "abab"

structurally marked LST's ω_τ of string "abab"

Greedy

$$\omega_{\tau_1} = {}_1\left({}_{1.1}\left({}_{1.1.1}a \right)_{1.1} {}_{1.1.2.1}b\ {}_{1.1.2.2}a \right)_1 {}_2 b$$

$$\omega_{\tau_2} = {}_1\left({}_{1.1.3.1}a\ {}_{1.1.3.2}b\ {}_{1.1.3.3}a \right)_1 {}_2 b \qquad POSIX$$

numerically marked LST's ω_τ of string "abab"

$$\omega_{\tau_1} = {}_1\left({}_2\left({}_3 a \right)_2 {}_4 b\ {}_5 a \right)_1 {}_9 b \qquad Greedy$$

$$\omega_{\tau_2} = {}_1\left({}_6 a\ {}_7 b\ {}_8 a \right)_1 {}_9 b \qquad POSIX$$

Greedy tree   *POSIX tree*

Fig. 1. Abstract syntax trees of the RE $e = \left((a)^+ \mid ba \mid aba \right)^* b$ (top); syntax trees and linearized trees of the ambiguous string $abab \in L(e)$ (bottom).

The structure of a RE e can be represented as an *abstract syntax tree* called *AST*, or as a *marked AST*, as shown in Fig. 1 (top left and right, respectively).

In the structurally marked form, each node subscript represents the path from the tree root (subscript Λ) to the node itself, in a dotted notation (the prefix "Λ" is always omitted). A *subexpression* (s.e.) is a well-formed substring of e that corresponds to a subtree in the whole AST of e.

Every string in the language $L(e)$ has (at least) one syntax tree, representable as an AST; if the string has two or more syntax trees, it is *ambiguous*, as well as the RE e itself. By a pre-order tree visit, we obtain a *linearized syntax tree* or LST. Two syntax trees for string $a\,b\,a\,b \in L(e)$ and the corresponding LST's are shown in Fig. 1 (bottom); more explanations come later in Sect. 5.

We have chosen two tree selection criteria often adopted: *Greedy* and *POSIX* [8]. Actually some versions of *Greedy* exist, e.g., the ones of [7] (we use this) and of the *Java* class "regex", which behave differently, whereas *POSIX* is a "de facto" standard. The next definition and example outline the two criteria.

Definition 1 (Tree selection criterion). *Given an ambiguous RE e, if in matching a string of $L(e)$, two or more s.e.'s are able to match in a string position, or one s.e. but in different ways, then the Greedy [7] and POSIX [8] selection criteria specify how to choose. Features common to both criteria: (i) among alternative s.e.'s, the leftmost s.e. has priority, unless overruled by (iv); (ii) if two iterative s.e.'s (Kleene star or cross) are nested, then maximizing the number of iterations of the inner s.e. has priority over the outer s.e.; (iii) if an iterative s.e. has a nullable argument unable to match a non-empty substring (in that string position), then it iterates once matching the empty string.*

POSIX additionally prescribes: (iv) for one s.e. the longest match has priority, and among alternative s.e.'s, a s.e. able to match a longer substring has priority over a s.e. able to match a shorter one, overruling (i) if necessary. □

Example 1. Figure 1 (bottom left) shows two trees for $a\,b\,a\,b$. *Greedy* [7] prefers the left tree, as s.e. $(a)^+$ is on the left of s.e. $a\,b\,a$, see (i). *POSIX* [8] prefers the right tree, as s.e. $a\,b\,a$ has a match longer than s.e. $(a)^+$, see (iv). RE $((a)^*)^*$ matches string $a\,a$ as $((a\,a))$ by both [7,8], see (ii), but as $((a)(a))$ by java.regex. □

For a language $L \subseteq \Sigma^+$, recall the classic definitions of the *initial, final* and *digram* (substring of length two) *sets*, as well as that of the *follow set*, of a terminal $a \in \Sigma$, which are respectively denoted by $Ini(L)$, $Fin(L)$, $Fol(L, a) \subseteq \Sigma$, and by $Dig(L) \subseteq \Sigma^2$, as follows:

$$\begin{aligned} Ini(L) &= \{\, a \mid a\,x \in L \,\} & Fin(L) &= \{\, b \mid x\,b \in L \,\} \\ Dig(L) &= \{\, a\,b \mid x\,a\,b\,y \in L \,\} & Fol(L, a) &= \{\, b \mid a\,b \in Dig(L) \,\} \end{aligned} \tag{1}$$

where $x, y \in \Sigma^*$. If it holds $L = L(e)$, we denote these sets by $Ini(e)$, etc. In Eq. (2) we recall the classic Definition [3] of *local* (or 2-*strictly locally testable*) language $L \subseteq \Sigma^+$. For some fixed sets Ini, Fin.and Dig, the strings of L start and end by a letter in Ini and Fin, respectively, and contain only digrams in Dig:

$$L = Ini \cdot \Sigma^* \,\cap\, \Sigma^* \cdot Fin - \left(\, \Sigma^* \cdot \left(\, \Sigma^2 - Dig \,\right) \cdot \Sigma^* \,\right) \tag{2}$$

It is well known [2,3] how to compute the sets Eq. (1) for a RE or subexpression, and how to construct a DFA for the local language defined by Eq. (2).

Definition 2 (Marked *RE*). *Let e be a RE over the terminal alphabet Σ. Let $\hat{e}$ be the string obtained from e by replacing each (terminal or meta) symbol $s \in \Omega$ that occurs in e at position h, with the marked symbol s_h. String $\hat{e}$ is called "fully marked RE". If only the symbols $a \in \Sigma$ are marked, the resulting RE is called "terminally marked" and is denoted by $\bar{e}$. A RE is said to be "marked" if it is fully or terminally marked. The set $\widehat{\Omega} = \widehat{\Sigma} \cup \widehat{M}$ of all the marked symbols in $\hat{e}$ is called "extended marked alphabet": it is the union of the marked terminals $\widehat{\Sigma}$ and marked metasymbols $\widehat{M}$. The marked alphabets depend on the RE e, but the dependence is left understood. The unmarking function unmark: $\widehat{\Omega} \to \Omega$ is defined as $s_h \xmapsto{unmark} s$ for any s, sssand extended to strings in the natural way.* $\square$

Marking yields a new *RE* $\hat{e}$ (or $\bar{e}$) from e, with no repeated marked symbols. Provided it is so, any rule that assigns marks to symbols is acceptable: the root-to-node structural rule or the left-to-right numerical one are just two options. To clarify Definition 2, we take the unmarked *RE* e Eq. (3) and we show how to *structurally* Eq. (4), *numerically* Eq. (5) and *terminally* Eq. (6) mark it (see also Fig. 1). Notice that both forms Eqs. (4) and (5) are fully marked:

$$e = \Big((a)^{+} \mid b\,a \mid a\,b\,a \Big)^{*} b \tag{3}$$

$$\hat{e} = {}_{1}\Big({}_{1.1.1}\big({}_{1.1.1.1}a \big)^{+}_{1.1.1} \mid_{1.1} {}_{1.1.2.1}b \cdot_{1.1.2} {}_{1.1.2.2}a \mid_{1.1} {}_{1.1.3.1}a \cdot_{1.1.3} {}_{1.1.3.2}b \cdot_{1.1.3} {}_{1.1.3.3}a \Big)^{*}_{1} \cdot_{\Lambda} {}_{2}b \tag{4}$$

$$\hat{e} = {}_{1}\Big({}_{2}(a_3)^{+}_{4} \mid {}_{5}b_6 \cdot_{7} a_8 \mid {}_{9}a_{10} \cdot_{11} b_{12} \cdot_{13} a_{14} \Big)^{*}_{15} b_{16} \tag{5}$$

$$\bar{e} = \Big((a_1)^{+} \mid b_2\,a_3 \mid a_4\,b_5\,a_6 \Big)^{*} b_7 \tag{6}$$

Remarks. The associative operators in the form Eq. (4) have an identical root-to-node subscript (mark) as they map to one tree node, the union and concatenation metasymbols are uniquely pinpointed by their adjacent marked symbols, and parentheses come in matching pairs. This permits to somewhat simplify form Eq. (4) and similarly Eq. (5): drop all the union and concatenation marks, and unify the two marks of matching parentheses. Thus we obtain a partially marked *RE*, which we may continue to call *fully marked*, e.g., from Eq. (5):

$$\hat{e} = {}_{1}\Big({}_{2}(a_3)^{+}_{2} \mid b_4\,a_5 \mid a_6\,b_7\,a_8 \Big)^{*}_{1} b_9 \tag{7}$$

In the next examples, for legibility we use such a simplified numerical marking. Also, as the end-of-text $\dashv$ occurs only once in a *RE*, it can be left unmarked.

Next we extend the sets *Ini*, *Fin*, *Dig* and *Fol* Eq. (1) to a fully marked *RE*. The key idea is to consider the metasymbols as terminal symbols. Thus a fully marked *RE* $\hat{e}$ is viewed as a well-formed string over the alphabet $\widehat{\Omega}$ (Definition 2), e.g., Eq. (7). Yet, $\hat{e}$ is not a valid *RE*, as the marked parentheses do not act as metasymbols any longer. So we put them back into the *RE*, but under another representation. The *reintroduction rules* Eq. (8) show how to rewrite a parenthesis pair in $\hat{e} \in \widehat{\Omega}^{+}$, as well as the empty set:

parenthesis pair	reintroduction rule - yields reparenthesized RE ĕ (8)
$_h(\hat{e})_h$ with no $*$ or $+$	$_h(\ [\hat{e}]\ )_h$ $\boxed{\text{and } 0_h \text{ is rewritten as } 0_h\,\emptyset}$
$_h(\hat{e})^*_h$ and $_h(\hat{e})^+_h$	$\left[\ _h\left(\ [\hat{e}]^+\ \right)_h\ \mid\ 1_h\ \right]$ and $_h\left(\ [\hat{e}]^+\ \right)_h$

In this *reparenthesized RE*, denoted by ĕ, it is the square brackets '[', ']' and the empty set $\emptyset$ that truly act as metasymbols. Notice the new marked meta-symbol 1_h added to $\widehat{M}$, to represent the empty string included in the Kleene star (the cross does not need it). Thus, the reintroduction rules Eq. (8) enlarge the alphabets $\widehat{M}$ and $\widehat{\Omega}$ (which remain finite). From now on, we assume that the alphabets $\widehat{M}$ and $\widehat{\Omega}$ are always completed in this way.

Example 2. Applying the reintroduction rules Eq. (8), *RE* $\hat{e}$ Eq. (7) becomes (Fig. 1):

$$\breve{e} = \left[\ _1\left(\ \left[\ _2(\ [a_3]^+\ )_2\ \mid\ b_4\,a_5\ \mid\ a_6\,b_7\,a_8\ \right]^+\ \right)_1\ \mid\ 1_1\ \right]\ b_9 \qquad (9)$$

The reparentesized *RE* ĕ has these sets of initials and digrams (not all shown):

$$Ini\,(\breve{e}) = \{_1(,\ 1_1\}\quad Dig\,(\breve{e}) = \{_1(_2(,\ _1(b_4,\ _1(a_6,\ 1_1b_9,\ _2(a_3,\ a_3a_3,\ a_3)_2,\ \dots\}\quad \square$$

Clearly the language $L\,(\breve{e})$ generated by the reparenthesized *RE* ĕ over the completed alphabet $\widehat{\Omega}$ is *local* Eqs. (1-2) and is defined by the sets *Ini, Fin, Dig* of ĕ.

3   Parser Construction

The classical Berry-Sethi algorithm builds a *DFA* of language $L\,(e)$. We recall this algorithm (described, e.g., in [2,3]), then we suitably extend it into a parser.

Classical Berry-Sethi Recognizer (BS). Given a *RE* e, the *BS* algorithm works as follows: it takes the terminally marked *RE* $\bar{e}$, it concatenates the end-of-text $\dashv$ to $\bar{e}$, and from $\bar{e}\dashv$ it builds a *DFA* for the language $L\,(e)$. Each *DFA* state q is identified by a set $TI(q) \subseteq \widehat{\Sigma}$ of marked letters a_h called *terminal items*. The initial state q_0 has the item set $TI(q_0) = Ini(\bar{e}\dashv)$, which is empty if, and only if (iff), it holds $L\,(e) = \emptyset$. No other state $q \neq q_0$ has an empty item set $TI(q)$. A state q is final iff its item set $TI(q)$ contains the end-of-text $\dashv$.

For an unmarked letter a (different from $\dashv$) and a state q, the subset $TI_a\,(q) \subseteq TI(q)$ contains the terminal items a_h such that it holds $a = unmark\,(a_h)$, which from now on are called "items of class a". For some pairs $(a,\,q)$ the set $TI_a\,(q)$ of items of class a may be empty. Else the *DFA* has an arc labeled a from q to the state q' identified by:

$$TI\,(q') = \{\ b_k\ \mid\ \exists\,a_h \in TI_a\,(q) \text{ such that } b_k\ \in\ Fol\,(\bar{e}\dashv,\ a_h)\ \} \qquad (10)$$

The state q' is identified by the follow sets of the terminal items of class a. The proof that this *BS DFA* recognizes the language $L\,(e)$ is in [2], simplified in [3].

BS Recognizer with Parser (BSP). Now we construct a DFA, called BSP, that not only recognizes the language $L(e)$, but also builds the parse tree(s). We enrich the previous BS DFA by adding to it syntactic information represented by marked metasymbols and by certain pointers. To this purpose, we extend the definitions of initial and follow sets to specify the metasymbols that precede a terminal symbol in a fully marked RE.

Definition 3 (Finished initial and follow sets). *A marked non-empty string $\zeta \in \widehat{\Omega}^+$ is called <u>finished</u> if it is the concatenation of a possibly empty prefix μ of metasymbols that does not contain either an empty set 0_l or two identically marked symbols 1_l, and of a marked terminal a_h. The formal definition of ζ is:*

$$\zeta = \mu\, a_h \quad with \quad \mu, a_h \in \widehat{M}^*, \widehat{\Sigma} \quad and \quad \forall 0_l, 1_l \in \widehat{M} \quad |\mu|_{0_l} = 0 \quad and \quad |\mu|_{1_l} \leq 1 \tag{11}$$

The marked terminal a_h is called the "end-letter" of ζ (a_h may be $\dashv$). The <u>finished initial set</u> of a RE e, denoted by Ini_{fin}, is the finite set of finished strings:

$$Ini_{fin}(e) = \{\, \mu\, a_h \mid \ Ini(\mu\, a_h) \subseteq Ini(\breve{e}) \ \wedge \ Dig(\mu\, a_h) \subseteq Dig(\breve{e}) \,\} \tag{12}$$

The <u>finished follow set</u> of a marked terminal symbol $a_h \in \widehat{\Sigma}$ (a_h may not be $\dashv$) in a RE e, denoted by Fol_{fin}, is the finite set of finished strings:

$$Fol_{fin}(e, a_h) = \{\, \mu\, b_k \mid \ Dig(a_h\, \mu\, b_k) \subseteq Dig(\breve{e}) \,\} \tag{13}$$

Both sets use the RE $\breve{e}$ Eq. (8). The set of all finished strings ζ is denoted by $\widehat{Z}$.
□

The prefix μ of a finished string is a possibly empty metasymbol sequence that encodes an acyclic path between the end-letters a_h and b_k in $\breve{e}$. The constraints stated in (11) prevent infinite string ambiguity: in particular, the null string ε defined by a Kleene star (or cross) with a nullable argument, such as for instance $(1)^*$, may not iterate two or more times; see Sect. 4 and Examples 5 and 7 for more discussion and details. For brevity we skip the computation of the number of finished strings for a given RE.

The new BSP we want to construct for a fully marked RE is a DFA $\mathcal{A}$ similar to the classical BS DFA $\mathcal{A}_{BS}$ for a terminally marked RE. Every state of $\mathcal{A}$ is identified by a set of items that enrich the terminal items with some syntactic information. Moreover, within a state each item is identified by an *item identifier* (e.g., an integer) to be later used for building the parse tree(s). The set of all the item identifiers is denoted by ID.

Definition 4 (Item). *An <u>item</u> is a pair $\iota = \langle \zeta, \Pi \rangle$, where $\zeta \in \widehat{Z}$ is a finished string and $\Pi \subseteq ID$ is a (possibly empty) set of item identifiers. An item is <u>final</u> if the end-letter of ζ is the end-of-text $\dashv$. An item set I is a binary relation included in the set product $\widehat{Z} \times \wp(ID)$.* □

Let X and $\mathcal{Y}$ be finite sets, and pose $\mathcal{W} = X \times \wp(\mathcal{Y})$. We define an operation Eq. (14) called *group-by$_X$* that collects all the pairs $(x, Y) \in \mathcal{W}$ that have the same left element $x \in X$. For a set $\mathcal{R} \subseteq \mathcal{W}$ of such pairs to be collected, define:

$$\mathcal{R} \xmapsto{\ group\text{-}by_X\ } \Big\{\, (\,x, Y_x\,) \mid Y_x = \{\, y \mid \exists\,(\,x, Y\,) \in \mathcal{R}\quad y \in Y \,\} \,\Big\} \qquad (14)$$

Algorithm 1. (***BS* recognizer with Parser - *BSP***) Construction of the the *DFA* $\mathcal{A}$ that recognizes a string in $L\,(e)$ and builds its parse tree(s).

Input: the sets Ini_{fin} and Fol_{fin} of a *RE* $e \dashv$ over a terminal alphabet Σ	
Output: a recognizer-parser *DFA* $\mathcal{A} = (\,\Sigma, Q, q_0, \delta, F\,)$ of the *RE* e	
$I\,(q_0) := \{\, \langle\, \zeta, \emptyset\,\rangle \mid \zeta \in Ini_{fin}\,(e \dashv)\,\}$ untagged	// initial state q_0 (12)
$Q := \{\, q_0\,\}\quad \delta := \emptyset$	// initialize state set Q and transition set δ
while $\exists$ state $q \in Q$ that is untagged **do**	// scan all states q
$\quad$ set state q tagged	// tag source state q
$\quad$ **foreach** unmarked letter $a \in \Sigma\ (\neq \dashv)$ **do**	// scan all letters $a\ (\neq \dashv)$
$\qquad I\,(q') := \emptyset$ untagged	// temporary target state q'
$\qquad$ **foreach** item $\iota = \langle\, \zeta, \Pi\,\rangle \in I_a\,(q)$ **do**	// scan items of class a
$\qquad\quad$ **foreach** $\begin{pmatrix} \text{finished string } \zeta' \text{ such that} \\ \zeta' \in Fol_{fin}\,(e \dashv, \text{end-letter } \zeta\,) \end{pmatrix}$ **do**	// scan the set of followers of end-letter of str. ζ (13)
$\qquad\qquad \iota' := \langle\, \zeta', \{\text{ item identifier of } \iota \text{ in } q\,\}\,\rangle$	// make new item ι'
$\qquad\qquad I\,(q') := I\,(q') \cup \{\, \iota'\,\}$	// add item ι' to state q'
$\qquad$ **if** $I\,(q') \neq \emptyset$ **then**	// state q' is not empty
$\qquad\quad I\,(q') := group\text{-}by_{\widehat{Z}}\,(I\,(q'))$	// group items of q' (14)
$\qquad\quad$ **if** $q' \notin Q$ **then**	// state q' is a new one
$\qquad\qquad Q := Q \cup \{\, q'\,\}$	// add new target state q'
$\qquad\quad \delta := \delta \cup \{\, q \xrightarrow{a} q'\,\}$	// add new a-transition
$F := \{\, q \in Q \mid \exists \text{ item } \iota \in I\,(q) \text{ that is final}\,\}$	// final state set F

The *BSP* of a *RE* e is a *DFA* $\mathcal{A} = (\,\Sigma, Q, q_0, \delta, F\,)$ constructed by Algorithm 1. The states of $\mathcal{A}$ extend those of the classical *BS*. Each state $q \in Q$ is identified by a set $I\,(q) \subseteq \widehat{Z} \times \wp\,(ID)$ of items grouped through the operation $group\text{-}by_{\widehat{Z}}$. For an unmarked letter a (different from $\dashv$) the subset $I_a\,(q) \subseteq I\,(q)$ contains the items $\iota = \langle\, \zeta, \Pi\,\rangle$ with $\zeta = \mu\, a_h$ such that it holds $a = unmark\,(a_h)$, called "items of class a" (as for *BS*).

The core of Algorithm 1 is the innermost *for loop* that checks the set Fol_{fin} and finds all the finished strings that follow the current one (Definition 3). The destination state q' of the a-arc $q \xrightarrow{a} q'$ is identified by the *finished follow sets* Eq. (13) of the *items* (Definition 4) of class a in the source state q. The new item ι' has syntactic information represented by the (prefix μ' of the) finished string ζ' and has a backward pointer to the source item ι. The two information pieces are unnecessary for string recognition, but specify the parse tree(s).

The states are tagged to avoid reexamining them. If it holds $L\,(e) = \emptyset$, then Algorithm 1 creates only the initial state q_0 with $I\,(q_0) = \emptyset$ and no arcs. Else, it creates the non-empty state q_0 and possibly more states, none empty, with the connecting arcs. Every state is both accessible and co-accessible by construction ($\mathcal{A}$ is trim). No item has a finished string that contains the empty set 0_h (see rules Eq. (8) and Definition 3). The items in q_0 do not have any backward link, whereas all the others are linked back to their source items.

Table 1. Finished initial and follow sets Ini_{fin} and Fol_{fin} of *RE* e Eq. (3), marking Eq. (7).

	Ini_{fin}	${}_1(\,{}_2(\,a_3$	${}_1(\,b_4$	${}_1(\,a_6$	$1_1\,b_9$	
Fol_{fin}	a_3	a_3	$)_2\,{}_2(\,a_3$	$)_2\,b_4$	$)_2\,a_6$	$)_2\,)_1\,b_9$
	b_4	a_5				
	a_5	${}_2(\,a_3$	b_4	a_6	$)_1\,b_9$	
	a_6	b_7				
	b_7	a_8				
	a_8	${}_2(\,a_3$	b_4	a_6	$)_1\,b_9$	
	b_9	$\dashv$				

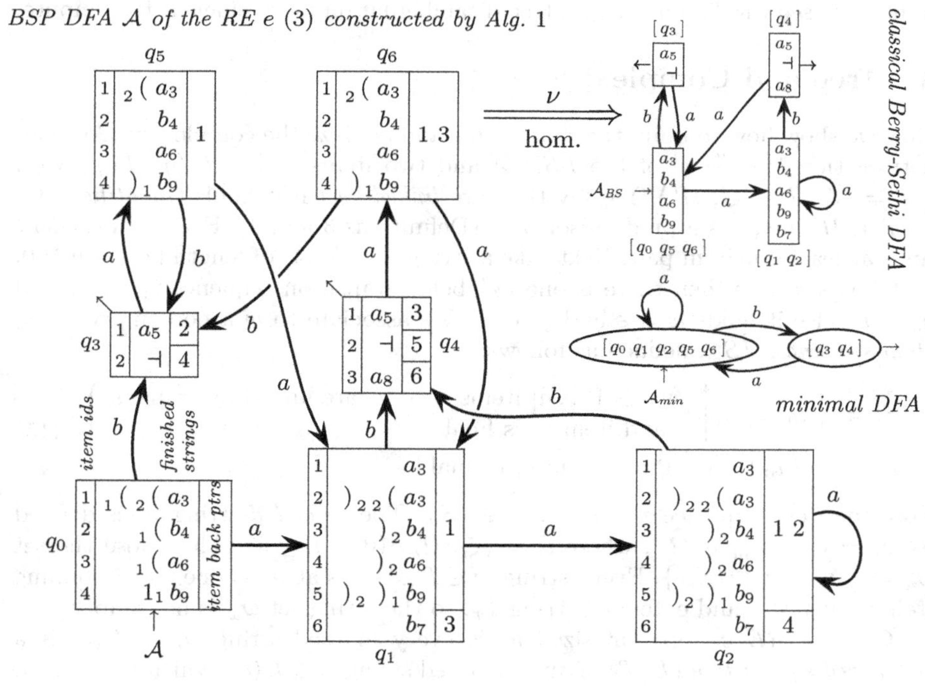

Fig. 2. *BSP DFA* $\mathcal{A}$ of the *RE* e Eq. (7), with the equivalent *BS* and minimal *DFA*'s.

The *BSP DFA* $\mathcal{A}$ recognizes the language $L(e)$ of *RE* e, since the *BS DFA* $\mathcal{A}_{BS} = (\Sigma, Q_{BS}, q_{BS_0}, \delta_{BS}, F_{BS})$ is a homomorphic image of $\mathcal{A}$, as it is proved in the following Proposition 1.

Proposition 1 (Equivalence of *DFA*'s). *Given a RE e, the DFA's $\mathcal{A}$ and $\mathcal{A}_{BS}$ of e are equivalent, i.e., they recognize the language $L(e)$.*   □

Sketch of Proof. There is a *DFA* homomorphism ν that maps the *BSP* $\mathcal{A}$ onto the pure recognizer $\mathcal{A}_{BS}$ by canceling the syntactic information represented by the metasymbols and the pointers in the (items of the) states of $\mathcal{A}$, whence the *DFA* $\mathcal{A}$ is equivalent to the *DFA* $\mathcal{A}_{BS}$. In general the *BSP* $\mathcal{A}$ is not minimal and has more states than $\mathcal{A}_{BS}$, which may be not minimal either.   □

Example 3. The sets Ini_{fin} and Fol_{fin} of the *RE* $e \dashv$ Eq. (3) with marking Eq. (7) are shown in Table 1 . The *BSP* $\mathcal{A}$ for e is shown in Fig. 2, along with the pure recognizer $\mathcal{A}_{BS} = \nu(\mathcal{A})$ (image of $\mathcal{A}$ through the homomorphism ν), and the minimal equivalent *DFA* $\mathcal{A}_{min}$. Although the recognizer $\mathcal{A}_{BS}$ has fewer states than the *BSP* $\mathcal{A}$ (4 vs 7), it is not minimal.   □

A general property of the *BSP* is that, for every state q and every item pair $\langle \zeta, \Pi \rangle$, $\langle \zeta', \Pi' \rangle$ in $I(q)$, the pointer sets Π and Π' are either identical or disjoint, see Fig. 2. This makes tree disambiguation more efficient to compute.

4   Tree and Complexity

Now we show how to build the syntax tree and analyze the complexity. Consider a transition $t: q \xrightarrow{a} q'$ of the *BSP* $\mathcal{A}$ and two items $\iota, \iota' \in I_a(q)$, $I(q')$ with $\iota, \iota' = \langle \zeta, \Pi \rangle$, $\langle \zeta', \Pi' \rangle$. We say they are *linked by t* if it holds (*identifier of ι in q*) $\in \Pi'$, i.e., ι is a predecessor of ι' (Definitions 3 and 4). Every transition t links at least one item pair. Next take a string $w \in L(\mathcal{A})$ of length $|w| = n \geq 0$. If it holds $w \neq \varepsilon$ then there is one w-labeled transition sequence $t_1 \ldots t_n$ and $q_n \in F$; else it holds $w = \varepsilon$ and $q_0 \in F$. We associate to w a non-empty *set of item sequences IS_w*, defined as follows:

$$IS_w = \left\{ \iota_0 \ldots \iota_n \;\middle|\; \begin{array}{l} \forall i \in [1 \ldots n] \text{ items } \iota_{i-1}, \iota_i \text{ are linked by trans. } t_i \\ \text{and item } \iota_n \text{ is final} \end{array} \right\} \tag{15}$$
$$\cup \; \{ \iota_0 \mid \; n = 0 \text{ and item } \iota_0 \text{ is final} \}$$

Pose the item sequence $\tau = \iota_0 \ldots \iota_n \in IS_w$. The *extended* string ω_τ is defined as $\omega_\tau = \zeta_0 \ldots \zeta_n \in \widehat{\Omega}^+$, where $\iota_i = \langle \zeta_i, \Pi_i \rangle$ $(0 \leq i \leq n)$. Then pose the set $\omega_w = \{ \omega_\tau \mid \tau \in IS_w \}$. Every string $w \in L(\mathcal{A})$ has at least one corresponding item sequence τ and extended string ω_τ, so the string set ω_w is not empty.

Given a *RE* e, reparentesized as $\check{e}$, every marked string $\omega_\tau \in L(\check{e})$ is a *linearized syntax tree (LST)* of an unmarked string $w \in L(e)$, which is obtained by unmarking string ω_τ and canceling the metasymbols. Conversely, every string $w \in L(e)$ has one *LST* $\omega_\tau \in L(\check{e})$, or more than one if it is ambiguous.

The important fact is that the language $L(\check{e})$ is *local*, see Eqs. (2) and (8) in Sect. 2, and that it contains the *LST*'s of language $L(e)$.

Example 4. Figure 1 (bottom right) shows the *LST*'s of the ambiguous string $a\,b\,a\,b$ for the *RE* e Eq. (3). Observe the structurally and numerically marked *LST*'s for the marked forms Eqs. (4) and (7) of e, respectively. See Example 1 for the distinction between the *Greedy* and *POSIX* forms of the syntax tree.   □

Definition 5 (Extended language). *Given a BSP* $\mathcal{A}$, *this language* $\widehat{L}(\mathcal{A})$:

$$\widehat{L}(\mathcal{A}) = \bigcup_{w \in L(\mathcal{A})} \omega_w \subseteq \widehat{\Omega}^+$$

is called the "extended language" of the BSP $\mathcal{A}$, *over the marked alphabet* $\widehat{\Omega}$. □

Notice that the extended language $\widehat{L}(\mathcal{A})$ contains the *LST*'s of language $L(\mathcal{A})$. The next proposition characterizes infinite ambiguity (the proof is omitted).

Proposition 2 (Infinite ambiguity). *A string* w *is infinitely ambiguous iff it has a LST* ω_τ *that contains a substring* $1_l \, \mu \, 1_l$, *for some mark* l *and* $\mu \in \widehat{M}^*$. □

Example 5. The *RE* $(1)^*$, marked $_1(1_2)_1^*$, has *LST* $_1(1_2 1_2)_1$. The *RE* $(1(a \mid 1))^*$, marked $_1(1_{2\,3}(a_4 \mid 1_5)_3)_1^*$, has *LST* $_1(1_{2\,3}(1_5)_3 1_{2\,3}(1_5)_3)_1$. □

We define the regular language $\widehat{N} = \widehat{\Omega}^* - \bigcup_{\forall \text{ null string } 1_l} \widehat{\Omega}^* 1_l \widehat{M}^* 1_l \widehat{\Omega}$ of all the strings not containing any substring $1_l \, \mu \, 1_l$, for any mark l and $\mu \in \widehat{M}^*$.

Theorem 1 (Linearized parse tree). *Let* e *be a RE and let* $\mathcal{A}$ *be the BSP of* e. *The extended language* $\widehat{L}(\mathcal{A})$ *coincides with the intersection* $L(\breve{e}) \cap \widehat{N}$. □

Proof. For each string $w \in L(\mathcal{A})$, the *BSP* $\mathcal{A}$ creates one or more item sequences $\tau = \iota_0 \ldots \iota_n \in IS_w$ Eq. (15) and as many extended strings $\omega_\tau = \zeta_0 \ldots \zeta_n$, where each finished string ζ_i ($0 \leq i \leq n = |w|$) is an element of the finished initial or follow set Ini_{fin} or Fol_{fin} (Definition 3) of the *RE* $e \dashv$. By (12-13) such sets have the same initial and digram sets Ini and Dig as the *RE* $\breve{e} \dashv$. Now let a_{h_i} be the end-letter of ζ_i. For $1 \leq i \leq n$, two strings ζ_{i-1} and ζ_i are concatenated in ω_τ if the digrams of $a_{h_{i-1}} \zeta_i$ are included in $Dig(\breve{e} \dashv)$, see Eq. (13). Hence the digrams of ω_τ originating from concatenation are the same as those of $\breve{e} \dashv$. Furthermore, since substring ζ_n is final ($n \geq 0$), the final set of ω_τ is $Fin = \{\dashv\}$. Whence the extended language $\widehat{L}(\mathcal{A})$ (disregarding the $\emptyset$ case) is defined by the sets Ini, $Fin = \{\dashv\}$ and Dig of $\breve{e} \dashv$, with the constraint of not containing any substring with two identically-marked null strings 1_l, as specified in Definition 3.

Language $L(\breve{e} \dashv)$ is defined by the sets Ini, $Fin = \{\dashv\}$ and Dig of the *RE* $\breve{e} \dashv$ (Sect. 3). Language $\widehat{L}(\mathcal{A})$ is defined by the same sets Ini, Fin and Dig of $\breve{e} \dashv$, and the constraint above is identical to Proposition 2, modeled by language $\widehat{N}$ (see above). Thus it follows that $\widehat{L}(\mathcal{A}) = L(\breve{e} \dashv) \cap \widehat{N}$, even without $\dashv$. □

Since by Proposition 1 it holds $L(\mathcal{A}) = L(e)$, Theorem 1 states that the *BSP* $\mathcal{A}$ specifies, by means of its extended language $\widehat{L}(\mathcal{A})$, the linearized syntax trees of language $L(e)$, except those that iterate a Kleene star (or cross) with a nullable argument twice or more times. By Proposition 2 such *LST*'s are those of the infinitely ambiguous strings of $L(e)$.

Example 6. The string $a\,a\,b \in L(\mathcal{A})$ of the running example (Fig. 1) has two item sequences: 1, 1, 5, 2 and 1, 2, 5, 2. Its extended strings ω_τ, i.e., *LST*'s, are immediately visible as link chains in Fig. 3, in the state sequence of $a\,a\,b$. □

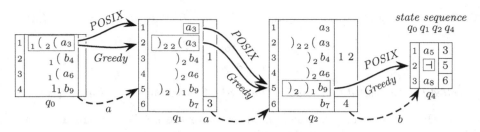

Fig. 3. The IS_{aab} of the ambiguous string $a\,a\,b \in L\left(\left((a)^+ \mid b\,a \mid a\,b\,a\right)^* b\right)$.

We examine the computational cost for Algorithm 1 to construct the *BSP* of a *RE* e, and the computational cost for the *BSP* of e to build a syntax tree for a string in the language $L(e)$. Clearly, as the *BSP* is a *DFA*, it recognizes the language $L(e)$ in real-time.

As the metasymbol prefix μ of a finished string $\zeta = \mu\,a_h$ does not contain any repeated identically-marked null strings 1_l (Definition 3), it does not contain any repeated open-closed parenthesis pairs either, else repeating a 1_l is unavoidable. Thus the number of items in a state of the *BSP* $\mathcal{A}$ is bounded from above by a constant $K = |\widehat{Z}| \times |ID|$ that depends only on the alphabet $\widehat{\Omega}$, and the size of $\mathcal{A}$ is bounded from above by 2^K. Hence the time complexity of Algorithm 1 is at least exponential in the size of $\widehat{\Omega}$.

The sets Ini_{fin} and Fol_{fin} (Definition 3) that drive Algorithm 1 have to be pre-computed from the sets Ini, $Fin = \{\dashv\}$ and Dig of $\breve{e} \dashv$. It suffices to find the acyclic paths in the *DFA* specified by such sets of $\breve{e} \dashv$ via Eq. (1), e.g., by means of the Dijkstra algorithm, in a time that depends only on $\widehat{\Omega}$.

If a string w of language $L(e)$ is not ambiguous, to build its unique *LST* ω_τ we trace back the item sequence IS_w by following the item links from the final one and concatenating the *LST* pieces on the way. The time complexity for building the *LST* is linear in the string length.

5 Disambiguation Criteria

First, we state conditions for string unambiguity, then we discuss how to select a syntax tree for an ambiguous string in accordance with the existing disambiguation criteria. Since [4] it has been known that the ambiguity of a *RE* is a decidable property, but it is interesting to state how the *RE* ambiguity is detected in the related *BSP*. The next proposition does so (proof omitted).

Proposition 3 (String and *RE* ambiguity). *A string $w \in L(e)$ is unambiguous if and only if, for every BSP state q that lies on the path of w, the next conditions are met:*

(a) for every item $\langle \zeta, \Pi \rangle \in q$, it holds $|\Pi| \leq 1$
(b) the last state contains only one final item

Furthermore, a RE e is unambiguous if and only if every state q of the BSP satisfies (a) and every final state of the BSP contains only one final item.   □

Clearly, applying Proposition 3 to a string or *RE* to decide if it is ambiguous, requires linear-time in the parser size (state number). This test eliminates any overhead needed for an ambiguous string when the string is not.

Example 7. In Fig. 2, the *RE* e Eq. (3) violates part (a) of Proposition 3 in the states q_2, q_6, and is (finitely) ambiguous (Fig. 1). The infinitely ambiguous *RE* $(1)^*$ marked as $_1(1_2)_1^*$ violates part (b), as its only state q_0 is identified by these final items: $\langle 1_1 \dashv, \emptyset \rangle$ and $\langle {}_1(1_2)_1 \dashv, \emptyset \rangle$; instead item $\langle {}_1(1_2 1_2)_1 \dashv, \emptyset \rangle$ is excluded, as it models two iterations of the Kleene star. For the infinitely ambiguous *RE*'s, Algorithm 1 only specifies the Kleene star iterations with $* = 0, 1$, which still suffice to verify the ambiguity of such *RE*'s.   □

Adding Disambiguation Criteria. When the user of a *RE* parser specifies a pattern to be matched by means of an ambiguous *RE*, he typically does not want to see all the syntax trees of a string, but just the one that he considers more relevant. Practical relevance is defined by either one of the two disambiguation criteria introduced in Definition 1. Notice that each of them assigns different priorities to almost identical comparisons. By exploiting the similarities, our tool is able to cover both criteria. When scanning an input string, the *BSP* traverses its state diagram from the initial to a final state and remembers the sequence of states encountered, unlike a pure recognizer. The result is a state sequence that contains items linked by identifiers (id's). Such items are the nodes of a directed graph, their id's are the edges, and the paths from final to initial items are the item sequences Eq. (15) of all the parse trees (*LST*'s). The tree builder algorithm visits such a graph from the final item(s) and marks all the touched items. Then it scans the marked items from the initial one(s): it considers the items that have two or more successors, selects one successor (or one of two initial items) according to the chosen disambiguation criterion, and discards all the others. Eventually it reaches a final item after building only one item sequence, which is the parse tree that fulfills the criterion.

Example 8. Figure 3 shows *Greedy / POSIX IS*'s of aab of $((a)^+ \mid ba \mid aba)^*b$.   □

We have implemented the *POSIX* criterion, as well as the *Greedy* one [7], by the Okui and Suzuki method [9] suitably modified. The method works by computing dynamic data to choose a prior path at each bifurcation point. Since the size of such data is determined by the number of id's in a *BSP* state, which is bounded by the maximum number of items and depends only on the *RE* complexity, each choice is taken in a constant time and the overall time complexity of tree building is linear in the input length. Notice that our approach allows us to use different disambiguation criteria seamlessly.

6   Conclusion

We have extended the classic approach by Berry and Sethi to construct a *DFA* for a *RE*, to obtain a parser and to check if the *RE* is ambiguous. The time complexities for string matching and tree building are linear in the input size. The existing *Greedy* and *POSIX* criteria to choose a syntax tree for an ambiguous string are easily incorporated into the parser. We have also implemented in *Java* both our parser and the parser by [9], to compare their performances: early measurements substantiate our expectation that *BSP* is significantly faster.

A future development is to implement the mixed parser/recognizer mode, in the sense that a user can specify the so-called *capturing* or *non-capturing* subexpressions for which he respectively wants, or does not, the syntax tree to be computed. For this, the parser generator has to leave the non-capturing subexpressions unmarked.

To demonstrate our method, an interactive *HTML*/JavaScript parser generator and parsing tool for *RE*'s is on *GitHub* (http://github.com/breveglieri/ebs).

References

1. Allauzen, C., Mohri, M.: A unified construction of the glushkov, follow, and antimirov automata. In: Královič, R., Urzyczyn, P. (eds.) MFCS 2006. LNCS, vol. 4162, pp. 110–121. Springer, Heidelberg (2006)
2. Berry, G., Sethi, R.: From regular expressions to deterministic automata. Theor. Comput. Sci. **48**(1), 117–126 (1986)
3. Berstel, J., Pin, J.E.: Local languages and the Berry-Sethi algorithm. Theor. Comput. Sci. **155**(2), 439–446 (1996)
4. Book, R., Even, S., Greibach, S., Ott, G.: Ambiguity in graphs and expressions. IEEE Trans. on Comp. C-**20(2)**, 149–153 (1971)
5. Breveglieri, L., Crespi Reghizzi, S., Morzenti, A.: Shift-reduce parsers for transition networks. In: Dediu, A.-H., Martín-Vide, C., Sierra-Rodríguez, J.-L., Truthe, B. (eds.) LATA 2014. LNCS, vol. 8370, pp. 222–235. Springer, Heidelberg (2014)
6. Dubè, D., Feeley, M.: Efficiently building a parse tree from a regular expression. Acta Inf. **37**(2), 121–144 (2000)
7. Frisch, A., Cardelli, L.: Greedy regular expression matching. In: Díaz, J., Karhumäki, J., Lepistö, A., Sannella, D. (eds.) ICALP 2004. LNCS, vol. 3142, pp. 618–629. Springer, Heidelberg (2004)
8. IEEE: std. 1003.2, POSIX, regular expression notation, section 2.8 (1992)
9. Okui, S., Suzuki, T.: Disambiguation in regular expression matching via position automata with augmented transitions. In: Domaratzki, M., Salomaa, K. (eds.) CIAA 2010. LNCS, vol. 6482, pp. 231–240. Springer, Heidelberg (2011)
10. Sakarovitch, J.: Elements of Automata Theory. Cambridge University Press, New York (2009)
11. Sulzmann, M., Lu, K.Z.M.: POSIX regular expression parsing with derivatives. In: Codish, M., Sumii, E. (eds.) FLOPS 2014. LNCS, vol. 8475, pp. 203–220. Springer, Heidelberg (2014)

Deciding Synchronous Kleene Algebra
with Derivatives

Sabine Broda$^{(\boxtimes)}$, Sílvia Cavadas, Miguel Ferreira, and Nelma Moreira

CMUP & DCC, Faculdade de Ciências da Universidade do Porto,
Rua do Campo Alegre, 4169-007 Porto, Portugal
{sbb,nam}@dcc.fc.up.pt, {silviacavadas,miguelferreira108}@gmail.com

Abstract. Synchronous Kleene algebra (SKA) is a decidable framework
that combines Kleene algebra (KA) with a synchrony model of concurrency. Elements of SKA can be seen as processes taking place within a
fixed discrete time frame and that, at each time step, may execute one
or more basic actions or then come to a halt. The synchronous Kleene
algebra with tests (SKAT) combines SKA with a Boolean algebra. Both
algebras were introduced by Prisacariu, who proved the decidability of
the equational theory, through a Kleene theorem based on the classical
Thompson ε-NFA construction. Using the notion of partial derivatives,
we present a new decision procedure for equivalence between SKA terms.
The results are extended for SKAT considering automata with transitions labeled by Boolean expressions instead of atoms. This work con-
tinous previous research done for KA and KAT, where derivative based
methods were used in feasible algorithms for testing terms equivalence.

Keywords: Synchronous Kleene algebra · Concurrency · Equivalence ·
Derivative

1   Introduction

Synchronous Kleene algebra (SKA) combines Kleene algebra (KA) with the
synchrony model of concurrency of Milner's Synchronous Calculus of Communication Systems (SCCS) [20]. Synchronous here means that two concurrent processes execute a single action simultaneously at each time instant of a
unique global clock. Although this synchrony model seems to be a very weak
model of concurrency when compared with asynchronous interleaving models,
its equational theory is powerful and the SCCS calculus includes the Calculus
of Communication Systems (CCS) as a sub-calculus. It also models the Esterel
programming language [5], a tool used by the industry [29].

This work was partially supported by CMUP (UID/MAT/00144/2013), which is
funded by FCT (Portugal) with national (MEC) and European structural funds
through the programs FEDER, under the partnership agreement PT2020, and
through the programme COMPETE and by the Portuguese Government through
the FCT under project FCOMP-01-0124-FEDER-020486.

© Springer International Publishing Switzerland 2015
F. Drewes (Ed.): CIAA 2015, LNCS 9223, pp. 49–62, 2015.
DOI: 10.1007/978-3-319-22360-5_5

SKA was introduced by Prisacariu [25]. It consists of a KA to which a synchrony operator and a notion of basic action are added. Using a Kleene's style theorem, Prisacariu proved the decidability of the equational theory. He also generalized Kleene algebra with tests (KAT) [15], an equational system that extends Kleene algebra with Boolean algebra. KAT is specially suited to capture and verify properties of simple imperative programs and, in particular, subsumes propositional Hoare logic [16]. For the resulting algebra, called synchronous Kleene algebra with tests (SKAT), the models considered were sets of guarded synchronous strings and decidability was also proved using the so called automata on guarded synchronous strings. SKAT can be seen as an alternative to Hoare logic for reasoning about parallel programs with shared variables in a synchronous system.

Decision procedures for Kleene algebra terms equivalence have been a subject of intense research in recent years [1,7,12,19,21,22,27]. This is partially motivated by the fact that regular expressions can be seen as a program logic that allows to express nondeterministic choice, sequence, and finite iteration of programs. Many proposed procedures decide equivalence based on the computation of a bisimulation (or a bisimulation up-to) between the two expressions [1,7,21,27]. Broda *et al.* studied the average size of derivative based automata both for KA and KAT [9]. For KAT terms, a coalgebraic decision procedure was presented by Kozen [18]. There, derivatives are considered with respect to symbols $v\sigma$ where σ is an action symbol but v corresponds to a valuation of the Boolean tests. This induces an exponential blow-up on the number of states or transitions of the automata and an accentuated exponential complexity when testing the equivalence of two KAT expressions (as noted in [3,23]). A. Silva [28] introduced a class of automata over guarded strings that avoids that blow-up. Broda *et al.* studied the average size of some automata of that class [9] and extended finite automata equivalence decision procedures to that class [10]. In this paper we continue this line of work and present new decision procedures for SKA and SKAT equivalence, based on the notion of partial derivatives. For SKA an ε-free NFA construction is presented which leads to smaller automata than the one given by Prisacariu. For SKAT we introduce a class of automata over guarded synchronous strings where transitions are labeled by Boolean expressions instead of valuations. This feature significally improves the performance of the associated methods. For both methods some experimental results are presented and discussed.

2 Deciding Synchronous Kleene Algebra

First we review some concepts related with SKA. A *Kleene algebra* (KA) is an algebraic structure $(\mathcal{A}, +, \cdot, ^*, 0, 1)$, where $+$ and $\cdot$ are binary operations on $\mathcal{A}$, * is a unary operation on $\mathcal{A}$, and 0 and 1 belong to $\mathcal{A}$, such that $(\mathcal{A}, +, \cdot, 0, 1)$ is an idempotent semiring, and * satisfies axioms (10)-(13) below. The natural order $\leq$ in $(\mathcal{A}, +, \cdot, 0, 1)$ is defined by $\alpha \leq \beta$ if and only if $\alpha + \beta = \beta$.

$$1 + \alpha\alpha^* \leq \alpha^* \qquad \alpha + \beta \cdot \gamma \leq \gamma \quad \rightarrow \quad \beta^* \cdot \alpha \leq \gamma$$
$$1 + \alpha^*\alpha \leq \alpha^* \qquad \alpha + \gamma \cdot \beta \leq \gamma \quad \rightarrow \quad \alpha \cdot \beta^* \leq \gamma$$

A *synchronous Kleene algebra* (SKA) over a finite set A_B is given by a structure $(\mathcal{A}, +, \cdot, \times, {}^*, 0, 1, A_B)$, where $A_B \subseteq \mathcal{A}$, $(\mathcal{A}, +, \cdot, {}^*, 0, 1)$ is a Kleene algebra, and $\times$ is a binary operator that is associative, commutative, distributive over $+$, with absorvent element 0 and identity 1. Furthermore, it satisfies $a \times a = a \quad \forall a \in A_B$, as well as the *synchrony axiom*

$$(\alpha^\times \cdot \alpha) \times (\beta^\times \cdot \beta) = (\alpha^\times \times \beta^\times) \cdot (\alpha \times \beta) \quad \forall \alpha^\times, \beta^\times \in A_B^\times,$$

where the set $A_B^\times$ is the smallest subset of $\mathcal{A}$ that contains A_B and is closed for $\times$. As usual, we will omit the operator $\cdot$ whenever it does not give rise to any ambiguity and use the following precedence over the operators: $+ < \cdot < \times < {}^*$.

We think of the elements of SKA as processes taking place within a fixed discrete time frame and that, at each time step, may execute one or more basic actions in A_B or then come to a halt.

The standard model of an SKA over A_B is the set of languages over the alphabet $\Sigma = \mathcal{P}(A_B) \setminus \{\emptyset\}$, which we will call *synchronous languages*. Each synchronous language represents a process described by its possible executions, which are given by the words over Σ, each one a sequence of sets of basic actions executed in a single time step. We call $\sigma \in \Sigma$ a *(synchronous) concurrent action*. The *synchronous product* of two words $x = \sigma_1 \cdots \sigma_m$ and $y = \tau_1 \cdots \tau_n$, with $n \geq m$, is defined by

$$x \times y = y \times x = (\sigma_1 \cup \tau_1 \cdots \sigma_m \cup \tau_m)\tau_{m+1} \cdots \tau_n.$$

In particular, the synchronous product of two letters in Σ is their union. The synchronous product of two languages L_1 and L_2 is defined by

$$L_1 \times L_2 = \{\, x \times y \,|\, x \in L_1, y \in L_2 \,\}.$$

It is clear that the synchronous regular languages over A_B contain the regular languages over Σ. It turns out that they are exactly the same set, i.e., the regular languages over Σ are also closed for $\times$. In [25], the classical Thompson construction for regular languages [30] is extended to build an automaton accepting the synchronous product of two languages given by their automata.

We now introduce the SKA analogue of the regular expressions. We denote by $\mathcal{T}_{\mathsf{SKA}}$ the set of SKA terms, containing 0 plus all terms generated by the grammar

$$\alpha \rightarrow 1 \mid a \mid \alpha + \alpha \mid \alpha \cdot \alpha \mid \alpha \times \alpha \mid \alpha^* \quad (a \in A_B). \tag{1}$$

Note that we do not include in $\mathcal{T}_{\mathsf{SKA}}$ compound expressions that have 0 as a subexpression. Given $\alpha \in \mathcal{T}_{\mathsf{SKA}}$, the language $\mathcal{L}(\alpha)$ denoted by α is inductively defined as follows, $\mathcal{L}(a) = \{\{a\}\}$, $\mathcal{L}(0) = \emptyset$, $\mathcal{L}(1) = \{\varepsilon\}$, $\mathcal{L}(\alpha^*) = \mathcal{L}(\alpha)^*$, $\mathcal{L}(\alpha + \beta) = \mathcal{L}(\alpha) \cup \mathcal{L}(\beta)$, $\mathcal{L}(\alpha\beta) = \mathcal{L}(\alpha)\mathcal{L}(\beta)$, $\mathcal{L}(\alpha \times \beta) = \mathcal{L}(\alpha) \times \mathcal{L}(\beta)$.

Example 1. Let $A_B = \{a, b\}$, hence $\Sigma = \{\{a\}, \{b\}, \{a, b\}\}$, and consider the SKA term $\alpha = (a(b + a)^*) \times (a + bb)^*$ over A_B. Then

$$\mathcal{L}(\alpha) = \{\{a\}, \{a\}\{a\}, \{a\}\{b\}, \ldots\} \times \{\varepsilon, \{a\}, \{a\}\{a\}, \{b\}\{b\}, \ldots\}$$
$$= \{\{a\}, \{a\}\{a\}, \{a\}\{b\}, \{a\}\{a, b\}, \{a, b\}\{b\}, \{a, b\}\{a, b\}, \ldots\}.$$

Given $\alpha, \beta \in \mathcal{T}_{\mathsf{SKA}}$, we say that they are *equivalent* if they denote the same language, i.e., $\mathcal{L}(\alpha) = \mathcal{L}(\beta)$. We also define $\varepsilon(\alpha) = 1$ if $\varepsilon \in \mathcal{L}(\alpha)$, and $\varepsilon(\alpha) = 0$ otherwise. A recursive definition of $\varepsilon : \mathcal{T}_{\mathsf{SKA}} \longrightarrow \{0, 1\}$ is given by the following, $\varepsilon(a) = \varepsilon(0) = 0$, $\varepsilon(1) = \varepsilon(\alpha^*) = 1$, $\varepsilon(\alpha + \beta) = \varepsilon(\alpha) + \varepsilon(\beta)$, and $\varepsilon(\alpha\beta) = \varepsilon(\alpha \times \beta) = \varepsilon(\alpha) \cdot \varepsilon(\beta)$. We generalize ε for sets $S \subseteq \mathcal{T}_{\mathsf{SKA}}$ by $\varepsilon(S) = \sum_{\alpha \in S} \varepsilon(\alpha)$.

2.1 Partial Derivative Automata for SKA

A *nondeterministic finite automaton* (NFA) is a tuple $\mathcal{A} = \langle S, \Sigma, S_0, \delta, F \rangle$, where S is a finite set of states, Σ is a finite alphabet, $S_0 \subseteq S$ a set of initial states, $\delta : S \times \Sigma \longrightarrow \mathcal{P}(S)$ the transition function, and $F \subseteq S$ a set of final states. The transition function δ is extended to words and sets of states in the natural way. A word $x \in \Sigma^*$ is *accepted* by $\mathcal{A}$ if and only if $\delta(S_0, x) \cap F \neq \emptyset$. The *language of* $\mathcal{A}$ is the set of words accepted by $\mathcal{A}$ and denoted by $\mathcal{L}(\mathcal{A})$.

In the context of SKA, we consider the alphabet $\Sigma = \mathcal{P}(\mathsf{A_B}) \setminus \{\emptyset\}$ and call the NFA a *nondeterministic automaton on synchronous strings*. Prisacariu presented a method of converting an SKA expression into an equivalent ε-NFA (in an ε-NFA transitions may be labelled by ε), based on the classical Thompson construction. Due to the local behaviour of the synchronization operator, in each step it is necessary to eliminate all ε-transitions except those entering the final state. The step for the synchronous product $\alpha \times \beta$ involves the construction of a classic product automaton from the automata corresponding to α and β, respectively. This leads easily to large automata for relatively small expressions. We present now a new method of converting of an SKA expression into an equivalent ε-free NFA. This method extends the classical partial derivative automata construction for regular expressions [4] and provides a new proof that the set of synchronous regular languages over $\mathsf{A_B}$ is precisely the set of regular languages over Σ.

As usual, the *left-quotient* of a synchronous language L w.r.t. a synchronous concurrent action σ is the set $\sigma^{-1}L = \{ x \mid \sigma x \in L \}$. The left quotient of L w.r.t. a word $x \in \Sigma^*$ is inductively defined by $\varepsilon^{-1}L = L$ and $(x\sigma)^{-1}L = \sigma^{-1}(x^{-1}L)$. Antimirov [4] introduced the notion of partial derivatives which we now generalize to the set $\mathcal{T}_{\mathsf{SKA}}$. Given sets $S, T \subseteq \mathcal{T}_{\mathsf{SKA}}$, let $S \odot T = \{ \alpha\beta \mid \alpha \in S \setminus \{0\}, \beta \in T \setminus \{0\} \}$ and $S \otimes T = \{ \alpha \times \beta \mid \alpha \in S \setminus \{0\}, \beta \in T \setminus \{0\} \}$. We consider $\alpha \odot S = \{\alpha\} \odot S$, and similarly for $S \odot \alpha$, $\alpha \otimes S$ and $S \otimes \alpha$. These definitions serve the following.

Definition 2. *The set of partial derivatives of a term* $\alpha \in \mathcal{T}_{\mathsf{SKA}}$ *w.r.t. the letter* $\sigma \in \Sigma$, *denoted by* $\partial_\sigma(\alpha)$, *is inductively defined by*

$$\partial_\sigma(0) = \partial_\sigma(1) = \emptyset \qquad\qquad \partial_\sigma(\alpha^*) = \partial_\sigma(\alpha) \odot \alpha^*$$

$$\partial_\sigma(a) = \begin{cases} \{1\} & \text{if } \sigma = \{a\} \\ \emptyset & \text{otherwise} \end{cases} \qquad \begin{aligned} \partial_\sigma(\alpha + \beta) &= \partial_\sigma(\alpha) \cup \partial_\sigma(\beta) \\ \partial_\sigma(\alpha\beta) &= \partial_\sigma(\alpha) \odot \beta \cup \varepsilon(\alpha) \odot \partial_\sigma(\beta) \end{aligned}$$

$$\partial_\sigma(\alpha \times \beta) = \left(\bigcup_{\sigma_1 \times \sigma_2 = \sigma} \partial_{\sigma_1}(\alpha) \otimes \partial_{\sigma_2}(\beta) \right) \cup \varepsilon(\alpha) \otimes \partial_\sigma(\beta) \cup \varepsilon(\beta) \otimes \partial_\sigma(\alpha).$$

The set of partial derivatives of $\alpha \in \mathcal{T}_{\mathsf{SKA}}$ *w.r.t. a word* $x \in \Sigma^*$ *is inductively defined by* $\partial_\varepsilon(\alpha) = \{\alpha\}$ *and* $\partial_{x\sigma}(\alpha) = \partial_\sigma(\partial_x(\alpha))$, *where, given a set* $S \subseteq \mathcal{T}_{SKA}$, $\partial_\sigma(S) = \bigcup_{\alpha \in S} \partial_\sigma(\alpha)$.

We denote by $\partial(\alpha)$ the set of all partial derivatives of α, $\partial(\alpha) = \bigcup_{x \in \Sigma^*} \partial_x(\alpha)$, and by $\partial^+(\alpha)$ the set of partial derivatives excluding the trivial derivative by ε, $\partial^+(\alpha) = \bigcup_{x \in \Sigma^+} \partial_x(\alpha)$. Given a set $S \subseteq \mathcal{T}_{\mathsf{SKA}}$, we define $\mathcal{L}(S) = \bigcup_{\alpha \in S} \mathcal{L}(\alpha)$. It is straightforward to show that for every $\mathcal{T}_{\mathsf{SKA}}$ term α and word x, $\mathcal{L}(\partial_x(\alpha)) = x^{-1}\mathcal{L}(\alpha)$. The following lemma will be used to show that $\partial(\alpha)$ is finite, as in the case for standard regular expressions.

Lemma 3. *The set $\partial^+(\alpha)$ satisfies the following.*

$$\partial^+(0) = \partial^+(1) = \emptyset \qquad \partial^+(\alpha + \beta) \subseteq \partial^+(\alpha) \cup \partial^+(\beta)$$
$$\partial^+(a) = \{1\} \quad (a \in \mathsf{A_B}) \qquad \partial^+(\alpha\beta) \subseteq \partial^+(\alpha) \odot \beta \cup \partial^+(\beta)$$
$$\partial^+(\alpha^*) \subseteq \partial^+(\alpha) \odot \alpha^* \qquad \partial^+(\alpha \times \beta) \subseteq \partial^+(\alpha) \otimes \partial^+(\beta) \cup \partial^+(\alpha) \cup \partial^+(\beta).$$

Proof. The proof proceeds by induction on the structure of α. It is clear that for $\partial^+(0)$, $\partial^+(1)$ and, for $\partial^+(a)$, $a \in \mathsf{A_B}$, the result is true. Now, suppose the claim is true for α and β, with $|\alpha|_{\mathsf{A_B}} \neq 0$ and $|\beta|_{\mathsf{A_B}} \neq 0$. Otherwise, one has $|\partial^+(\alpha)| = 0$ and/or $|\partial^+(\beta)| = 0$, simplifying the arguments below. There are four induction cases to consider, in which we will make use of the fact that, for any SKA expression γ and letter $\sigma \in \Sigma$, the set $\partial^+(\gamma)$ is closed for taking derivatives w.r.t. σ, i.e., $\partial_\sigma(\partial^+(\gamma)) \subseteq \partial^+(\gamma)$.

i. One can check by induction on the length of x that, for $x \in \Sigma^+$, $\partial_x(\alpha + \beta) = \partial_x(\alpha) \cup \partial_x(\beta)$. Hence, $\partial^+(\alpha + \beta) = \partial^+(\alpha) \cup \partial^+(\beta)$.

ii. We will prove by induction on the length of x that $\partial_x(\alpha\beta) \subseteq \partial^+(\alpha) \odot \beta \cup \partial^+(\beta)$ for every word $x \in \Sigma^+$. The claim is true for $\sigma \in \Sigma$ since $\partial_\sigma(\alpha\beta) = \partial_\sigma(\alpha) \odot \beta \cup \varepsilon(\alpha) \odot \partial_\sigma(\beta)$. Assuming it is true for x, $\partial_{x\sigma}(\alpha\beta) = \partial_\sigma(\partial_x(\alpha\beta)) \subseteq \partial_\sigma(\partial^+(\alpha) \odot \beta \cup \partial^+(\beta)) \subseteq \partial_\sigma(\partial^+(\alpha)) \odot \beta \cup \partial_\sigma(\beta) \cup \partial_\sigma(\partial^+(\beta)) \subseteq \partial^+(\alpha) \odot \beta \cup \partial^+(\beta)$.

iii. We prove by induction on the length of x that, for every word $x \in \Sigma^+$, $\partial_x(\alpha \times \beta) \subseteq \partial^+(\alpha) \otimes \partial^+(\beta) \cup \partial^+(\alpha) \cup \partial^+(\beta)$. The claim is true for $\sigma \in \Sigma$ because $\partial_\sigma(\alpha \times \beta) = \bigcup_{\sigma_1 \times \sigma_2 = \sigma} \partial_{\sigma_1}(\alpha) \otimes \partial_{\sigma_2}(\beta) \cup \varepsilon(\alpha) \otimes \partial_\sigma(\beta) \cup \varepsilon(\beta) \otimes \partial_\sigma(\alpha)$; supposing it is true for x, $\partial_{x\sigma}(\alpha \times \beta) = \partial_\sigma(\partial_x(\alpha \times \beta)) \subseteq \partial_\sigma(\partial^+(\alpha) \otimes \partial^+(\beta) \cup \partial^+(\alpha) \cup \partial^+(\beta)) \subseteq (\bigcup_{\sigma_1 \times \sigma_2 = \sigma} \partial_{\sigma_1}(\partial^+(\alpha)) \otimes \partial_{\sigma_2}(\partial^+(\beta))) \cup \partial_\sigma(\partial^+(\alpha)) \cup \partial_\sigma(\partial^+(\beta)) \subseteq \partial^+(\alpha) \otimes \partial^+(\beta) \cup \partial^+(\alpha) \cup \partial^+(\beta)$.

iv. We show by induction on the length of x that $\partial_x(\alpha^*) \subseteq \partial^+(\alpha) \odot \alpha^*$ for $x \in \Sigma^+$. It is true for $\sigma \in \Sigma$ because $\partial_\sigma(\alpha^*) = \partial_\sigma(\alpha) \odot \alpha^*$; supposing the claim true for x, $\partial_{\sigma x}(\alpha^*) = \partial_\sigma(\partial_x(\alpha^*)) \subseteq \partial_\sigma(\partial^+(\alpha) \odot \alpha^*) \subseteq \partial_\sigma(\partial^+(\alpha)) \odot \alpha^* \cup \partial_\sigma(\alpha^*) \subseteq \partial^+(\alpha) \odot \alpha^* \cup \partial_\sigma(\alpha) \odot \alpha^* \subseteq \partial^+(\alpha) \odot \alpha^*$. $\qquad\square$

Now, it is easy to obtain the following upper bound for the size of $\partial^+(\alpha)$.

Proposition 4. *Given $\alpha \in \mathcal{T}_{\mathsf{SKA}}$, $|\partial^+(\alpha)| \leq 2^{|\alpha|_{\mathsf{A_B}}} - 1$, where $|\alpha|_{\mathsf{A_B}}$ denotes the number of occurrences of elements of $\mathsf{A_B}$ in α. Thus, $|\partial(\alpha)| \leq 2^{|\alpha|_{\mathsf{A_B}}}$.*

We note that this upper bound is exactly the same obtained for the number of partial derivatives for regular expressions with the shuffle operator [11]. In the latter case, however, the correspondent version of Lemma 3 establishes equalities instead of inclusions.

We extend to SKA terms the standard Antimirov automaton or partial derivative automaton. Given $\alpha \in \mathcal{T}_{SKA}$, we define the *partial derivative automaton* associated to α by $\mathcal{A}(\alpha) = \langle \partial(\alpha), \Sigma, \{\alpha\}, \delta_\alpha, F_\alpha \rangle$, where $F_\alpha = \{ \gamma \in \partial(\alpha) \mid \varepsilon(\gamma) = 1 \}$ and $\delta_\alpha(\gamma, \sigma) = \partial_\sigma(\gamma)$. Then, it is easy to see that $\mathcal{L}(\mathcal{A}(\alpha)) = \mathcal{L}(\alpha)$.

Example 5. Consider again the expression α from Example 1 and let $\beta = (b + a)^*$ and $\gamma = (a + bb)^*$, i.e. $\alpha = (a\beta) \times \gamma$. Furthermore, let $\alpha_0 = \alpha$, $\alpha_1 = \beta \times \gamma$, $\alpha_2 = \beta \times (b\gamma)$, $\alpha_3 = \beta$, $\alpha_4 = b\gamma$, and $\alpha_5 = \gamma$. The nonempty sets of partial derivatives of α are the following: $\partial_{\{a\}}(\alpha_0) = \{\alpha_1, \alpha_3\}$, $\partial_{\{a,b\}}(\alpha_0) = \{\alpha_2\}$, $\partial_{\{a\}}(\alpha_1) = \{\alpha_1, \alpha_3, \alpha_5\}$, $\partial_{\{b\}}(\alpha_1) = \{\alpha_2, \alpha_3, \alpha_4\}$, $\partial_{\{a,b\}}(\alpha_1) = \{\alpha_1, \alpha_2\}$, $\partial_{\{b\}}(\alpha_2) = \{\alpha_1, \alpha_5\}$, $\partial_{\{a,b\}}(\alpha_2) = \{\alpha_1\}$, $\partial_{\{a\}}(\alpha_3) = \partial_{\{b\}}(\alpha_3) = \{\alpha_3\}$, $\partial_{\{b\}}(\alpha_4) = \partial_{\{a\}}(\alpha_5) = \{\alpha_5\}$, $\partial_{\{b\}}(\alpha_5) = \{\alpha_4\}$. Then, $\mathcal{A}(\alpha)$ is the following.

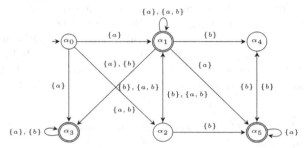

It is worthwhile to note that this automaton has 6 states and 19 transitions, while the one obtained using Prisacariu's Thompson-based construction has 16 states and 73 transitions, even after some necessary ε-transition eliminations.

2.2   Equivalence of SKA Expressions

We are interested in an algorithm that decides whether or not two SKA terms represent the same regular language. Since we already know how to construct an NFA that accepts a given SKA term, the problem is tantamount to deciding the language equivalence of two automata. One possible approach is to search for the existence of a bisimulation in the determinized NFAs (DFAs), as presented by Hopcroft and Karp [14]. This algorithm can be easily extended to NFAs as in Almeida *et al.* [1]. A presentation of this algorithm and an improved variant, together with proofs of correctness, can be found in Bonchi and Pous [6].

2.3   Implementation and Experimental Results

A Python module for manipulating SKA terms and automata over synchronous strings was implemented within the FAdo library [26], which includes several algorithms for regular expressions and finite automata. For the efficient computation of the set of partial derivatives of a term w.r.t. a symbol, in FAdo a function is used, that given an expression α computes the set of pairs $(\sigma, \partial_\sigma(\alpha))$ with $\sigma \in \Sigma$ [4]. We extended it to the synchronous product,

$$f : \mathcal{T}_{\mathsf{SKA}} \; \rightarrow \; \mathcal{P}(\Sigma \times \mathcal{P}(\mathcal{T}_{\mathsf{SKA}}))$$

$$f(0) = f(1) = \emptyset \qquad\qquad f(a) = \{(\{a\}, \varepsilon)\} \quad f(\alpha^*) = f(\alpha) \cdot \alpha^*$$

$$f(\alpha + \beta) = f(\alpha) \cup f(\beta) \qquad\qquad f(\alpha\beta) = f(\alpha) \odot \beta \cup f(\beta) \odot \varepsilon(\alpha)$$

$$f(\alpha \times \beta) = \{\, (\sigma_1 \cup \sigma_2, \alpha_1 \times \alpha_2) \mid (\sigma_1, \alpha_1) \in f(\alpha), (\sigma_2, \alpha_2) \in f(\beta) \,\}$$
$$\cup\, f(\beta) \odot \varepsilon(\alpha) \cup f(\alpha) \odot \varepsilon(\beta)$$

where, as before, for $\alpha \neq 1$, $\Gamma \odot \alpha = \{\, (\sigma, \alpha'\alpha) \mid (\sigma, \alpha') \in \Gamma \,\}$.

For running some experiments we uniformly random generated SKA terms. The FAdo random generator has as input a grammar, the size k of the alphabet, and the size n of the words to be generated. A prefix notation version of the grammar (1) was used in order to obtain terms, uniformly generated in the size $|\alpha|$ of the syntactic tree (i.e. parentheses not counted). For each size, $n = |\alpha|$ and $k = |A_\mathsf{B}|$, samples of 1000 terms were generated. We compared the sizes of the partial derivative automata $\mathcal{A}(\alpha)$ and the automata proposed by Prisacariu, a variant of the Thompson construction, $(S_{\mathrm{tho}}, \Sigma, I_{\mathrm{tho}}, \delta_{\mathrm{tho}}, F_{\mathrm{tho}})$. Table 1 presents average values obtained for $n \in \{50, 100\}$ and $k \in \{5, 10, 20\}$.

Table 1. Experimental results for uniform random generated $\mathcal{T}_{\mathsf{SKA}}$ expressions

k	$|\alpha|$	$|S_{\mathrm{tho}}|$	$|\delta_{\mathrm{tho}}|$	$|\partial(\alpha)|$	$|\delta_\alpha|$	$\frac{|\partial(\alpha)|}{|S_{\mathrm{tho}}|}$	$\frac{|\delta_\alpha|}{|\delta_{\mathrm{tho}}|}$
5	50	59	496	23	159	0.389	0.321
5	100	491	47288	128	4133	0.261	0.087
10	50	49	271	18	97	0.364	0.358
10	100	358	15096	96	1691	0.268	0.112
20	50	44	165	16	69	0.364	0.418
20	100	194	2126	60	559	0.309	0.263

Analyzing the table, it seems that the partial derivative automaton is always smaller than the Thompson-like construction, and that the exponential blow up of the automaton size may not occur on average. For regular expressions it is known that after eliminating ε-transitions from the Thompson automaton one obtains the Glushkov automaton [13], of which the partial derivative automaton is a quotient. Asymptotically and on average the size of the partial derivative automaton is half the size of the Glushkov automaton [8], which on the other hand is linear on the size of the expression. As noticed before, for the synchronous product the Thompson construction considers a product automaton and thus a quadratic number of transitions is expected. We also note that for every synchronisation ε-transitions are eliminated, reducing the size of the resulting automata that otherwise should be much larger. No such procedures are needed for the partial derivative automata. For testing the equivalence of SKA terms we can use one of the algorithms mentioned above.

3   Deciding Synchronous Kleene Algebra with Tests

Synchronous Kleene algebra with tests (SKAT) was also introduced by Prisacariu as a natural extension of the Kleene algebra with tests to the synchronous setting. The SKA axiomatization was extended to SKAT, whose standard models are sets of guarded synchronous strings. Prisacariu defined automata over guarded synchronous strings that were based on the ones considered by Kozen for guarded strings [17]. In the synchronous case, automata were built in two layers: one that processed a synchronous string and another to represent the valuations of the boolean tests (called atoms, as defined below). Our contribution in this section is to consider a much simpler notion of automata and to show that the derivative based methods developed in the previous section for SKA can be extended to SKAT. We use standard finite automata where transitions are labeled both with action symbols and boolean tests (instead of atoms). This kind of automata for KAT terms were introduced by Silva [28] and Broda *et al.* [9,10]. In the next subsection, we revise the notions of SKAT and guarded synchronous strings.

3.1   SKAT and Guarded Synchronous Strings

Formally, an SKAT is a structure $(\mathcal{A}, \mathcal{B}, +, \cdot, \times, *, \neg, 0, 1, A_B, T)$, where $T \subseteq \mathcal{B} \subseteq \mathcal{A}$ and A_B and T are disjoint finite sets, $(\mathcal{A}, +, \cdot, \times, *, 0, 1, A_B \cup T)$ is an SKA, $(\mathcal{B}, +, \cdot, \neg, 0, 1)$ and $(\mathcal{B}, +, \times, \neg, 0, 1)$ are Boolean algebras, and $(\mathcal{B}, +, \cdot, \times, 0, 1)$ is a subalgebra of $(\mathcal{A}, +, \cdot, \times, 0, 1)$.

Similar to what was done for SKA, we consider the set $\mathcal{B}_{\mathsf{SKAT}}$ of boolean expressions and the set $\mathcal{T}_{\mathsf{SKAT}}$ of SKAT expressions over $A_B \cup T$. $\mathcal{B}_{\mathsf{SKAT}}$ is the set of terms finitely generated from $T \cup \{0, 1\}$ and operators $+, \cdot, \times, \neg$, while $\mathcal{T}_{\mathsf{SKAT}}$ denotes the set of terms finitely generated from $A_B \cup \mathcal{B}_{\mathsf{SKAT}}$ and operators $+, \cdot, \times, *$. Elements of $\mathcal{B}_{\mathsf{SKAT}}$ and $\mathcal{T}_{\mathsf{SKAT}}$ will be denoted by $b, b_1, \ldots$ and $\alpha, \beta, \alpha_1, \ldots$, respectively, and are generated by the following grammar

$$b \to 0 \mid 1 \mid t \mid b + b \mid b \cdot b \mid b \times b \mid \neg b \quad (t \in T),$$
$$\alpha \to a \mid b \mid \alpha + \alpha \mid \alpha \cdot \alpha \mid \alpha \times \alpha \mid \alpha^* \quad (a \in A_B).$$

The set At of *atoms* over $T = \{t_0, \ldots, t_{l-1}\}$, with $l \geq 1$, is the set of all boolean assignments to all elements of T, i.e. $\mathsf{At} = \{\ x_0 \cdots x_{l-1} \mid x_i \in \{t_i, \bar{t}_i\},\ t_i \in T\ \}$. We denote elements of At by v, v_1, etc. Note that each atom $v \in \mathsf{At}$ has associated a binary word of l bits $(w_0 \cdots w_{l-1})$ where $w_i = 0$ if $\bar{t}_i \in v$, and $w_i = 1$ if $t_i \in v$. The standard model of SKAT consists of the sets of guarded synchronous strings. The set of guarded synchronous strings over $A_B \cup T$ is $\mathsf{GSS} = (\mathsf{At} \cdot \Sigma)^* \cdot \mathsf{At}$, where, as before, $\Sigma = \mathcal{P}(A_B) \setminus \{\emptyset\}$. For $x = v_0 \sigma_1 \cdots \sigma_m v_m$ and $y = v_0' \sigma_1' \cdots \sigma_n' v_n' \in \mathsf{GSS}$, where $m, n \geq 0$, $v_i, v_j' \in \mathsf{At}$ and $\sigma_i, \sigma_j' \in \Sigma$, we define the *fusion product* $x \diamond y = v_0 \sigma_1 \cdots \sigma_m v_m \sigma_1' \cdots \sigma_n' v_n'$, if $v_m = v_0'$, leaving it undefined otherwise. Similarly, for $m \leq n$ the product $x \times y = y \times x$ is defined only if $v_0 = v_0', \ldots, v_m = v_m'$ by $x \times y = v_0 (\sigma_1 \cup \sigma_1') \cdots (\sigma_m \cup \sigma_m') v_m \sigma_{m+1}' \cdots \sigma_n' v_n'$. For sets $X, Y \subseteq \mathsf{GSS}$, $X \diamond Y = \{\ x \diamond y \mid x \in X, y \in Y, x \diamond y \text{ exists}\ \}$ and

$X \times Y = \{ x \times y \mid x \in X, y \in Y, x \times y \text{ exists } \}$. Finally, let $X^0 = \mathsf{At}$ and $X^{n+1} = X \diamond X^n$, for $n \geq 0$, and define $X^* = \bigcup_{n \geq 0} X^n$.

Given a SKAT expression α, we define $\mathsf{GSS}(\alpha) \subseteq \mathsf{GSS}$ inductively as follows,

$$
\begin{aligned}
\mathsf{GSS}(a) &= \{ \mathrm{v}_1\{a\}\mathrm{v}_2 \mid \mathrm{v}_1, \mathrm{v}_2 \in \mathsf{At} \} & \mathsf{GSS}(\alpha \cdot \beta) &= \mathsf{GSS}(\alpha) \diamond \mathsf{GSS}(\beta) \\
\mathsf{GSS}(b) &= \{ \mathrm{v} \mid \mathrm{v} \in \mathsf{At} \wedge \mathrm{v} \leq b \} & \mathsf{GSS}(\alpha \times \beta) &= \mathsf{GSS}(\alpha) \times \mathsf{GSS}(\beta) \\
\mathsf{GSS}(\alpha + \beta) &= \mathsf{GSS}(\alpha) \cup \mathsf{GSS}(\beta) & \mathsf{GSS}(\alpha^*) &= \mathsf{GSS}(\alpha)^*,
\end{aligned}
$$

where $\mathrm{v} \leq b$ if $\mathrm{v} \to b$ is a propositional tautology. For $T \subseteq \mathcal{T}_{\mathsf{SKAT}}$, let $\mathsf{GSS}(T) = \bigcup_{\alpha \in T} \mathsf{GSS}(\alpha)$. Given two $\mathcal{T}_{\mathsf{SKAT}}$ expressions α and β, we say that they are *equivalent* if $\mathsf{GSS}(\alpha) = \mathsf{GSS}(\beta)$.

3.2 Automata for Guarded Synchronous Strings

We extend to for guarded synchronous strings the automata defined for KAT in [9,10,28]. Besides their simplicity when compared with the two-level automata of Prisacariu, their transitions are labeled with tests instead of atoms, avoiding in this way the inevitable exponential blow-up on the size of the automata induced by the number of valuations of tests.

A *(nondeterministic) automaton with tests* (NTA) over the alphabets Σ and T is a tuple $\mathcal{A} = \langle S, s_0, o, \delta \rangle$, where S is a finite set of states, $s_0 \in S$ is the initial state, $o : S \to \mathcal{B}_{\mathsf{SKAT}}$ is the output function, and $\delta \subseteq \mathcal{P}(S \times (\mathcal{B}_{\mathsf{SKAT}} \times \Sigma) \times S)$ is the transition relation. A synchronous guarded string $\mathrm{v}_0 \sigma_1 \ldots \sigma_n \mathrm{v}_n$, with $n \geq 0$, is accepted by the automaton $\mathcal{A}$ if and only if there is a sequence of states $s_0, s_1, \ldots, s_n \in S$, where s_0 is the initial state, and, for $i = 0, \ldots, n-1$, one has $\mathrm{v}_i \leq b_i$ for some $(s_i, (b_i, \sigma_{i+1}), s_{i+1}) \in \delta$, and $\mathrm{v}_n \leq o(s_n)$. The set of all guarded strings accepted by $\mathcal{A}$ is denoted by $\mathsf{GSS}(\mathcal{A})$. We say that an SKAT expression $\alpha \in \mathcal{T}_{\mathsf{SKAT}}$ is *equivalent* to an automaton $\mathcal{A}$, and write $\alpha = \mathcal{A}$, if $\mathsf{GSS}(\mathcal{A}) = \mathsf{GSS}(\alpha)$.

3.3 Partial Derivatives for SKAT

In the following, we extend the notion of partial derivative, previously defined in [10] for KAT, to SKAT expressions. The main novelty of the approach in [10] is that derivatives are considered only w.r.t. action symbols σ instead of all combinations $\mathrm{v}\sigma$ for $\mathrm{v} \in \mathsf{At}$ and $\sigma \in \Sigma$.

Definition 6. *For $\alpha \in \mathcal{T}_{\mathsf{SKAT}}$ and $\sigma \in \Sigma$, the set $\partial_\sigma(\alpha)$ of partial derivatives of α w.r.t. σ is a subset of $\mathcal{B}_{\mathsf{SKAT}} \times \mathcal{T}_{\mathsf{SKAT}}$ inductively defined as follows,*

$$
\begin{aligned}
\partial_\sigma(a) &= \begin{cases} \{(1,1)\} & \text{if } \sigma = \{a\} \\ \emptyset & \text{otherwise} \end{cases} & \partial_\sigma(\alpha^*) &= \partial_\sigma(\alpha) \odot \alpha^* \\
& & \partial_\sigma(\alpha + \beta) &= \partial_\sigma(\alpha) \cup \partial_\sigma(\beta) \\
\partial_\sigma(b) &= \emptyset & \partial_\sigma(\alpha\beta) &= \partial_\sigma(\alpha) \odot \beta \cup \mathsf{out}(\alpha) \odot \partial_\sigma(\beta)
\end{aligned}
$$

$$
\partial_\sigma(\alpha \times \beta) = \left(\bigcup_{\sigma_1 \times \sigma_2 = \sigma} \partial_{\sigma_1}(\alpha) \otimes \partial_{\sigma_2}(\beta) \right) \cup \mathsf{out}(\alpha) \otimes \partial_\sigma(\beta) \cup \mathsf{out}(\beta) \otimes \partial_\sigma(\alpha),
$$

where $\mathsf{out} : \mathcal{T}_{\mathsf{SKAT}} \longrightarrow \mathcal{B}_{\mathsf{SKAT}}$ is defined by

$$
\begin{aligned}
\mathsf{out}(a) &= 0 & \mathsf{out}(b) &= b & \mathsf{out}(\alpha^*) &= 1 & \mathsf{out}(\alpha + \beta) &= \mathsf{out}(\alpha) + \mathsf{out}(\beta) \\
\mathsf{out}(\alpha \cdot \beta) &= \mathsf{out}(\alpha) \cdot \mathsf{out}(\beta) & & & \mathsf{out}(\alpha \times \beta) &= \mathsf{out}(\alpha) \times \mathsf{out}(\beta),
\end{aligned}
$$

and for $S, T \subseteq \mathcal{B}_{SKAT} \times \mathcal{T}_{SKAT}$, $\alpha \neq 0$ in $\mathcal{T}_{SKAT}$, and $b \neq 0$ in $\mathcal{B}_{SKAT}$, $S \odot \alpha = \{ (b', \alpha'\alpha) \mid (b', \alpha') \in S, \alpha' \neq 0 \}$, $b \odot S = \{ (b \cdot b', \alpha') \mid (b', \alpha') \in S, b' \neq 0 \}$, $S \odot 0 = 0 \odot S = \emptyset$ and $S \otimes T = \{ (b \times b', \alpha \times \alpha') \mid (b, \alpha) \in S, (b', \alpha') \in T, b, b', \alpha, \alpha' \neq 0 \}$. Given $\alpha \in \mathcal{T}_{SKAT}$ and $\sigma \in \Sigma$ we define the set of expressions derived from α w.r.t. a letter σ by $\Delta_\sigma(\alpha) = \{ \alpha' \mid (b, \alpha') \in \partial_\sigma(\alpha) \text{ for some } b \}$. The functions ∂_σ, out, and Δ_σ are naturally extended to sets of SKAT expressions and words $\in \Sigma^\star$.

Let $\Delta(\alpha) = \bigcup_{x \in \Sigma^*} \Delta_x(\alpha)$. Given $\alpha \in \mathcal{T}_{SKAT}$, we define the partial derivative automaton associated to α by $\mathcal{A}(\alpha) = \langle \Delta(\alpha), \alpha, \text{out}, \delta_\alpha \rangle$, where

$$\delta_\alpha = \{ (\gamma, (b, \sigma), \gamma') \mid \gamma \in \Delta(\alpha), (b, \gamma') \in \partial_\sigma(\gamma) \}.$$

In order to justify the correctness of the partial derivative automaton, i.e., to show that $\text{GSS}(\mathcal{A}(\alpha)) = \text{GSS}(\alpha)$, we first note that, using an almost identical proof as for Proposition 4 in Sect. 2, one can show by induction on the structure of $\alpha \in \mathcal{T}_{SKAT}$ that $|\Delta^+(\alpha)| \leq 2^{|\alpha|_{A_B}} - 1$, where again $\Delta^+(\alpha)$ is the set of expressions derived from α excluding the trivial derivation w.r.t. the empty word ε. Thus, $\Delta(\alpha)$ is finite. Finally, the correctness of the partial derivative automaton is guaranteed by the following result.

Proposition 7. Let $\gamma \in SKAT$ and $x \in (\text{At} \times \Sigma)^* \cdot \text{At}$. If $x = \mathbf{v}$, then $x \in \text{GSS}(\gamma)$ if and only if $\mathbf{v} \leq \text{out}(\gamma)$. Furthermore, if $x = \mathbf{v}\sigma x'$, then $x \in \text{GSS}(\gamma)$ if and only if there is some $(b, \gamma') \in \partial_\sigma(\gamma)$, such that $\mathbf{v} \leq b$ and $x' \in \text{GSS}(\gamma')$.

Proof. The proof is by induction on the structure of γ. We only present for *ii.* the cases for $\gamma = \alpha\beta$ and $\gamma = \alpha \times \beta$. Let $\gamma = \alpha\beta$ and $x = \mathbf{v}\sigma x'$. One has $x \in \text{GSS}(\alpha\beta)$ iff $x \in \text{GSS}(\alpha) \diamond \text{GSS}(\beta)$. This means that either, $\mathbf{v} \in \text{GSS}(\alpha)$ and $x \in \text{GSS}(\beta)$, or or $x' = x_1 \diamond x_2$, with $\mathbf{v}\sigma x_1 \in \text{GSS}(\alpha)$ and $x_2 \in \text{GSS}(\beta)$. The former is equivalent to $\mathbf{v} \leq \text{out}(\alpha)$, $\mathbf{v} \leq b$ and $x' \in \text{GSS}(\gamma')$ for some $(b, \gamma') \in \partial_\sigma(\beta)$, i.e. to $\mathbf{v} \leq \text{out}(\alpha)b$ and $x' \in \text{GSS}(\gamma')$ for some $(\text{out}(\alpha)b, \gamma') \in \partial_\sigma(\alpha\beta)$. The latter is equivalent to $\mathbf{v} \leq b$, $x_1 \in \text{GSS}(\gamma')$ and $x_2 \in \text{GSS}(\beta)$ for some $(b, \gamma') \in \partial_\sigma(\alpha)$, i.e. to $\mathbf{v} \leq b$ and $x' = x_1 \diamond x_2 \in \text{GSS}(\gamma') \diamond \text{GSS}(\beta) = \text{GSS}(\gamma'\beta)$ for some $(b, \gamma'\beta) \in \partial_\sigma(\alpha\beta)$.

Consider $\gamma = \alpha \times \beta$ and $x = \mathbf{v}\sigma x'$. One has $x \in \text{GSS}(\alpha \times \beta)$ iff $x \in \text{GSS}(\alpha) \times \text{GSS}(\beta)$. This means that either, $x = (\mathbf{v}\sigma_1 x_1) \times (\mathbf{v}\sigma_2 x_2)$ for some $\mathbf{v}\sigma_1 x_1 \in \text{GSS}(\alpha)$, $\mathbf{v}\sigma_2 x_2 \in \text{GSS}(\beta)$ such that $\sigma = \sigma_1 \cup \sigma_2$ and $x' = x_1 \times x_2$, or $\mathbf{v} \in \text{GSS}(\alpha)$ and $x \in \text{GSS}(\beta)$, or $\mathbf{v} \in \text{GSS}(\beta)$ and $x \in \text{GSS}(\alpha)$. The proof for the two last cases are analogous to the first case for the concatenation. On the other hand, $\mathbf{v}\sigma_1 x_1 \in \text{GSS}(\alpha)$ and $\mathbf{v}\sigma_2 x_2 \in \text{GSS}(\beta)$ is equivalent to $\mathbf{v} \leq b_1$, $x_1 \in \text{GSS}(\gamma_1')$, $\mathbf{v} \leq b_2$ and $x_2 \in \text{GSS}(\gamma_2')$ for some $(b_1, \gamma_1') \in \partial_{\sigma_1}(\alpha)$ and $(b_2, \gamma_2') \in \partial_{\sigma_2}(\beta)$, i.e. to $\mathbf{v} \leq b_1 \times b_2$, $x' = x_1 \times x_2 \in \text{GSS}(\gamma_1') \times \text{GSS}(\gamma_2') = \text{GSS}(\gamma_1' \times \gamma_2')$ for some $(b_1 \times b_2, \gamma_1' \times \gamma_2') \in \partial_\sigma(\alpha \times \beta)$. □

Example 8. Consider the expressions $\alpha = (t_1 p)^\star \neg t_1$ and $\beta = (t_2 pq + \neg t_2 q)$, which represent the programs while t_1 do p and if t_2 then $p; q$ else q, respectively. The partial derivative automaton for $\alpha \times \beta$, corresponding to the synchronous execution of both programs is the following.

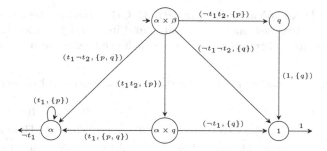

To test the equivalence of SKAT terms we can consider the algorithm that tests the equivalence of NTAs as presented by Broda *et al.* [10], and implicitly use the definition of the partial derivative automaton associated to an SKAT expression.

4   Experimental Results

We implemented (in Python) the algorithm by Broda *et al.* for testing NTAs equivalence and performed some experiments[1]. The implementation uses BDDs (binary decision diagrams) for dealing with boolean functions. To compare the performance of the new NTAs with respect to the ones that use explicitly derivatives w.r.t $v\sigma \in \text{At}\Sigma$ we considered the same experiments as in Almeida [2, Sect. 3.5.2]. Each sample has 10000 KAT expressions generated uniformly at random of a given size. For each sample we performed two experiments: (1) we tested the equivalence of each KAT expression against itself; (2) we tested the equivalence of two consecutively generated KAT expressions. For each pair of KAT expressions we measured: the number of pairs of derivatives generated (H), the number of iterations (it), which gives an estimate of the boolean assignments that must be tested for each program symbol, and the number ($|\alpha|_T$) of tests of T in each expression. Table 2 summarizes both the results obtained and the ones obtained by Almeida. Each row corresponds to a sample, where the three first columns characterize the sample, respectively, $|A_B|$ (k), $|T|$ (l), and the length of each KAT expression generated. Rows *a.* to *e.* contain our results, and corresponding ones obtained by Almeida are listed in rows *f.* to *j.*. Column four has the number of elements of T in each expression ($|\alpha|_T$). Columns five and seven give the average size of H in the experiment (1) and (2), respectively. Columns six and eight have the number of iterations. These two columns have no entries for Almeida's results, as in that algorithm all assignments of $|At|$ are considered for each symbol of $|A_B|$. Finally, the last two columns are the average times, in seconds, of each experiment. For Almeida's results the implementation was in Ocaml and the values where obtained with an Intel® Xeon® 5140 at 2.33 GHz with 4 GB of RAM, whereas the new values were obtained with an AMD®

[1] Source code at http://www.dcc.fc.up.pt/~nam/web/resources/katexp.tgz.

Phenom(tm)® II X4 955 ar 3.20 GHz with 32 GB of RAM. The most significative cases are the two last ones, in d. and e. and in i. and j. respectively, where a substantial performance improvement was achieved with the new algorithm.

Table 2. Experimental results for uniformly random generated KAT expressions

	1	2	3	4	5	6	7	8	9	10
	k	l	$|\alpha|$	$|\alpha|_\mathsf{T}$	$H(1)$	$it(1)$	$H(2)$	$it(2)$	Time(1)	Time(2)
a	5	5	50	10.33	9.21	149	0.49	0.19	0.08552	0.00148
b	5	5	100	19.55	15.74	2854	0.66	0.88	2.7256	0.00278
c	10	10	50	10.32	11.61	59.56	0.30	0.03	0.05424	0.0035
d	10	10	100	19.89	20.87	516	0.35	0.09	1.1969	0.01274
e	15	15	50	10.31	12.78	50.7	0.25	0.013	0.0616	0.00738
f	5	5	50	9.98	7.35	n.a	0.53	n.a	0.0097	0.00087
g	5	5	100	19.71	15.74	n.a	0.76	n.a	0.0875	0.00223
h	10	10	50	11.12	8.30	n.a	0.50	n.a	0.5050	0.30963
i	10	10	100	21.93	16.78	n.a	0.67	n.a	20.45	1.31263
j	15	15	50	11.57	8.47	n.a	0.47	n.a	6.4578	55.22

Damien Pous developed an equivalence test for symbolic automata [31] and performed some tests for KAT terms [24]. To ensure equivalence of a pair of KAT terms (α_1, α_2) he added $\mathsf{A_B}^\star$ to each term. We ran a similar test considering a sample of 10000 pairs of KAT terms with $k = 7$, $l = 7$ and $|\alpha| = 100$. The values obtained were $|\alpha|_\mathsf{T} = 18.58$, $H = 41.34$, $it = 1745$ and $Time = 1.9456$, which are competitive with the ones in [24] (for Antimirov's algorithm).

5   Conclusion

In this paper we extended the notion of derivative to sets of (guarded) synchronous strings and showed that the methods based on derivatives lead to simple and elegant decision procedures for testing SKA and SKAT expressions equivalence. Based on our experiments, it may be worthwhile to study the average-case size of the SKA automata, in the analytic combinatorics framework. We also implemented the new class of SKAT automata based on NTAs automata. As the performance of testing NTA equivalence seems competitive we believe that our extension to SKAT automata is also much more efficient than the one proposed by Prisacariu.

References

1. Almeida, M., Moreira, N., Reis, R.: Testing regular languages equivalence. J. Automata Lang. Comb. **15**(1/2), 7–25 (2010)

2. Almeida, R.: Decision algorithms for Kleene algebra with tests and Hoare logic. Master's thesis, Faculdade de Ciências da Universidade do Porto, July 2012. http://www.dcc.fc.up.pt/~nam/web/resources/docs/thesisRA.pdf

3. Almeida, R., Broda, S., Moreira, N.: Deciding KAT and Hoare logic with derivatives. In: Faella, M., Murano, A. (eds.) 3rd GANDALF. EPTCS, vol. 96, pp. 127–140 (2012)

4. Antimirov, V.M.: Partial derivatives of regular expressions and finite automaton constructions. Theoret. Comput. Sci. **155**(2), 291–319 (1996)

5. Berry, G., Gonthier, G.: The Esterel synchronous programming language: design, semantics, implementation. Sci. Comput. Program. **19**(2), 87–152 (1992)

6. Bonchi, F., Pous, D.: Checking NFA equivalence with bisimulations up to congruence. In: Giacobazzi, R., Cousot, R. (eds.) POPL 2013, pp. 457–468. ACM (2013)

7. Braibant, T., Pous, D.: Deciding Kleene algebras in Coq. Log. Methods Comput. Sci. **8**(1), 1–42 (2012)

8. Broda, S., Machiavelo, A., Moreira, N., Reis, R.: On the average size of Glushkov and partial derivative automata. Int. J. Found. Comput. Sci. **23**(5), 969–984 (2012)

9. Broda, S., Machiavelo, A., Moreira, N., Reis, R.: On the average size of Glushkov and equation automata for KAT expressions. In: Gąsieniec, L., Wolter, F. (eds.) FCT 2013. LNCS, vol. 8070, pp. 72–83. Springer, Heidelberg (2013)

10. Broda, S., Machiavelo, A., Moreira, N., Reis, R.: On the equivalence of automata for KAT-expressions. In: Beckmann, A., Csuhaj-Varjú, E., Meer, K. (eds.) CiE 2014. LNCS, vol. 8493, pp. 73–83. Springer, Heidelberg (2014)

11. Broda, S., Machiavelo, A., Moreira, N., Reis, R.: Partial derivative automaton for regular expressions with shuffle. In: Shallit, J., Okhotin, A. (eds.) DCFS 2015. LNCS, vol. 9118, pp. 21–32. Springer, Heidelberg (2015)

12. Coquand, T., Siles, V.: A decision procedure for regular expression equivalence in type theory. In: Jouannaud, J.-P., Shao, Z. (eds.) CPP 2011. LNCS, vol. 7086, pp. 119–134. Springer, Heidelberg (2011)

13. Glushkov, V.M.: The abstract theory of automata. Russ. Math. Surv. **16**, 1–53 (1961)

14. Hopcroft, J., Karp, R.M.: A linear algorithm for testing equivalence of finite automata. Technical report TR 71-114, University of California, Berkeley, California (1971)

15. Kozen, D.: Kleene algebra with tests. Trans. Prog. Lang. Syst. **19**(3), 427–443 (1997)

16. Kozen, D.: On Hoare logic and Kleene algebra with tests. ACM Trans. Comput. Log. **1**(1), 60–76 (2000)

17. Kozen, D.: Automata on guarded strings and applications. Matématica Contemporânea **24**, 117–139 (2003)

18. Kozen, D.: On the coalgebraic theory of Kleene algebra with tests. Technical report, Cornell University (2008). http://hdl.handle.net/1813/10173

19. Krauss, A., Nipkow, T.: Proof pearl: regular expression equivalence and relation algebra. J. Autom. Reasoning **49**, 95–109 (2011)

20. Milner, R.: Communication and concurrency. PHI Series in computer science. Prentice Hall, Upper Saddle River (1989)

21. Moreira, N., Pereira, D., Melo de Sousa, S.: Deciding regular expressions (in-) equivalence in Coq. In: Kahl, W., Griffin, T.G. (eds.) RAMICS 2012. LNCS, vol. 7560, pp. 98–113. Springer, Heidelberg (2012)

22. Nipkow, T., Traytel, D.: Unified decision procedures for regular expression equivalence. In: Klein, G., Gamboa, R. (eds.) ITP 2014. LNCS, vol. 8558, pp. 450–466. Springer, Heidelberg (2014). Archive of Formal Proofs 2014

23. Pereira, D.: Towards certified program logics for the verification of imperative programs. Ph.D. thesis, University of Porto (2013)
24. Pous, D.: Symbolic algorithms for language equivalence and Kleene algebra with tests. In: Rajamani, S.K., Walker, D. (eds.) 42nd POPL 2015, pp. 357–368. ACM (2015)
25. Prisacariu, C.: Synchronous Kleene algebra. J. Log. Algebr. Program. **79**(7), 608–635 (2010)
26. Project FAdo: FAdo: tools for formal languages manipulation. http://fado.dcc.fc.up.pt/. (Accessed on 01 April 2015)
27. Rot, J., Bonsangue, M., Rutten, J.: Coinductive proof techniques for language equivalence. In: Dediu, A.-H., Martín-Vide, C., Truthe, B. (eds.) LATA 2013. LNCS, vol. 7810, pp. 480–492. Springer, Heidelberg (2013)
28. Silva, A.: Position automata for Kleene algebra with tests. Sci. Ann. Comp. Sci. **22**(2), 367–394 (2012)
29. Synopsys: Esterel studio. http://www.synopsys.com/home.aspx
30. Thompson, K.: Regular expression search algorithm. Commun. ACM **11**(6), 410–422 (1968)
31. Veanes, M.: Applications of symbolic finite automata. In: Konstantinidis, S. (ed.) CIAA 2013. LNCS, vol. 7982, pp. 16–23. Springer, Heidelberg (2013)

On the Hierarchy of Block Deterministic Languages

Pascal Caron, Ludovic Mignot$^{(\boxtimes)}$, and Clément Miklarz

LITIS, Université de Rouen, 76801 Saint-Étienne du Rouvray Cedex, France
{pascal.caron,ludovic.mignot,clement.miklarz1}@univ-rouen.fr

Abstract. A regular language is k-block deterministic if it is specified by a k-block deterministic regular expression. This subclass of regular languages has been introduced by Giammarresi *et al.* as a possible extension of one-unambiguous regular languages defined and characterized by Brüggemann-Klein and Wood. We first show that each k-block deterministic regular language is the alphabetic image of some one-unambiguous regular language. Moreover, we show that the conversion from a minimal DFA of a k-block deterministic regular language to a k-block deterministic automaton not only requires state elimination, and that the proof given by Han and Wood of a proper hierarchy in k-block deterministic languages based on this result is erroneous. Despite these results, we show by giving a parameterized family that there is a proper hierarchy in k-block deterministic regular languages.

1 Introduction

A Document Type Definition (DTD) containing a grammar is used to know whether an XML file fits some specification. These grammars are made of rules whose right-hand part is a restricted regular expression. Brüggemann-Klein and Wood have formalized these regular expressions and have shown that the set of languages specified is strictly included in the set of regular ones. The distinctive aspect of such expressions is the one-to-one correspondence between each letter of the input word and a unique position in them. The resulting Glushkov automaton is deterministic. The languages specified are called one-unambiguous regular languages.

Several extensions of one-unambiguous expressions have been considered:

- k-block deterministic regular expressions [4] are such that while reading an input word, there is a one-to-one correspondence between the next at most k input symbols and the same number of symbols of the expression. These expressions have particular Glushkov automata. The transitions of these automata can be labeled by words of length at most k and for every couple of words labeling two output transitions of a single state, these words are not prefix from each other.
- k-lookahead regular expressions form another generalization. This time, the reading of the next k symbols of the input word allows one to know the next position in the expression. This extension has been proposed in [6].

© Springer International Publishing Switzerland 2015
F. Drewes (Ed.): CIAA 2015, LNCS 9223, pp. 63–75, 2015.
DOI: 10.1007/978-3-319-22360-5_6

– (k, l)-unambiguous regular expressions [3] is another extension of one-unambiguity, where the next k symbols may induce several paths, but with at most
one common state.

These three families of expressions fit together as families of languages in the
way that a language is k-block deterministic (resp. k-lookahead deterministic,
(k, l)-unambiguous) if there exists a k-block deterministic (resp. k-lookahead
deterministic, (k, l)-unambiguous) expression to represent it.

Preliminaries are gathered in Sect. 2. In Sect. 3, we recall several results
from [4,6] on which we question their truthfulness. Indeed, we show in Sect. 4
that, due to an erroneous statement of Lemma 4, the witness family given as
a proof of Theorem 3 is invalid; and present an alternative family, proving the
infinite hierarchy of k-block deterministic regular languages w.r.t. k.

2   Preliminaries

2.1   Languages and Automata Basics

Let Σ be a non-empty finite *alphabet*. A *word* w over Σ is a finite sequence of
symbols from Σ. The *length* of a word w is denoted by $|w|$, and the *empty word*
is denoted by ε. The word x is a *prefix* of w if there exists a word u such that
$w = xu$. The set of all prefixes of w is denoted by $\mathrm{Pref}(w)$.

Let Σ^* denote the set of all words over Σ. A *language over* Σ is a subset of
Σ^*. Let L and L' be two languages over Σ. The following operations are defined:

– *the union*: $L \cup L' = \{w \mid w \in L \vee w \in L'\}$
– *the concatenation*: $L \cdot L' = \{w \cdot w' \mid w \in L \wedge w' \in L'\}$
– *the Kleene star*: $L^* = \bigcup_{k \in \mathbb{N}} L^k$ with $L^0 = \{\varepsilon\}$ and $L^{k+1} = L \cdot L^k$

A *regular expression over* Σ is built from $\emptyset$ (the empty set), ε, and symbols
in Σ using the binary operators $+$ and $\cdot$, and the unary operator *. The *language*
$\mathrm{L}(E)$ *specified by a regular expression* E is defined as follows:

$$\mathrm{L}(\emptyset) = \emptyset, \qquad\qquad \mathrm{L}(\varepsilon) = \{\varepsilon\}, \qquad\qquad \mathrm{L}(a) = \{a\},$$
$$\mathrm{L}(F + G) = \mathrm{L}(F) \cup \mathrm{L}(G), \quad \mathrm{L}(F \cdot G) = \mathrm{L}(F) \cdot \mathrm{L}(G), \quad \mathrm{L}(F^*) = \mathrm{L}(F)^*,$$

with $a \in \Sigma$, and F, G some regular expressions over Σ. Given a language L,
if there exists a regular expression E such that $\mathrm{L}(E) = L$, then L is a *regular
language*.

A *finite automaton* A is a 5-tuple $(\Sigma, Q, I, F, \delta)$ where: Q is a finite set of
states, $I \subset Q$ is the set of initial states, $F \subset Q$ is the set of final states, and
$\delta \subset Q \times \Sigma \times Q$ is a set of transitions. The set δ is equivalent to a function
of $Q \times \Sigma \rightarrow 2^Q$: $(p, a, q) \in \delta \iff q \in \delta(p, a)$. This function can be extended
to $2^Q \times \Sigma^* \rightarrow 2^Q$ as follows: for any subset $Q' \subset Q$, for any symbol $a \in \Sigma$,
for any word $w \in \Sigma^*$: $\delta(Q', \varepsilon) = Q'$, $\delta(Q', a) = \bigcup_{q \in Q'} \delta(q, a)$, $\delta(Q', a \cdot w) =
\delta(\delta(Q', a), w)$; finally, we set $\delta(q, w) = \delta(\{q\}, w)$. The *language* $\mathrm{L}(A)$ *recognized*
by A is the set $\{w \in \Sigma^* \mid \delta(I, w) \cap F \neq \emptyset\}$. Two automata are *equivalent* if they

recognize the same language. The *right language of a state q of A* is denoted by $L_q(A) = \{w \in \Sigma^* \mid \delta(q,w) \cap F \neq \emptyset\}$. Two states are *equivalent* if they have the same right language.

An automaton $A = (\Sigma, Q, I, F, \delta)$ is *standard* if $|I| = 1$ and $\forall q \in Q, \forall a \in \Sigma, \delta(q,a) \cap I = \emptyset$. If A is not a standard automaton, then it is possible to compute an equivalent standard automaton $(\Sigma, Q_s, I_s, F_s, \delta_s)$ as follows:

- $Q_s = Q \cup \{i_s\}$ with $i_s \notin Q$
- $I_s = \{i_s\}$
- $F_s = F \cup \{i_s\}$ if $I \cap F \neq \emptyset$, F otherwise
- $\delta_s = \delta \cup \{(i_s, a, q) \mid \exists i \in I, (i, a, q) \in \delta\}$

This operation is called *standardization*.

An automaton $A = (\Sigma, Q, I, F, \delta)$ is *deterministic* if $|I| = 1$ and $\forall t_1 = (p, a, q_1)$, $t_2 = (p, b, q_2) \in \delta, (t_1 \neq t_2) \implies (a \neq b)$. If A is not deterministic, it is possible to compute an equivalent deterministic automaton by using the powerset construction described in [10].

A deterministic automaton $A = (\Sigma, Q_A, \{i_A\}, F_A, \delta_A)$ is *minimal* if there is no equivalent deterministic automaton $B = (\Sigma, Q_B, \{i_B\}, F_B, \delta_B)$ such that $|Q_B| < |Q_A|$. If A is not minimal, it is possible to compute an equivalent minimal deterministic automaton by merging equivalent states [7,9]. Notice that two equivalent minimal deterministic automata are isomorphic.

Kleene's Theorem [8] asserts that the set of the languages specified by regular expressions is the same as the set of languages recognized by finite automata. The conversion of regular expressions into automata has been deeply studied, *e.g.* by Glushkov [5]. To differentiate each occurence of the same symbol in a regular expression, a *marking* of all the symbols of the alphabet is performed by indexing them with their relative position in the expression. The marking of a regular expression E produces a new regular expression denoted by $E^\sharp$ over the alphabet of indexed symbols denoted by Π_E where each indexed symbol occurs at most once in $E^\sharp$. The reverse of marking is the *dropping* of subscripts, denoted by $\natural$, such that if $x \in \Pi_E$ and $x = a_k$, then $x^\natural = a$.

Let E be a regular expression over an alphabet Σ. The following functions are defined:

- $\mathrm{Null}(E) = \{\varepsilon\}$ if $\varepsilon \in L(E)$, $\emptyset$ otherwise
- $\mathrm{First}(E) = \{x \in \Sigma \mid \exists w \in \Sigma^*, xw \in L(E)\}$
- $\mathrm{Last}(E) = \{x \in \Sigma \mid \exists w \in \Sigma^*, wx \in L(E)\}$
- $\mathrm{Follow}(E, x) = \{y \in \Sigma \mid \exists u, v \in \Sigma^*, uxyv \in L(E)\}, \forall x \in \Sigma$

From these functions, an automaton recognizing $L(E)$ can be computed:

Definition 1. *The* Glushkov *automaton of a regular expression E over an alphabet Σ is denoted by $G_E = (\Sigma, Q_E, I_E, F_E, \delta_E)$ with:*

- $Q_E = \Pi_E \cup \{i\}$
- $I_E = \{i\}$
- $F_E = \mathrm{Last}(E^\sharp) \cup \{i\}$ *if* $\mathrm{Null}(E^\sharp) = \{\varepsilon\}$, $\mathrm{Last}(E^\sharp)$ *otherwise*

- $\delta_E = \{(x,a,y) \in \Pi_E \times \Sigma \times \Pi_E \mid y \in \text{Follow}(E^\sharp, x) \wedge a = y^\natural\}$
  $\cup\{(i,a,y) \in \{i\} \times \Sigma \times \Pi_E \mid y \in \text{First}(E^\sharp) \wedge a = y^\natural\}$

Finally, an automaton is a *Glushkov automaton* if it is the Glushkov automaton of a regular expression E.

Example 1. Let $E = (a + b)^*a + \varepsilon$. Then $E^\sharp = (a_1 + b_2)^*a_3 + \varepsilon$ with $\Pi_E = \{a_1, b_2, a_3\}$, and G_E is given in Fig. 1.

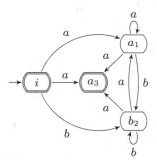

Fig. 1. The Glushkov automaton G_E of $E = (a + b)^*a + \varepsilon$

2.2   One-Unambiguous Regular Languages

We present the notion of one-unambiguity introduced in [1].

Definition 2. *A regular expression E is* one-unambiguous *if G_E is deterministic. A regular language is* one-unambiguous *if it is specified by some one-unambiguous regular expression.*

Brüggemann-Klein and Wood showed that the one-unambiguity of a regular language is stucturally decidable over its minimal DFA. This decision procedure is related to the strongly connected components of the underlying graph and to their links with the remaining parts.

Let $A = (\Sigma, Q, I, F, \delta)$ be a deterministic automaton. A set $O \subset Q$ is called an *orbit* if it is a strongly connected component. An orbit is *trivial* if it consists of only one state and there is no transition from it to itself in A. *The orbit of a state q*, denoted by $O(q)$ is the orbit to which q belongs. The set of orbits of A is denoted by $\mathcal{O}_A$. Let $O \in \mathcal{O}_A$ be an orbit and $p \in O$ be a state. The state p is a *gate of O* if $(p \in F) \vee (\exists a \in \Sigma, \exists q \in (Q \setminus O), q \in \delta(p, a))$. The set of gates of O is denoted by $G(O)$. The automaton A has the *orbit property* if all the gates of each orbit have identical connections to the outside. More formally:

Definition 3. *An automaton $A = (\Sigma, Q, I, F, \delta)$ has the* orbit property *if, for any orbit O in $\mathcal{O}_A$, for any two states (p, q) in $G(O)$, the two following conditions are satisfied:*

- $p \in F \Longrightarrow q \in F$,
- $\forall r \in (Q \setminus O), \forall a \in \Sigma, r \in \delta(p, a) \Longrightarrow r \in \delta(q, a)$.

Let $q \in Q$ be a state. The *orbit automaton* A_q *of the state* q *in* A is the automaton obtained by restricting the states and the transitions of A to $O(q)$ with initial state q and final states $G(O(q))$. For any state $q \in Q$, the languages $L(A_q)$ are called the *orbit languages of* A. A symbol $a \in \Sigma$ is A-*consistent* if there exists a state $q_a \in Q$ such that all final states of A have a transition labelled by a to q_a. A set S of symbols is A-*consistent* if each symbol in S is A-consistent. The S-*cut* A_S of A is constructed from A by removing, for each $a \in S$, all transitions labelled by a that leave a final state of A. All these notions can be used to characterize one-unambiguous regular languages:

Theorem 1 ([1]). *Let* M *be a minimal deterministic automaton and* S *be a* M-*consistent set of symbols. Then,* $L(M)$ *is one-unambiguous if and only if:*

1. *the* S-*cut* M_S *of* M *has the orbit property*
2. *all orbit languages of* M_S *are one-unambiguous.*

Furthermore, if M *consists of a single non-trivial orbit and* $L(M)$ *is one-unambiguous,* M *has at least one* M-*consistent symbol.*

This theorem suggests an inductive algorithm to decide, given a minimal deterministic automaton M whether $L(M)$ is one-unambiguous: the *BKW test*. Furthermore, the theorem defines a sufficient condition over non-minimal deterministic automaton:

Lemma 1 ([1]). *Let* A *be a deterministic automaton and* M *be its equivalent minimal deterministic automaton.*

1. *If* A *has the orbit property, then so does* M
2. *If all orbit languages of* A *are one-unambiguous, then so are all orbit languages of* M.

Consequently, the BKW test is extended to deterministic automata which are not minimal. Reinterpreting the results in [1], it can be shown that

Lemma 2. *The Glushkov automaton of a one-unambiguous regular expression passes the BKW test.*

2.3   Block Deterministic Regular Languages

We present the notion of block determinism introduced in [4].

Let Σ be an alphabet and k be an integer. The *set of blocks* $B_{\Sigma,k}$ is the set $\{w \mid w \in \Sigma^* \wedge 1 \leq |w| \leq k\}$. The notions of regular expression and automaton can be extended to ones over set of blocks. Let E be a regular expression over Γ and $A = (\Gamma, Q, I, F, \delta)$ be an automaton. Let Σ be an alphabet and k be an integer, if $\Gamma \subset B_{\Sigma,k}$ then E and A are (Σ, k)-*block*. And since $\Gamma \subset B_{\Sigma,k} \subset \Sigma^*$, a language over Γ is also a language over Σ. To distinguish blocks as syntactic components in a regular expression, we write them between square brackets. Those are omitted for one letter blocks. The notion of determinism can be extended to block-determinism.

Definition 4. *An automaton $A = (\Gamma, Q, I, F, \delta)$ is k-block deterministic if the following conditions hold:*

- *there exists an alphabet Σ such that A is (Σ, k)-block,*
- $|I| = 1$,
- $\forall t_1 = (p, b_1, q_1), t_2 = (p, b_2, q_2) \in \delta, (t_1 \neq t_2) \Longrightarrow (b_1 \notin \mathrm{Pref}(b_2))$.

Since $\Sigma = B_{\Sigma,1}$, regular expressions and automata can be considered as ones over a set of blocks. Moreover, the blocks can be treated as single symbols, as we do when we refer to the elements of an alphabet. With this assumption, the marking of block regular expressions induces the construction of a Glushkov automaton from a block regular expression, and the usual automaton transformations such as determinization and minimization can be easily performed.

Example 2. Let $E = [aa]^*([ab]b + ba)b^*$. Then $E^\sharp = [aa]_1^*([ab]_2 b_3 + b_4 a_5)b_6^*$, and G_E is given in Fig. 2.

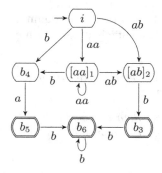

Fig. 2. The $(\Sigma, 2)$-block Glushkov automaton G_E

Finally, the block determinism of a Glushkov automaton can be used to extend the block determinism to block expression:

Definition 5. *A block regular expression E is k-block deterministic if G_E is k-block deterministic. A regular language is k-block deterministic if it is specified by some k-block deterministic regular expressions.*

Example 3. Since the Glushkov automaton in Fig. 2 is 2-block deterministic, $L([aa]^*([ab]b + ba)b^*)$ is 2-block deterministic.

Let $A = (\Sigma, Q, I, F, \delta)$ be an automaton and Γ be a set. Then the automaton $B = (\Gamma, Q, I, F, \delta')$ is an *alphabetic image* of A if there exists an injection ϕ from Σ to Γ such that $\delta' = \{(p, \phi(a), q) \mid (p, a, q) \in \delta\}$. In this case, we set $B = \phi(A)$. Caron and Ziadi showed in [2] that an automaton is a Glushkov one if and only if the two conditions hold:

- it is homogeneous (for any state q, for any two transitions (p, a, q) and (r, b, q), the symbols a and b are the same);
- it satisfies some structural properties over the transition structure.

One can check that any injection ϕ from Σ to Γ preserves such conditions, since the alphabetical image preserves the transition structure by only changing the symbol labeling a transition. Therefore

Lemma 3. *The alphabetic image of an automaton A is a Glushkov automaton if and only if A is a Glushkov automaton.*

Let us show that the BKW test can be used to characterize the k-block determinism of a regular language:

Theorem 2. *A regular language L is k-block deterministic if and only if it is recognized by a k-block deterministic automaton K such that K is the alphabetic image of a deterministic automaton which passes the BKW test.*

Proof. Let us show the double implication.

1. Let L be a k-block deterministic regular language over Σ. Then there exists a k-block deterministic Glushkov automaton $K = (B_{\Sigma,k}, Q, \{i\}, F, \delta_K)$ that recognizes L. Let $\Pi = \{[b] \mid b \in B_{\Sigma,k}\}$ be an alphabet, $\varphi : \Pi \to B_{\Sigma,k}$ be the bijection such that for every $[b] \in \Pi, \varphi([b]) = b$. Let $A = (\Pi, Q, \{i\}, F, \delta_A)$ be a Glushkov automaton such that $K = \varphi(A)$. Let us suppose that A is not deterministic. Then, there exist two transitions $(p, a, q), (p, a, r) \in \delta_A$ such that $q \neq r$. Thus, $(p, \varphi(a), q), (p, \varphi(a), r) \in \delta_K$, which contradicts the fact that K is k-block deterministic. So, A is a deterministic Glushkov automaton, and therefore passes the BKW test following Lemma 2.
2. Let $A = (\Pi, Q_A, \{i_A\}, F_A, \delta_A)$ be a deterministic automaton which passes the BKW test, $K = \{\Gamma, Q_A, \{i_A\}, F_A, \delta_K\}$ be a k-block deterministic automaton, and $\varphi : \Pi \to \Gamma$ be an injection such that $K = \varphi(A)$. Now, $\varphi : \Pi \to \Gamma$ is extended into the morphism $\varphi : \Pi^* \to \Gamma^*$ such that for every letter $a \in \Pi$ and every word $w \in \Pi^*$ we have $\varphi(a \cdot w) = \varphi(a) \cdot \varphi(w)$ and $\varphi(\varepsilon) = \varepsilon$. In this case, $L(K) = \varphi(L(A))$. Since A passes the BKW test, there exists an equivalent deterministic Glushkov automaton $G = (\Pi, Q_G, \{i_G\}, F_G, \delta_G)$. Following Lemma 3, there also exists a Glushkov automaton $H = (\Gamma, Q_G, \{i_G\}, F_G, \delta_H)$ such that $H = \varphi(G)$ and $L(H) = \varphi(L(G))$. Since A and G are equivalent deterministic automata, $\varphi(L(G)) = \varphi(L(A))$. And so $L(H) = L(K)$. Let us suppose that H is not k-block deterministic, then there exist two transitions $(p_H, \varphi(a), q_H), (p_H, \varphi(b), r_H) \in \delta_H$ such that either $(\varphi(a) = \varphi(b)) \wedge (q_H \neq r_H)$

or $(\varphi(a) \neq \varphi(b)) \wedge (\varphi(a) \in \mathrm{Pref}(\varphi(b)))$. By definition, (p_H, a, q_H), $(p_H, b, r_H) \in$ δ_G. But since G and A are equivalent deterministic automata, there exist two transitions (p_A, a, q_A), $(p_A, b, r_A) \in \delta_A$, and by definition, $(p_A, \varphi(a), q_A)$, $(p_A, \varphi(b), r_A) \in \delta_K$. Let us suppose that $(\varphi(a) = \varphi(b)) \wedge (q_h \neq r_h)$. Since φ is an injection, $(a = b) \wedge (q_h \neq r_h)$, which contradicts the fact that G is deterministic. So let us suppose that $(\varphi(a) \neq \varphi(b)) \wedge (\varphi(a) \in \mathrm{Pref}(\varphi(b)))$, it contradicts the fact that K is k-block deterministic. Therefore, H is a k-block deterministic Glushkov automaton, and $\mathrm{L}(K)$ is k-block deterministic.   □

It has been proved that one-unambiguous regular languages are a proper subfamily of k-block deterministic regular languages. Therefore one can wonder whether there exists an infinite hierarchy in k-block deterministic regular languages regarding k. That has been achieved by Han and Wood [6], but with an invalid assumption.

3   Previous Results on Block-Deterministic Languages

In [4], a method is presented for creating from a block automaton an equivalent block automaton with larger blocks by eliminating states while preserving the right language of every other states.

Let $A = (\Gamma, Q, I, F, \delta)$ be a block automaton. The *state elimination of q in A* creates a new block automaton, denoted by $\mathcal{S}(A, q)$, computed as follows: first, the state q and all transitions going in and out of it are removed; second, for every two transitions (r, u, q) and (q, v, s) in δ, the transition (r, uv, s) is added. This transformation is illustrated in Fig. 3.

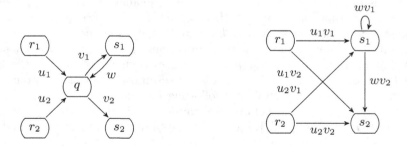

Fig. 3. The state elimination of the state q

Definition 6. *Let $A = (\Gamma, Q, I, F, \delta)$ be a block automaton. A state $q \in Q$ satisfies the* state elimination precondition *if it is neither an initial state nor a final state and it has no self-loops.*

The state elimination is extended to a set $S \subset Q$ of states if every state in S satisfies the state elimination precondition, and the subgraph induced by S is acyclic. In this case, we can eliminate the states in S in any order. Giammarresi *et al.* [4] suggest that state elimination is sufficient to decide the k-block determinism of a regular language.

Lemma 4 ([4,6]). *Let M be a minimal deterministic automaton of a k-block-deterministic regular language. We can transform M to a k-block deterministic automaton that satisfies the orbit property using state elimination.*

Using this lemma, Han and Wood stated that:

Theorem 3 ([6]). *There is a proper hierarchy in k-block-deterministic regular languages.*

Proof. Han and Wood exhibited the family of languages L_k specified by regular expressions $E_k = ([a^k])^*([a^{k-1}b]b + ba)b^*$ whose minimal deterministic automata M_k are represented in Fig. 4. Following Lemma 4, there is no other choice but to eliminate states q_1 to q_{k-1}, in any order, to have the orbit property. Thus, L_k is k-block deterministic and not $(k-1)$-block deterministic. ☐

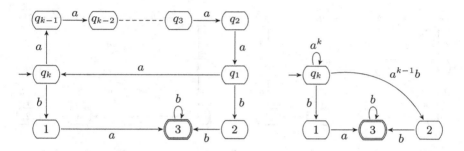

Fig. 4. The minimal deterministic automaton M_k and its equivalent k-block deterministic automaton after having eliminated states q_1 to q_{k-1}

4   A Witness for the Infinite Hierarchy

In this section, we exhibit a counter-example for Lemma 4. We can find a k-block deterministic language with a minimal deterministic automaton from which we cannot get any k-block deterministic automaton that satisfies the orbit property. In Fig. 5, the leftmost automaton is minimal and none of its states can be eliminated. However, by applying standardization, we create an equivalent deterministic automaton from which we can eliminate the state i to get the rightmost equivalent 2-block deterministic automaton.

This clearly shows that the only action of state elimination is not enough to decide whether a language is k-block deterministic. Using this operation, we show that:

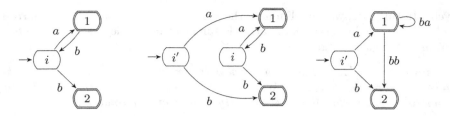

Fig. 5. The counter-example

Proposition 1. $\forall k \in \mathbb{N} \setminus \{0\}$, *the language L_k is 2-block deterministic.*

Proof. As shown in Fig. 6, we can always standardize M_k, proceed to the state elimination of q_k and get a 2-block deterministic automaton which respects the conditions stated in Theorem 2. Thus, L_k is 2-block deterministic and is specified by the regular expressions $F_k = (a^{k-1}([aa]a^{k-2})^*([ab]a + bb) + ba)b^*$.   □

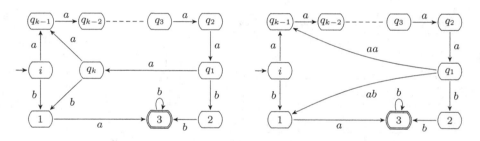

Fig. 6. The standardization of M_k followed by the state elimination of q_k

However, Theorem 3 is still correct since we can give proper details about the proof with our own parameterized family of languages. Let $k \in \mathbb{N} \setminus \{0\}$ be an integer and $A_k = (\Sigma, Q_k, I_k, F_k, \delta_k)$ be the automaton (given in Fig. 7) such that:

- $\Sigma = \{a, b, c\}$
- $Q_k = \{f\} \cup \{\alpha_j, \beta_j \mid 1 \leq j \leq k\}$
- $I_k = \{\beta_k\}$
- $F_k = \{f\} \cup \{\alpha_k, \beta_k\}$
- $\delta_k = \Delta_k \cup \Gamma_k$ with:
  - $\Delta_k = \{(\beta_k, a, \alpha_k), (\beta_1, b, f), (\alpha_k, a, \alpha_k), (\alpha_1, b, f), (\alpha_1, c, \beta_k)\}$
  - $\Gamma_k = \{(\alpha_j, b, \alpha_{j-1}), (\beta_j, b, \beta_{j-1}) \mid 2 \leq j \leq k\}$

First of all, let us notice that the word $b^j \in L(A_k)$ if and only if $j = k$. Thus, for all $k \neq k'$, $L(A_k) \neq L(A_{k'})$. Furthermore,

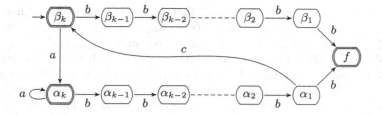

Fig. 7. The k-block deterministic automaton A_k

Proposition 2. $\forall k \in \mathbb{N} \setminus \{0\}$, $\mathrm{L}(A_k)$ *is k-block deterministic.*

Proof. By construction, for all k, A_k is trimmed and deterministic. So, any automaton that we can get from eliminating states such that the state elimination precondition is respected is a block deterministic automaton.

For any integer k in $\mathbb{N} \setminus \{0\}$, we can eliminate the set of states $\{\alpha_j, \beta_j \mid 1 \leq j \leq k - 1\}$ because none of these states are initial or final and their induced subgraph is acyclic. Thus, we can get a k-block deterministic automaton B_k, such that $\mathrm{L}(B_k) = \mathrm{L}(A_k)$, shown in Fig. 8. Obviously B_k respects the conditions stated in Theorem 2, so $\mathrm{L}(A_k)$ is k-block deterministic. Furthermore, it can be checked that $\mathrm{L}(A_k)$ is specified by the k-block deterministic regular expression $(a(\varepsilon + [b^{k-1}c]))^*(\varepsilon + [b^k])$. □

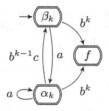

Fig. 8. The k-block deterministic automaton B_k

Finally, let us show that the index cannot be reduced:

Proposition 3. $\forall k \in \mathbb{N} \setminus \{0\}$, $\mathrm{L}(A_k)$ *is not $(k-1)$-block deterministic.*

Proof. Let $B = (B_{\Sigma,k-1}, Q_B, \{i_B\}, F_B, \delta_B)$ be a $(k-1)$-block deterministic automaton equivalent to A_k.

We first show that there exists a non-trivial orbit $O \subset Q_B$ and two states $\alpha, \beta \in O$ such that $\mathrm{L}_\alpha(B) = \mathrm{L}_{\alpha_k}(A_k)$ and $\mathrm{L}_\beta(B) = \mathrm{L}_{\beta_k}(A_k)$. Let us consider the following state sequences: $(\alpha_{k,j})_{j \in \mathbb{N}} \subset F_B$ and $(\beta_{k,j})_{j \in \mathbb{N}} \subset F_B$, such that $\beta_{k,0} = i_B$, $\delta_B(\beta_{k,j}, a) = \alpha_{k,j}$ and $\delta_B(\alpha_{k,j}, b^{k-1}c) = \beta_{k,j+1}$. It follows that $\delta_B(i_B, (ab^{k-1}c)^j) = \beta_{k,j}$ and $\delta_B(i_B, (ab^{k-1}c)^j a) = \alpha_{k,j}$. Notice that the existence of $\alpha_{k,j}$ and $\beta_{k,j}$ is ensured by the fact that $\mathrm{L}(B) = \mathrm{L}(A_k)$. Let us suppose that there exists $j \in \mathbb{N}$ such that $L_{\beta_{k,j}}(B) \neq L_{\beta_k}(A_k)$. Then there exists $w \in \Sigma^*$ such that $w \in L_{\beta_{k,j}}(B) \triangle L_{\beta_k}(A_k)$, where for any two sets X and Y, $X \triangle Y = (X \setminus Y) \cup (Y \setminus X)$. And since $\delta_k(\beta_k, (ab^{k-1}c)^j) = \beta_k$, $(ab^{k-1}c)^j \cdot w \in L(B) \triangle L(A_k)$. Thus, $L(B) \neq L(A_k)$ which is contradictory. So, for every $j \in \mathbb{N}$, we have $L_{\beta_{k,j}}(B) = L_{\beta_k}(A_k)$. The proof that for every $j \in \mathbb{N}$, we have $L_{\alpha_{k,j}}(B) = L_{\alpha_k}(A_k)$, is done in the same way. Now, let us suppose that for every $j \neq j' \in \mathbb{N}$, we have $\alpha_{k,j} \neq \alpha_{k,j'}$ and $\beta_{k,j} \neq \beta_{k,j'}$. Then Q_B would be infinite, which would contradict the fact that B is a finite automaton. So, there exist $j < j' \in \mathbb{N}$ such that $\alpha_{k,j} = \alpha_{k,j'}$ or $\beta_{k,j} = \beta_{k,j'}$. Thus, either there exists a path going from $\beta_{k,j}$ to $\alpha_{k,j}$ and a path going from $\alpha_{k,j}$ to $\beta_{k,j'} = \beta_{k,j}$, and $\beta_{k,j}$ and $\alpha_{k,j}$ belong to the same orbit; or there exists a path going from $\alpha_{k,j}$ to $\beta_{k,j+1}$ and a path going from $\beta_{k,j+1}$ to $\alpha_{k,j'} = \alpha_{k,j}$, and $\alpha_{k,j}$ and $\beta_{k,j+1}$ belong to the same orbit.

Finally, let us focus on such an orbit O with two gates α and β such that $L_\alpha(B) = L_{\alpha_k}(A_k)$ and $L_\beta(B) = L_{\beta_k}(A_k)$. We know that for every $i \in \mathbb{N}$ such that $1 \leq i < k$, we have $\delta_k(\beta_k, b^i) = \beta_{k-i}$ with $|L_{\beta_{k-i}}(A_k)| < \infty$. Since $L_\beta(B) = L_{\beta_k}(A_k)$ and B is a $(k-1)$-block deterministic, there exist $j \in \mathbb{N}$ and $p \in Q_B$ such that $1 \leq j < k$, $\delta_B(\beta, [b^j]) = p$ and $L_p(B) = L_{\beta_{k-j}}(A_k)$. This means that $|L_p(B)| < \infty$, so $p \notin O$. Now, if there does not exist a state $q \in Q_B$ such that $\delta_B(\alpha, [b^j]) = q$, then B does not have the orbit property. So, let us suppose that such a state exists. We know that for every $i \in \mathbb{N}$ such that $1 \leq i < k$, we have $\delta_k(\alpha_k, b^i) = \alpha_{k-i}$ with $|L_{\alpha_{k-i}}(A_k)| = \infty$. Since $L_\alpha(B) = L_{\alpha_k}(A_k)$, we have $L_q(B) = L_{\alpha_{k-j}}(A_k)$ and $|L_q(B)| = \infty$. So $p \neq q$ and B does not have the orbit property.

Since $L(A_k)$ cannot be recognized by a $(k-1)$-block deterministic alphabetic image of an automaton passing the BKW test, following Theorem 2 it holds that $L(A_k)$ is not $(k-1)$-block deterministic.   □

References

1. Brüggemann-Klein, A., Wood, D.: One-unambiguous regular languages. Inf. Comput. **140**(2), 229–253 (1998). http://dx.doi.org/10.1006/inco.1997.2688
2. Caron, P., Ziadi, D.: Characterization of Glushkov automata. Theoret. Comput. Sci. **233**(1–2), 75–90 (2000)
3. Caron, P., Flouret, M., Mignot, L.: (k,l)-unambiguity and quasi-deterministic structures: an alternative for the determinization. In: Dediu, A.-H., Martín-Vide, C., Sierra-Rodríguez, J.-L., Truthe, B. (eds.) LATA 2014. LNCS, vol. 8370, pp. 260–272. Springer, Heidelberg (2014)
4. Giammarresi, D., Montalbano, R., Wood, D.: Block-deterministic regular languages. In: Restivo, A., Ronchi Della Rocca, S., Roversi, L. (eds.) ICTCS 2001. LNCS, vol. 2202, pp. 184–196. Springer, Heidelberg (2001)
5. Glushkov, V.M.: The abstract theory of automata. Russ. Math. Surv. **16**, 1–53 (1961)
6. Han, Y.S., Wood, D.: Generalizations of 1-deterministic regular languages. Inf. Comput. **206**(9–10), 1117–1125 (2008)

7. Hopcroft, J.E.: An $n \log n$ algorithm for minimizing the states in a finite automaton. In: Kohavi, Z. (ed.) The Theory of Machines and Computations, pp. 189–196. Academic Press, New York (1971)
8. Kleene, S.: Representation of events in nerve nets and finite automata. In: Shannon, C., McCarthy, J. (eds.) Automata Studies, pp. 3–41. Princeton University Press, Princeton (1956). Annals of Mathematics Studies 34
9. Moore, E.F.: Gedanken experiments on sequential machines. In: Shannon, C., McCarthy, J. (eds.) Automata Studies, pp. 129–153. Princeton University Press, Princeton (1956)
10. Rabin, M.O., Scott, D.: Finite automata and their decision problems. IBM J. Res. 3(2), 115–125 (1959)

Security of Numerical Sensors in Automata

Zhe Dang, Dmitry Dementyev, Thomas R. Fischer,
and William J. Hutton III[✉]

School of Electrical Engineering and Computer Science,
Washington State University, Pullman, WA 99164, USA
william.hutton@gmail.com

Abstract. Numerical sensors are numerical functions applied on memory contents. We study the computability of the mutual information rate between two sensors in various forms of automata, including nondeterministic pushdown automata augmented with reversal-bounded counters as well as discrete timed automata. The computed mutual information rate can be used to determine whether it is the case that there is essentially no information flow between a low sensor and a high sensor and hence could provide a way to quantitatively and algorithmically analyze some covert channels.

1 Introduction

An automaton is an abstract computing device that works on a memory and interacts with its environment (through input and/or output). A large part of traditional automata theory focuses on the language aspect of automata; e.g., what kind of languages an automaton accepts when it runs and what kind of properties the languages have (for instance; decidability of emptiness, equivalence, etc.). In this paper, we investigate the security aspect of automata with a focus on confidentiality. Automata are a fundamental model for all modern programs and software systems. This paper will provide insight for practitioners working on software and systems security.

In our setting, a sensor is a (many-to-one) function applied to memory contents. When the range of the function is vectors of numbers, we call the sensor numerical. For instance, for a pushdown automaton, a sensor that maps a stack content to its height is obviously a numerical sensor. Suppose that we are given an automaton M and two sensors low and $high$. For every reachable configuration C of M, we may obtain a pair of measurements, ($low(C)$ and $high(C)$). We can then define an (often infinite) set of measurements containing all such pairs. We seek to compute the maximal mutual information rate between the two sensors from the measurements set. The rate characterizes the amount of information (bit rate) that flows between the two sensors. We borrow the names of the sensors from software security [19,21], where a high variable and a low variable are often used to identify memory contents that are private and public, respectively, so that the information flow between the two is expected to be almost zero. Herein, we generalize the concept from variables to sensors in the

© Springer International Publishing Switzerland 2015
F. Drewes (Ed.): CIAA 2015, LNCS 9223, pp. 76–88, 2015.
DOI: 10.1007/978-3-319-22360-5_7

sense that the content $high(C)$ to be kept private may be an abstract form (e.g., the sum $x_1 + x_2$ of two counter values in the automaton) of C instead of being part (e.g., x_1) of C.

It turns out that computing the mutual information rate from a given measurements set is a difficult problem since the set can be infinite (e.g., for an infinite state automaton). Our approach is as follows. When the measurements set is finite, we can treat it as a bipartite graph G which also defines a channel. We find that the maximal mutual information between the sender and the receiver in the channel is related to, but different from the traditional channel capacity studied in information theory [6]. To compute the maximal mutual information, we show two fundamental results. First, the (estimated) maximal mutual information can be approximated with a small error by $\log \frac{N_{left} N_{right}}{E}$, where N_{left} and N_{right} are respectively the numbers of left nodes and the number of right nodes in the bipartite graph G and $E \geq \max(N_{left}, N_{right})$ is the number of edges in G. This approximation holds with high probability for a random G (i.e. the probability goes to 1 as G gets infinitely large). Second, the exact value of the maximal information can be computed as the logarithm of the size of maximum matching in G. Then, for an infinite set of measurements, the mutual information rate can be computed asymptotically using the estimated maximal mutual information. Notice that a computable mutual information rate directly leads to an algorithmic solution to the problem whether low and $high$ sensors are secure (in the sense of information secrecy) in an automaton. We shall point out that our results do not depend on a predefined probability model, which, in practice, is hard to obtain.

We show an intimate relationship between mutual information rate (between two numerical sensors) and information rate of a formal language [5, 7, 8, 18, 20]. Using this result, we show that the mutual information rate is computable for a variety of infinite state automata including reversal-bounded multicounter machines [15] with linear numerical sensors (over the counter values). In particular, we show that our definition of the numerical sensors is also flexible enough to be used on an abstract "memory"such as time. To this end, we show that the mutual information rate is also computable in discrete timed automata [1], when the sensors are durations of certain events. In this way, we give a formal and quantified treatment of a time-based covert channel embedded in a real-time system. We believe that the approach can also be used in fundamental studies for other forms of covert channels.

In the context of software security [19, 21], traditional approaches [2, 3, 22] in software security use static code analysis and/or an explicit probability model to calculate information dependency of variables in a program. As stated previously, such a probability is usually very hard to obtain in practice. Some work also does not depend on an explicit probability model (e.g., min-entropy in [26]). Our work handles asymptotic mutual information and does not assume a uniform distribution on the low and/or $high$ sensors. Also, unlike our recent work [11, 20] that studies a long-term mutual information rate (which is measured over a sequence of activities in a program that extends to infinity), we focus on short term mutual information rate in this paper.

2   Preliminaries

Let $\mathbb{N}$ be the set of nonnegative integers. $\mathbf{V} \subseteq \mathbb{N}^k$ is a *linear set* if $\mathbf{V} = \{\mathbf{v} : \mathbf{v} = \mathbf{v}_0 + t_1 \mathbf{v}_1 + \cdots + t_i \mathbf{v}_m, t_1, \cdots, t_m \geq 0\}$ where $\mathbf{v}_0, \cdots, \mathbf{v}_m$ are constant k-arity vectors in $\mathbb{N}^k$, for some $m \geq 0$. $\mathbf{V}$ is a *semilinear set* if it is the union of finitely many linear sets. Every finite subset of $\mathbb{N}^k$ is semilinear – it is a finite union of linear sets whose generators are constant vectors. Clearly, semilinear sets are closed under union and projection. It is also known that semilinear sets are closed under intersection and complementation.

Let Y be a finite set of integer variables. An atomic Presburger formula on Y is either a linear constraint $\sum_{y \in Y} a_y y < b$, or a mod constraint $x \equiv_d c$, where a_y, b, c and d are integers with $0 \leq c < d$. A Presburger formula can always be constructed from atomic Presburger formulas using $\neg$ and $\wedge$. Presburger formulas are closed under quantification. Let S be a set of k-tuples in N^k. S is Presburger definable if there is a Presburger formula $P(y_1, \cdots, y_k)$ such that the set of non-negative integer solutions is exactly S. It is well-known that S is a semilinear set iff S is Presburger definable. It can be shown that $\mathbf{V} \subseteq \mathbb{Z}^k$ is a semilinear set iff $\mathbf{V}$ is Presburger definable.

Let $\Sigma = \{a_1, \cdots, a_k\}$ be an alphabet. For each word $\alpha \in \Sigma^*$, define the Parikh map of α to be the vector $\#(\alpha) = (\#_{a_1}(\alpha), \cdots, \#_{a_k}(\alpha))$, where each symbol count $\#_{a_i}(\alpha)$ denotes the number of symbol a_i's in α. For a language $L \subseteq \Sigma^*$, the Parikh map of L is the set $\#(L) = \{\#(\alpha) : \alpha \in L\}$. The language L is semilinear if $\#(L)$ is a semilinear set.

A counter is a non-negative integer variable that can be incremented by 1, decremented by 1, or remain unchanged. Additionally, a counter can be tested for equality with 0. A *reversal-bounded NCM M* is a one-way non-deterministic finite automaton augmented with k (for some k) reversal-bounded counters. A counter is reversal-bounded if it makes at most r (a fixed constant like 3) alternations between non-decreasing and non-increasing modes in any computation. It is known that the language $L(M)$ accepted by M is semilinear [15]. The semilinearity remains when M is a reversal-bounded NPCM [15], that is a nondeterministic pushdown automaton M augmented with a number of reversal-bounded counters. (A reversal-bounded NCM is a special case of a reversal-bounded NPCM.)

Reversal-bounded NCMs have been extensively studied since their introduction in 1978 [15]; many generalizations are identified; e.g., multiple tapes, two-way tapes, stacks, etc. In particular, reversal-bounded NCMs have found applications in areas like Alur and Dill's [1] time-automata [9,10], Paun's [23] membrane computing systems [16], and Diophantine equations [27].

3   Maximal Mutual Information

A multi-bipartite graph $\hat{G}$ is a bipartite graph where an edge between node i and node j is labeled by a multiplicity $k(i,j) \geq 0$. In this case, the edge is counted as $k(i,j)$ edges. Suppose that $\hat{G}$ has N_{left} left nodes and N_{right} right nodes, and $\hat{E}$ edges (the sum of all multiplicities). Let X and Y be random variables on the left and on the right nodes, respectively. The joint distribution $p(X, Y)$ is uniform on all the $\hat{E}$ edges; i.e. $p(X, Y) = \frac{k(X,Y)}{\hat{E}}$. We require that $\hat{E} \geq \max(N_{left}, N_{right})$. From this definition of joint distribution, we can compute the entropy $H(X)$, the entropy $H(Y)$, the joint entropy $H(X, Y)$, and the mutual information $I(X; Y) = H(X) + H(Y) - H(X, Y)$, denoted by $H_L(\hat{G}), H_R(\hat{G}), H(\hat{G})$ and $I(\hat{G})$, respectively. Finally, we define $\lambda_{\hat{G}} = \log \frac{N_{left} N_{right}}{\hat{E}}$.

The multi-bipartite graph $\hat{G}$ induced a bipartite graph G obviously: in G, (i, j) is an edge iff $k(i, j) > 0$, where, the i and the j ranges over the left and the right nodes, respectively. Suppose that the bipartite graph G has E edges. Recall that $\lambda_G = \log \frac{N_{left} N_{right}}{E}$. The joint distribution $p(X, Y)$ of the $\hat{G}$ naturally induces a joint distribution for G. From this latter joint distribution, we can define the mutual information of G (induced from $\hat{G}$), denoted $I_{\hat{G}}(G)$ and have

$$I_{\hat{G}}(G) = I(\hat{G}). \tag{1}$$

Let $\hat{G}$ be a "random" (defined in a moment) multi-bipartite graph with N_{left} left nodes, N_{right} right nodes, and $N_{left} N_{right} \geq \hat{E} \geq N = \max(N_{left}, N_{right})$ edges. Herein, N_{left}, N_{right} and $\hat{E}$ are constants, instead of random variables. Suppose that the left nodes are $l_1, \cdots, l_{N_{left}}$ and the right nodes are $r_1, \cdots, r_{N_{right}}$. Define $U = \{(l_i, r_j) : 1 \leq i \leq N_{left}, 1 \leq j \leq N_{right}\}$ to be the set of all possible edges between left nodes and right nodes. We now uniformly and independently select $\hat{E}$ edges from the U (each edge is selected with probability $\frac{1}{N_{left} N_{right}}$). The result forms a random multi-bipartite graph $\hat{G}$. (An edge in U can be selected multiple times, so we end up with a multi graph instead of a graph). We now state our main result so far as follows.

Theorem 1. *With high probability (w.h.p, i.e. the probability goes to 1 as the value $\min(N_{left}, N_{right}) \to \infty$),*

$$|I(\hat{G}) - \lambda_{\hat{G}}| = O(\log \log N) \text{ and } |I_{\hat{G}}(G) - \lambda_G| = O(\log \log N), \tag{2}$$

where the $O(\log \log N)$ terms, as usual, is bounded by $c \log \log N$ for some constant c.

Let G be a bipartite graph where all the E edges are given. We can now treat the G as a channel. Now, we define $I(G)$ as $\max I(X; Y)$ where the max is over all the probability distributions over the E edges of the graph G. We use M to denote the size of maximum matching in G (i.e. the number of pairs in the matching).

Theorem 2. *For any bipartite graph G, $I(G) = \log M$.*

4   Secure Numerical Sensing in Automata

An automaton is a device that is equipped with a piece of (potentially unbounded) memory. After reading an input symbol, an automaton may make a move. The result of this move may update the content of the automaton's memory. The simplest class of automata are finite automata. Finite memory stores one control state. In the more complex case of pushdown automata, memory stores both a control state and the content of the stack. By convention, control states are finitely many, which correspond to line labels in a modern program.

Therefore, a run of automaton $\mathbf{M}$ is a sequence $C_0 \xrightarrow{a_1} \cdots \xrightarrow{a_n} C_n$, for some n, where the input symbols read so far are $w = a_1 \cdots a_n$, and each C_i is the content of memory at step i, and C_0 is the initial memory content, which is given usually in the definition of $\mathbf{M}$. In this case, we say that C_n is reachable.

At a high level, the memory contents can be understood as a value of one or more data structures. For instance, in case of a pushdown automaton, the memory would be a dynamic array where its first element stores a control state and its second element stores the bottom symbol of the stack and the rest of the array stores the stack content with the top symbol of the stack being at the end of the array.

Let $\mathcal{C}$ be the set of all the reachable memory contents and $High$ and Low be two sets. We are now given two functions $low : \mathcal{C} \to Low$ and $high : \mathcal{C} \to High$, which are called *sensors*. For a $C \in \mathcal{C}$, the results $low(C)$ and $high(C)$ are called *measurements*.

In practice, low (respectively, $high$) is to specify certain information about the memory that is public (respectively, private). For instance, in a pushdown automaton, $low(C)$ can be defined as the top symbol of the stack specified in the $C \in \mathcal{C}$, while $high(C)$ can be defined as the control state specified in the $C \in \mathcal{C}$. Following these two definitions, an attacker therefore can only observe the top symbol of the stack in a C.

We look at another example. Consider a multicounter automaton $\mathbf{M}$. A reachable memory content C therefore contains s (the control state) and $v_1, \cdots, v_k$ (nonnegative integer values for counters $x_1, \cdots, x_k$). We can define $low(C)$ to be $\sum_{i=0}^{k} v_i$ and $high(C) = (v_1, \cdots, v_k)$. From these definitions, it is interesting to see, when an attacker can measure the sum of all the counter values at a moment during a run of $\mathbf{M}$, whether the attacker can deduce some information of the individual counter values $v_1, \cdots, v_k$. In particular, we would like to compute the amount of the information; i.e. the mutual information between the two sensors.

Notice that the attacker can measure $low(C)$ without directly observing the C (this is why the function is called a sensor). For each $C \in \mathcal{C}$, we use $|low(C)|$ and $|high(C)|$ to denote the sizes (in bits or another given unit) of $low(C)$ and $high(C)$. For a given size n, we use $\mathcal{C}_n$ to denote all the C's in $\mathcal{C}$ such that both $|low(C)|$ and $|high(C)|$ are at most n. Now, for the $\mathcal{C}_n$, we construct a bipartite graph G_n where, for each $C \in \mathcal{C}_n$, there is an edge from the left node $l_{low(C)}$ to the right node $r_{high(C)}$. Notice that the number of edges E in G_n is the cardinality of the following set

$$R_n = \{(low(C), high(C)) : C \in \mathcal{C}_n\}. \tag{3}$$

We use N_{left} and N_{right} to denote the number of left nodes and right nodes in G_n, respectively, with $N = \max(N_{left}, N_{right})$. Recall that

$$\lambda_{G_n} = \log \frac{N_{left} N_{right}}{E}. \tag{4}$$

We now assume that $\min(N_{left}, N_{right})$ goes to ∞ as $n \to \infty$. That is, both $Low = \{low(C) : C \in \mathcal{C}\}$ and $High = \{high(C) : C \in \mathcal{C}\}$ are an infinite set. In this case, because of (2), λ_{G_n} indeed provides a good estimation of the mutual information between the left node $low(C)$ and the right node $high(C)$, when C is drawn from $\mathcal{C}_n$, with at most $O(\log \log N)$ bits of error. In fact, if we consider *estimated mutual information rate* $\frac{\lambda_{G_n}}{n}$, then the error goes to 0 asymptotically. Hence, we now define

$$I(low; high) = \limsup \frac{\lambda_{G_n}}{n} \tag{5}$$

and called it (estimated) mutual information rate between the two sensors *low* and *high*.

However, when one or both of *Low* and *High* are finite sets, one can verify that $I(low; high)$ is still defined and is simply 0. This is consistent with our intuition since now the mutual information between the left node $low(C)$ and the right node $high(C)$ can not be larger than $\log \min(|Low|, |High|)$ which is a finite constant. Hence, the mutual information rate has to be zero asymptotically.

Recall that we used $I(G_n)$ to denote the maximal mutual information in G_n, which, according to Theorem 2, equals $\log M$ where M is the size of maximum matching in G_n. When one or both of *Low* and *High* are finite sets, $\max_n I(G_n)$ is a finite value. In this case, we define $C(low; high)$ as the maximal mutual information between the two sensors *low* and *high*. In this case, once the definitions of *low* and *high* are computed from the definition of automaton $\mathbf{M}$, the maximal mutual information can be precisely computed using Theorem 2.

In the rest of the paper, we focus on how to compute $I(low; high)$, which is a much harder problem (because *Low* and *High* are infinite sets). In reality, it is assumed that a *low* sensor should not deduce any nontrivial amount of information in the *high* sensor. Clearly, the computability of $I(low; high)$ and $C(low; high)$ leads directly to algorithmic solutions of the problem, regardless of if this assumption is true or false.

4.1 Secure Numerical Sensing w.r.t. Estimated Mutual Information Rate

For a language L, we use $S_n(L)$ to denote the number of words with length n in L, λ_L, to denote the information rate of L, and $\limsup \frac{\log S_n(L)}{n}$ to denote the limit superior, which always exists. Whenever an actual limit (instead of lim sup) exists, we say that L is *converging*.

The definition of information rate comes from Shannon [25] in describing a channel capacity. Before proceeding further, we must explain the intuition behind Shannon's definition. $\frac{\log S_n(L)}{n}$ specifies the average number of bits needed per symbol (i.e. bit rate) if a word of length n in L is losslessly compressed. The information rate λ_L is simply the asymptotic bit rate. In other words, λ_L is the average amount of information per symbol contained in a word in L.

The following theorem is a fundamental result:

Theorem 3. *The information rate of a regular language is computable [5].*

As pointed out in [5], the information rate can actually be efficiently computed using a matrix algorithm. A language is suffix-closed if, for every word w in the language, there is a word $w' = wa$, for some symbol a, in the language.

Theorem 4. *A suffix-closed regular language is converging.*

We now prove a fundamental result. Let k be a constant. For a nonnegative integer $v < 2^n$, we use $[v, n]$ to denote its n-bit binary representation (in reverse order). For instance, when v is 4, $[v; 6]$ is 001000 (it is the reverse of 000100 = 4). For a vector of k nonnegative integers $\mathbf{v} = (v_1, \cdots, v_k)$, we say that it is 2^n-bounded if each $v_i < 2^n$. In this case, we use $[v_1, \cdots, v_k; n]$ to denote its n-bit representation, which is a k-track word, where, for each $1 \leq i \leq k$, the i-th track is $[v_i; n]$. Let $\mathbf{V}$ be a set of vectors $\mathbf{v} = (v_1, \cdots, v_k)$ of k nonnegative integers. We use $[\mathbf{V}; n]$ to denote the set of n-bit representations of its 2^n-bounded elements; i.e. $[\mathbf{V}; n] = \{[\mathbf{v}; n] : \mathbf{v} \text{ is } 2^n\text{-bounded}, \mathbf{v} \in \mathbf{V}\}$. Notice that $[\mathbf{V}; n]$ is a language of k-track words on bits with length n. We use $[\mathbf{V}]$ to denote $\cup_n [\mathbf{V}; n]$. Now we can show that:

Theorem 5. *For a semilinear set $\mathbf{V}$, $[\mathbf{V}]$ is a converging regular language.*

We now consider *numerical sensors*. That is, the sensors *low* and *high* return nonnegative integer vectors as their measurements. In this case, the measurements set $\{(low(C), high(C)) : C \in \mathcal{C}\}$ is simply a set of vectors, denoted by $\mathbf{V}_{low,high}$.

Theorem 6. *For numerical sensors low and high, when their set of measurements $\mathbf{V}_{low,high}$ is effectively a semilinear set (i.e. the description of the semilinear set can be computed), the mutual information rate $I(low; high)$ is computable.*

When a numerical sensor returns integer (instead of nonnegative integer) vectors as measurements, we add a sign bit to each measurement; e.g., -5 is treated as $(1, 5)$ while $+5$ is treated as $(0, 5)$. In this way, such an integer numerical sensor can be considered a numerical sensor returning nonnegative integer vectors. More precisely, an integer vector $\mathbf{v} \in \mathbb{Z}^k$ is now encoded as a nonnegative integer vector $\hat{\mathbf{v}}$ in $\mathbb{N}^{2k}$, after adding k sign bits. For a set $\mathbf{V} \subseteq \mathbb{Z}^k$, we call it an *integer semilinear set* when $\{\hat{\mathbf{v}} : \mathbf{v} \in \mathbf{V}\}$ is a semilinear set in $\mathbb{N}^{2k}$. The mapping between $\mathbf{v}$ and $\hat{\mathbf{v}}$ is one-to-one and length preserving. Therefore, Theorem 6 can be straightforwardly generalized.

Theorem 7. *For integer numerical sensors low and high, when their measurements set $\mathbf{V}_{low,high}$ is effectively Presburger definable (i.e. the description of the Presburger formula can be computed), the mutual information rate $I(low; high)$ is computable.*

4.2   Secure Numerical Sensing in Automata

Below, we will show how to use Theorem 6 to establish some decidability results on secure sensing of various computational models.

We consider a reversal-bounded NPCM M (with input tape and with reversal-bounded counters $x_1, \cdots, x_k$). A *linear numerical sensor* is an integer linear combination of the counters $x_1, \cdots, x_k$; e.g., $2x_1 - 3x_2 + 4x_3$. Such a sensor is also an integer numerical sensor. Suppose that *low* and *high* are linear numerical sensors. Now, the measurements set $\mathbf{V}_{low,high}$ is the set of all $(low(x_1, \cdots, x_k), high(x_1, \cdots, x_k))$ whenever counter values $x_1, \cdots, x_k$ are reachable in M. Since it can be shown that the measurements set $\mathbf{V}_{low,high}$ is an integer semilinear set, from Theorem 7, the following result is immediate:

Theorem 8. *Suppose that low and high are linear numerical sensors in a reversal-bounded NPCM M. Then, the mutual information rate $I(low; high)$ is computable.*

For a word w of even length, *low* observes the number of a's in the first half while *high* observes the number of b's in the second half. When w is drawn from a language L accepted by an NPCM (e.g., L is a context-free language), the mutual information rate $I(low; high)$ is computable, using Theorem 8. Notice that for the context-free language $\{a^n b^n : n \geq 0\}$, the rate is 1 bit (i.e. *low* and *high* share the complete information), while for the context-free language $\{a^m c^k b^n : m, k, n \geq 0, m + k = n\}$, the rate is 0 bit (asymptotically, *low* and *high* share no information).

We now consider an automaton working on multiple input tapes. In practice, such tapes can be used to model a system's input, internal states, or even its output. For instance, consider a printer that accepts user input and prints out a file. Herein, the user input and the output file can be modeled as two input tapes to an abstract automaton. In this case, an output (write) action of the printer is simulated by an input (read) action of the automaton.

Let M be a 2-tape NFA which has two input tapes. Let *low* and *high* be numerical sensors on first and the second tapes respectively. We define $R(M) = \{x \heartsuit y : M$ on input (x, y) accepts$\}$, where $\heartsuit$ is a delimiter (a new symbol). At each step, the transition of M is of the form $q : (a, b) \rightarrow p$, where $a, b \in \Sigma \cup \{\epsilon\}$ (ϵ is the null symbol). The transition means that M in state q reading a and b on the two tapes enters state p.

Let the numerical sensors *low* and *high* return the Parikh maps of the two inputs; i.e. $low(x) = \#(x)$ and $high(y) = \#(y)$. In this case, the measurements set $\mathbf{V}_{low,high}$ for the M is defined as $\mathbf{V}_{low,high} = \{(low(x), high(y)) : x \heartsuit y \in R(M)\}$. We can show that $R(M)$ is a semilinear set. Hence:

Theorem 9. *For a 2-tape NFA M, its mutual information rate $I(low; high)$ is computable when the numerical sensors low and high are defined above.*

The above can further be generalized to k-tape NFA M with reversal-bounded counters in a straightforward way. In this case, M can perform counter operations while reading input symbols from the tapes. Let *low* and *high* be Parikh maps of some of the tapes (e.g., *low* is for the first $\lceil\frac{k}{2}\rceil$ tapes while *high* for the rest). Then, Theorem 9 still holds for k-tape NFA:

Theorem 10. *For a k-tape NFA M augmented with reversal-bounded counters, its mutual information rate $I(low; high)$ is computable when the numerical sensors low and high are defined above.*

Notice that $R(M)$ is quite complex for a multi-tape NFA M. For instance, the language $L = \{x\heartsuit y\heartsuit xy : x, y \in (a+b)^*\}$ is the $R(M)$ of a 3-tape DFA, which is not even a context-free language. One can slightly generalize the L as follows: $L' = \{x\heartsuit y\heartsuit xy : x, y \in (a + b)^*, x \in L_1, y \in L_2, xy \in L_3, \#_a(x) = \#_b(y)\}$ where L_1, L_2, L_3 are some given regular languages ($\#_a(x)$ is the number of symbol a's in word x). In this case, L' is still the $R(M')$ for some 3-tape NFA augmented with reversal-bounded counters. Suppose that, for a word $w = x\heartsuit y\heartsuit xy \in L'$, $low(w) = (|x|, \#_a(x))$ and $high(w) = (|y|, \#_b(y))$. Then, from Theorem 10, the mutual information rate $I(low; high)$ of L' can be computed.

There are cases when the mutual information rate $I(low; high)$ is not computable. For example, a 2-head DFA M is a DFA with two one-way heads. The move of the machine depends on the state and the symbols scanned by the two heads. In a move, the machine changes state and moves each head (independently) at most one cell to the right. For a word $x \in L(M)$, we define $low(x) = \#_a(x)$ and $high(x) = \#_b(x)$. In this case, the measurements set $\mathbf{V}_{low,high}$ for the M is defined as $\mathbf{V}_{low,high} = \{(low(x), high(y)) : x \in L(M)\}$.

Proposition 1. *The mutual information rate $I(low; high)$ for a 2-head DFA is not computable when the numerical sensors low and high are defined above.*

As mentioned earlier, the two sensors *low* and *high* establish an abstract communication channel and the mutual information rate $I(low; high)$ measures the amount of information "flowing" over the channel. Communication is mostly direct; i.e. sending a message from one party to another. However, in the setting of a covert channel, a party can indirectly send a "message" using an unexpected channel (for instance, file sizes and time durations can be used to encode some information). Herein, numerical sensors are measured over abstract "memory" such as time. This is interesting, since time is often used as a media for covert channel that leaks information in a clever way.

"A covert channel is a path of communication that was not designed to be used for communication [4]." Covert channels can include both explicit and implicit communication. Deliberate communication between two or more entities is explicit. Implicit communication can occur when an entity is not aware that they are communicating information with one or more entities. The use of a common resource, such as the Linux */tmp* directory, which is usually both readable and writable by all users of the system, is an example of explicit communication. That communication becomes covert when the entities use the resource to communicate in a way that is was not designed for. For example, sending a message one character at a time by converting each character to its UTF-8 decimal value, then creating a file of corresponding size using random data. The file creation date of each file is used to determine the order to convert the file sizes back into UTF-8 characters to read the message. If the entities synchronize their communication using a simple protocol, they could use a single file, changing its size after the previous character was acknowledged by the receiver. Of course the previously mentioned *low(C)* and *high(C)* sensors could easily detect this covert channel by observing the rate of change of the file's modification timestamp.

The case of implicit communication is also interesting. Consider a user that is using a web site to generate a strong password. The user refreshes (reloads) the page to generate a new password until he sees one he likes, then writes it down. In an effort to obfuscate the fact that the last page requested contains his password, he refreshes the web page a few more times. The *low(C)* and *high(C)* sensors could be used to compare the time between configuration changes to infer which of the received password was the selected password.

We now investigate numerical sensors for time durations in a real-time systems. When a real-time system is modeled as a timed automaton, the mutual information rate of some time intervals (between events) can be algorithmically computed, as we will show. This opens the possibility for automatically analyzing covert time channels in such systems. But first, some definitions are required.

A timed automaton [1] is a finite state transition system augmented with a number of real-valued clocks. All the clocks progress synchronously with rate 1, except that a clock can be reset to 0 at some transition. A discrete timed automaton is the case when all the clocks take nonnegative integer values.

A *clock constraint* is a Boolean combination of *atomic clock constraints*, each of which is either in the form of $x \# c$ or in the form of $x - y \# c$, where $\#$ denotes $\leq, \geq, <, >$, or $=$, c is an integer, x, y are clocks. That is, clock constraints allow us to compare a single clock or the difference of two clocks against an integer constant.

A discrete timed automaton M is specified by (Σ, Q, X, T), where Σ is a finite set of observable event alphabet. We use τ to denote an unobservable event (i.e. silent event), which corresponds to ϵ (empty symbol) in automata theory. The Q is a finite set of *(control) states* with a designated initial state and the X is a finite set of clocks $x_1, \cdots, x_k$, for some k. Finally, $T \subseteq Q \times (\Sigma \cup \{\tau\}) \times 2^X \times \mathcal{C}_X \times Q$ is a finite set of *transitions*, where $\mathcal{C}_X$ is the set of clock constraints over clocks $x_1, \cdots, x_k$. Each transition $\langle q, a, r, \delta, q' \rangle$ leads from state q to state q' with event a (either observable or unobservable), the enabling condition δ, which is a clock constraint, and clock reset set $r \subseteq X$.

A configuration α is a tuple of a state and k clock values. For a transition $t = \langle q, a, r, \delta, q' \rangle$, if α is at state q, and the enabling condition δ is satisfied by the clock values specified in α, we use $\alpha \to^t \beta$, called one-step transition, to denote the fact that the transition sends configuration from α to β, where the state in β is q', event a is observed if $a \in \Sigma$, and the new clock values are in β: if $r = \emptyset$ (no clock resets on the transition), then every old clock value in α is incremented by 1; if $r \neq \emptyset$, then every old clock value in α is unchanged except for those clocks in r, which are reset to value 0.

A timestamp is a value of the *now* clock that never resets. We assume that M contains such a clock. When this is the case, a timed word [1] is used to record, when it runs, a sequence of observable events along with the timestamps when the events are observed. We only consider finite timed words. For example, the timed word $(a, 7)(b, 8)(a, 12)$ says that events a, b, a are observed at times $7, 8, 12$, respectively. More precisely, let $\hat{w}$ be a timed word. We say that configuration α can reach configuration β through time word $\hat{w}$, written $\alpha \rightsquigarrow_M^{\hat{w}} \beta$, if there are $\alpha = \alpha_0, \cdots, \alpha_n = \beta$, for some n, such that $\alpha_i \rightarrow^{t_i} \alpha_{i+1}$ holds for every $0 \leq i < n$ and some transition t_i. In particular, the observable events along with their timestamps collected from $t_0, \cdots, t_{n-1}$ (in this order) are exactly $\hat{w}$.

For a timed word $\hat{w}$, we use $D(\hat{w}, a)$ to denote the time elapse (duration) between the first a-event and the last a-event in $\hat{w}$ (when there is no such event, $D(\hat{w}, a) = 0$.). For instance, for the aforementioned timed word $\hat{w} = (a, 7)(b, 8)(a, 12)$, $D(\hat{w}, a) = 5$ and $D(\hat{w}, b) = 0$. Suppose that we are interested in two sets of events, say $\{l_1, \cdots, l_u\}$ and $\{h_1, \cdots, h_v\}$. For a timed word $\hat{w}$, the *low* sensor is defined as $low(\hat{w}) = (D(\hat{w}, l_1), \cdots, D(\hat{w}, l_u))$ and the *high* sensor is defined similarly as $high(\hat{w}) = (D(\hat{w}, h_1), \cdots, D(\hat{w}, h_v))$. Such sensors are called *duration sensors*. M can also generalize to discrete pushdown timed automaton in a straightforward way [9,10]. We are now given two sets of configurations A and B and the set of measurements

$$\mathbf{V}_{low,high} = \{(low(\hat{w}), high(\hat{w})) : \alpha \rightsquigarrow_M^{\hat{w}} \beta, \alpha \in A \text{ and } \beta \in B\}. \tag{6}$$

Theorem 11. *The mutual information rate $I(low; high)$ for $\mathbf{V}_{low,high}$ defined in (6), where A and B are Presburger-definable and low and high are duration sensors, is computable for a discrete pushdown timed automaton M.*

For example, let P_l and P_h be two real-time processes that access a finite shared memory C, which is the critical region. We are interested in read events in P_l and write events in P_h. A common setup would be the party P_h leaks some information to the party P_l through the content of the critical region that P_h writes (and later read by P_l). However, P_h could set up a time-based covert channel with P_l by controlling the duration x_{ab} between two special events, say, write(a) and write(b). In the meanwhile, the party P_l uses the duration y_{ab} between two events, say, read(a) and read(b), to recover the part or all of the information "sent" by P_h through x_{ab}. When the concurrent real-time system of P_l and P_h is specified by a timed automaton (or even discrete pushdown timed automaton), the information rate that P_h sends to P_l through the time covert channel can be computed, using Theorem 11 (after renaming events properly).

5   Conclusions

Numerical sensors are numerical functions applied to memory contents. We study the computability of the mutual information rate between two sensors in various forms of automata, including nondeterministic pushdown automata augmented with reversal-bounded counters as well as discrete timed automata. The computed mutual information rate can be used to determine whether it is the case that there is essentially no information flow between a low sensor and a high sensor and hence could provide a way to quantitatively and algorithmically analyze some type of covert channels.

References

1. Alur, R., Dill, D.L.: A theory of timed automata. Theoret. Comput. Sci. **126**(2), 183–235 (1994)
2. Alvim, M.. Andrs, M., Palamidessi, C.: Probabilistic information flow. In: LICS 2010, pp. 314–321
3. Backes, M., Berg, M., Köpf, B.: Non-uniform distributions in quantitative information-flow. In: ASIACCS 2011, pp. 367–375
4. Bishop, M.: Introduction to Computer Security. Addison-Wesley, Reading (2011)
5. Chomsky, N., Miller, G.A.: Finite state languages. Inf. Control **1**, 91–112 (1958)
6. Cover, T.M., Thomas, J.A.: Elements of Information Theory, 2nd edn. Wiley-Interscience, New York (2006)
7. Cui, C., Dang, Z., Fischer, V, Ibarra, O.H.: Execution information rate for some classes of automata. Information and Computation (accepted)
8. Cui, C., Dang, Z., Fischer, T.R., Ibarra, O.H.: Information Rate of Some Classes of Nonregular Languages: An Automata-Theoretic Approach, Information and Computation (conditionally accepted)
9. Dang, Z., Ibarra, O.H., Bultan, T., Kemmerer, R.A., Su, J.: Binary reachability analysis of discrete pushdown timed automata. In: Emerson, E.A., Sistla, A.P. (eds.) CAV 2000. LNCS, vol. 1855, pp. 65–84. Springer, Heidelberg (2000)
10. Dang, Z.: Pushdown timed automata: a binary reachability characterization and safety verification. Theoret. Comput. Sci. **302**(13), 93–121 (2003)
11. Dang, Z., Fischer, T., Hutton, W., Ibarra, O., Li, Q.: Quantifying communication in synchronized languages. In: COCOON 2015 (to appear)
12. Flajolet, P., Sedgewick, R.: Analytic Combinatorics. Cambridge University Press, Cambridge (2009)
13. Gonnet, G.H.: Expected length of the longest probe sequence in hash code searching. J. ACM **28**, 289–304 (1981)
14. Hopcroft, J.E., Ullman, J.D.: Introduction to Automata Theory, Languages and Computation. Addison-Wesley Publishing Company, Reading (1979)
15. Ibarra, O.H.: Reversal-bounded multicounter machines and their decision problems. J. ACM **25**(1), 116–133 (1978)
16. Ibarra, O.H., Dang, Z., Egecioglu, O., Saxena, G.: Characterizations of catalytic membrane computing systems. In: Rovan, B., Vojtáš, P. (eds.) MFCS 2003. LNCS, vol. 2747, pp. 480–489. Springer, Heidelberg (2003)
17. Kaminger, F.P.: The noncomputability of the channel capacity of context-sensitive languages. Inf. Comput. **17**(2), 175–182 (1970)
18. Kuich, W.: On the entropy of context-free languages. Inf. Control **16**(2), 173–200 (1970)
19. Lanotte, R., Maggiolo-Schettini, A., Troina, A.: Time and probability-based information flow analysis. IEEE TSE **36**(5), 719–734 (2010)
20. Li, Q., Dang, Z.: Sampling automata and programs. Theoret. Comput. Sci. **577**, 125–140 (2015)

21. Lowe, G.: Defining information flow quantity. J. Comput. Secur. **12**(3–4), 619–653 (2004)
22. Mu, C., Clark, D.: Quantitative analysis of secure information flow via probabilistic semantics. In: ARES 2009, pp. 49–57
23. Paun, G.: Membrane Computing: An Introduction. Springer, Berlin (2000)
24. Raab, M., Steger, A.: "Balls into Bins" - a simple and tight analysis. In: Rolim, J.D.P., Serna, M., Luby, M. (eds.) RANDOM 1998. LNCS, vol. 1518, p. 159. Springer, Heidelberg (1998)
25. Shannon, C.E., Weaver, W.: The Mathematical Theory of Communication. University of Illinois Press, Champaign (1949)
26. Smith, G.: On the foundations of quantitative information flow. In: de Alfaro, L. (ed.) FOSSACS 2009. LNCS, vol. 5504, pp. 288–302. Springer, Heidelberg (2009)
27. Xie, G., Dang, Z., Ibarra, O.: A solvable class of quadratic Diophantine equations with applications to verification of infinite-state systems. In: Baeten, J.C.M., Lenstra, J.K., Parrow, J., Woeginger, G.J. (eds.) ICALP 2003. LNCS, vol. 2719. Springer, Heidelberg (2003)

Jumping Finite Automata: Characterizations and Complexity

Henning Fernau, Meenakshi Paramasivan[✉], and Markus L. Schmid

Fachbereich 4 – Abteilung Informatik, Universität Trier, 54286 Trier, Germany
{Fernau,Paramasivan,MSchmid}@uni-trier.de

Abstract. We characterize the class of languages described by jumping finite automata (i. e., finite automata, for which the input head after reading (and consuming) a symbol, can jump to an arbitrary position of the remaining input) in terms of special shuffle expressions. We can characterize some interesting subclasses of this language class. The complexity of parsing these languages is also investigated.

1 Introduction

Throughout the history of automata theory, the classical finite automaton has been extended in many different ways: two-way automata, multi-head automata, automata with additional resources (counters, stacks, etc.) and so on. However, for all these variants, it is always the case that the input is read in a continuous fashion. On the other hand, there exist models that are closer to the classical model in terms of computational resources, but that differ in how the input is processed (e. g., restarting automata [17] and biautomata [13]). One such model that has drawn comparatively little attention are the jumping finite automata (JFA) introduced by Meduna and Zemek [15,16], which are like classical finite automata with the only difference that after reading (i. e., consuming) a symbol and changing in a new state, the input head can jump to an arbitrary position of the remaining input.

We provide a characterization of the JFA-languages in terms of expressions using shuffle, union and iterated shuffle, which enables us to put them into the context of classical formal language results from around 1980. This also resolves an open problem in [15]. By showing that any such expression is equivalent to one with a star-height (with respect to iterated shuffle) of at most 1, we obtain a normal form for this language class. If we interpret *general* finite automata, i. e., finite automata the transitions of which can be labeled by words instead of single symbols, as jumping automata, then we obtain a much more powerful model. This is demonstrated by showing that the universal word problem for JFA can be solved in polynomial time (for fixed alphabets), whereas it is NP-complete for general JFA, even for finite languages over a fixed binary alphabet.

Due to space restrictions, results marked with (∗) are not proven here.

Jumping Finite Automata. Following Meduna and Zemek, we denote a *general finite machine* as $M = (Q, \Sigma, R, s, F)$, where Q is a finite set of *states*, Σ

© Springer International Publishing Switzerland 2015
F. Drewes (Ed.): CIAA 2015, LNCS 9223, pp. 89–101, 2015.
DOI: 10.1007/978-3-319-22360-5_8

is the *input alphabet*, $\Sigma \cap Q = \emptyset$, R is a finite set of *rules* of the form $py \to q$ where $p, q \in Q$ and $y \in \Sigma^*$, $s \in Q$ is the *start state* and $F \subseteq Q$ is a set of *final states*. If all rules $py \to q \in R$ satisfy $|y| \le 1$, then M is a *finite machine*.

We interpret M in two ways:

– As a (general) finite automaton, a *configuration* of M is any string in $Q\Sigma^*$. The binary *move relation* on $Q\Sigma^*$, written as $\Rightarrow$, is defined as follows:

$$pw \Rightarrow qz : \iff \exists\, py \to q \in R : w = yz.$$

– As a (general) jumping finite automaton, a *configuration* of M is any string in $\Sigma^* Q \Sigma^*$. The binary *jumping relation* on $\Sigma^* Q \Sigma^*$, written as $\curvearrowright$, satisfies:

$$vpw \curvearrowright v'qz' : \iff \exists\, py \to q \in R\ \exists\, z \in \Sigma^* : w = yz \,\wedge\, vz = v'z'.$$

If M is a (general) finite machine, we can hence obtain the following languages:

$$L_{\mathrm{FA}}(M) = \{w \in \Sigma^* : \exists\, f \in F : sw \Rightarrow^* f\} \text{ and}$$
$$L_{\mathrm{JFA}}(M) = \{w \in \Sigma^* : \exists\, u, v \in \Sigma^*\ \exists\, f \in F : w = uv \wedge usv \curvearrowright^* f\}.$$

This defines us the language classes $\mathcal{REG}$ (accepted by (generalized) finite automata), $\mathcal{JFA}$ (accepted by jumping finite automata, or JFAs for short) and $\mathcal{GJFA}$ (accepted by general jumping finite automata, or GJFAs for short). $\mathcal{CFL}$ denotes the class of context-free languages.

2 Operations on Languages and Their Properties

The reader is assumed to be familiar with the standard operations on formal languages, like catenation, union and iterated catenation, aka Kleene star.

Definition 1. *Let $u, v \in \Sigma^*$, the shuffle operation, denoted by $\sqcup\!\sqcup$, is a binary operation on words, described by $u \sqcup\!\sqcup v = \{x_1 y_1 x_2 y_2 \ldots x_n y_n : u = x_1 x_2 \ldots x_n, v = y_1 y_2 \ldots y_n, x_i, y_i \in \Sigma^*, 1 \le i \le n, n \ge 1\}$. It is extended on languages in the natural way: for $L_1, L_2 \subseteq \Sigma^*$, $L_1 \sqcup\!\sqcup L_2 := \{z : z \in x \sqcup\!\sqcup y, x \in L_1, y \in L_2\}$.*

Definition 2. *For $L \subseteq \Sigma^*$, the iterated shuffle of L is defined by:*

$$L^{\sqcup\!\sqcup,*} := \bigcup_{n=0}^{\infty} L^{\sqcup\!\sqcup,n} \quad \text{where } L^{\sqcup\!\sqcup,0} = \{\varepsilon\} \text{ and } L^{\sqcup\!\sqcup,i} := L^{\sqcup\!\sqcup,i-1} \sqcup\!\sqcup L.$$

Let us now recall the following computation rules from [8].

Proposition 1. *Let M_1, M_2, M_3 be arbitrary languages.*

1. $M_1 \sqcup\!\sqcup M_2 = M_2 \sqcup\!\sqcup M_1$ *(commutative law)*
2. $(M_1 \sqcup\!\sqcup M_2) \sqcup\!\sqcup M_3 = M_1 \sqcup\!\sqcup (M_2 \sqcup\!\sqcup M_3)$ *(associative law)*
3. $M_1 \sqcup\!\sqcup (M_2 \cup M_3) = M_1 \sqcup\!\sqcup M_2 \cup M_1 \sqcup\!\sqcup M_3$ *(distributive law)*
4. $(M_1 \cup M_2)^{\sqcup\!\sqcup,*} = (M_1)^{\sqcup\!\sqcup,*} \sqcup\!\sqcup (M_2)^{\sqcup\!\sqcup,*}$

5. $(M_1^{\sqcup,*})^{\sqcup,*} = (M_1)^{\sqcup,*}$
6. $(M_1 \sqcup M_2^{\sqcup,*})^{\sqcup,*} = (M_1 \sqcup (M_1 \cup M_2)^{\sqcup,*}) \cup \{\varepsilon\}$

The second, third and fifth rule are also true when you consider (iterated) catenation instead of (iterated) shuffle. This is no coincidence, as we will see. The sixth rule will play a crucial rôle in the proof of our main normal form result.

We can deduce from the first three computation rules the following:

Proposition 2. (*) $(2^{\Sigma^*}, \cup, \sqcup, \emptyset, \{\varepsilon\})$ is a commutative semiring.

Definition 3. The set of all permutations of w, $\mathrm{perm}(w)$, is defined as follows:

$$\mathrm{perm}(w) = \begin{cases} \{\varepsilon\}, & |w| = 0 \\ \{a\} \sqcup \mathrm{perm}(u), & w = a \cdot u, a \in \Sigma, u \in \Sigma^* \end{cases}$$

For $L \subseteq \Sigma^*$, $\mathrm{perm}(L) = \bigcup_{w \in L} \mathrm{perm}(w)$.

We summarize two important properties of perm in the following two lemmas.

Lemma 1. $\mathrm{perm} : 2^{\Sigma^*} \to 2^{\Sigma^*}$ is a hull operator, i.e., it is extensive, (monotone) increasing and idempotent.

By the well-known correspondence between hull operators and (systems of) closed sets, we will also speak of *perm-closed languages* in the following, i.e., languages L satisfying $\mathrm{perm}(L) = L$. Such languages are also called *commutative*.

Lemma 2. The set $\{\mathrm{perm}(w) : w \in \Sigma^*\}$ is a partition of Σ^*. There is a natural bijection between this partition and the set of functions $\mathbb{N}^\Sigma$, given by the Parikh mapping $\pi_\Sigma : \Sigma^* \to \mathbb{N}^\Sigma, w \mapsto (a \mapsto |w|_a)$, where $|w|_a$ is the number of occurrences of a in w. Namely, $\{\mathrm{perm}(w) : w \in \Sigma^*\} = \pi_\Sigma^-(\pi_\Sigma(w))$.

Due to Lemma 2, we conclude:

Proposition 3. For $L_1, L_2 \subseteq \Sigma^*$, $\mathrm{perm}(L_1) = \mathrm{perm}(L_2)$ iff $\pi_\Sigma(L_1) = \pi_\Sigma(L_2)$.

By the definition of the work of a jumping finite automaton M, it is clear that $w \in L_{\mathrm{JFA}}(M)$ implies that $\mathrm{perm}(w) \subseteq L_{\mathrm{JFA}}(M)$, i.e., $\mathrm{perm}(L_{\mathrm{JFA}}(M)) \subseteq L_{\mathrm{JFA}}(M)$. Since perm is extensive as a hull operator, we can conclude:

Corollary 1. If $L \in \mathcal{JFA}$, then L is perm-closed.

This also follows by results in [15]. In particular, we mention the following important characterization theorem from [16], that we enrich by combining it with the well-known theorem of Parikh [18], using Proposition 3.

Theorem 1. $\mathcal{JFA} = \mathrm{perm}(\mathcal{REG}) = \mathrm{perm}(\mathcal{CFL})$.

This theorem also generalizes the main result of [14]. According to the analysis indicated in [5], Parikh's original proof would produce, starting from a context-free grammar G with n variables, a regular expression E of length $O\left(2^{2^{n^2}}\right)$ such that $\mathrm{perm}(L(G)) = \mathrm{perm}(L(E))$, whose corresponding NFA is even bigger, while the construction of [5] results in an NFA A with only 4^n states, satisfying $\mathrm{perm}(L(G)) = \mathrm{perm}(L(A))$.

Corollary 2. *Let L be a finite language. Then, $L \in \mathcal{JFA}$ iff L is perm-closed.*

This also shows that all finite $\mathcal{JFA}$ languages are so-called commutative regular languages as studied by Ehrenfeucht, Haussler and Rozenberg in [4]. We will come back to this issue later.

The relation between (iterated) catenation and (iterated) shuffle can now be neatly expressed as follows.

Theorem 2. *(*) perm : $2^{\Sigma^*} \to 2^{\Sigma^*}$ is a semiring morphism from the semiring $(2^{\Sigma^*}, \cup, \cdot, \emptyset, \{\varepsilon\})$ to the semiring $(2^{\Sigma^*}, \cup, \sqcup\!\sqcup, \emptyset, \{\varepsilon\})$ that also respects the iterated catenation resp. shuffle operation.*

Clearly, perm cannot be an isomorphism, as the catenation semiring is not commutative, while the shuffle semiring is, see Proposition 2.

3   Alphabetic Shuffle Expressions

Shuffle expressions and variants thereof have been an active field of study over decades; we only point the reader to [9–11]. Here, we describe one special variant tightly linked to jumping finite automata. We hence give an inductive definition of what we call *alphabetic shuffle expressions*, or α-SHUF expressions for short, in the following.

Definition 4. *The symbols $\emptyset, \varepsilon$ and each $a \in \Sigma$ are α-SHUF expressions (base case). If S_1, S_2 are α-SHUF expressions, then $(S_1 + S_2), (S_1 \sqcup\!\sqcup S_2)$ and $S_1^{\sqcup\!\sqcup,*}$ are α-SHUF expressions.*

The semantics of α-SHUF expressions is defined in the expected way. For instance, $L((a + b)^{\sqcup\!\sqcup,*}) = \{a, b\}^{\sqcup\!\sqcup,*}$. The corresponding class of languages was termed $\mathcal{L}_3$ in [7]. If S_1, S_2 are two expressions, then $S_1 \equiv S_2$ means that they are equivalent, i.e., they describe the same language, or, more formally, $L(S_1) = L(S_2)$. Sometimes, to avoid confusion with arithmetics, we also write $\cup$ in expressions instead of $+$.

Notice that we could introduce (classical) regular expressions in the very same way. Clearly, these characterize the regular languages.

Definition 5. *The symbols $\emptyset, \varepsilon$ and each $a \in \Sigma$ are regular expressions. If S_1, S_2 are regular expressions, then $(S_1 + S_2), (S_1 \cdot S_2)$ and S_1^* are regular expressions.*

Lemma 3. *Let R' be a regular expression. Let the α-SHUF expression R be obtained from R' by consequently replacing all $\cdot$ by $\sqcup\!\sqcup$, and all $*$ by $\sqcup\!\sqcup,*$ in R'. Then, $\mathrm{perm}(L(R')) = L(R)$.*

Proof. Let R' be a regular expression. By definition, this means that $L(R') = K$, where K is some expression over the languages $\emptyset$, $\{\varepsilon\}$ and $\{a\}$, $a \in \Sigma$, using only union, catenation and Kleene-star. By Theorem 2, $\mathrm{perm}(K)$ can be transformed into an equivalent expression K' using only union, shuffle and iterated shuffle. Furthermore, in K', the operation perm only applies to languages of the form $\emptyset$, $\{\varepsilon\}$ and $\{a\}$, $a \in \Sigma$, which means that by simply removing all perm operators, we obtain an equivalent expression K'' of languages $\emptyset$, $\{\varepsilon\}$ and $\{a\}$, $a \in \Sigma$, using only union, shuffle and iterated shuffle. This expression directly translates into the α-SHUF expression R with $L(R) = \mathrm{perm}(L(R'))$.   □

We are now ready to prove our characterization theorem for $\mathcal{JFA}$.

Theorem 3. *A language $L \subseteq \Sigma^*$ is in $\mathcal{JFA}$ if and only if there is some α-SHUF expression R such that $L = L(R)$.*

Proof. If $L \in \mathcal{JFA}$, then there exists a regular language L' such that $L = \mathrm{perm}(L')$ by Theorem 1. L' can be described by some regular expression R'. By Lemma 3, we find an α-SHUF expression R such that $L = \mathrm{perm}(L(R')) = L(R)$.

Conversely, if L is described by some α-SHUF expression R, i.e., $L = L(R)$, then construct the regular expression R' by consequently replacing all $\sqcup\!\sqcup$ by $\cdot$ and all $\sqcup\!\sqcup^*$ by * in R. Clearly, we face the situation described in Lemma 3, so that we conclude that $\mathrm{perm}(L(R')) = L(R) = L$. As $L(R')$ is a regular language, $\mathrm{perm}(L(R')) = L \in \mathcal{JFA}$ by Theorem 1.   □

Since α-SHUF languages are closed under iterated shuffle, we obtain the following corollary as a consequence of Theorem 3, adding to the list of closure properties given in [15].

Corollary 3. *$\mathcal{JFA}$ is closed under iterated shuffle.*

Let us finally mention a second characterization (recall the first characterization from Corollary 2) of the finite perm-closed sets.

Proposition 4. *Let L be some language. Then, L is finite and perm-closed if and only if there is an α-SHUF expression R, with $L = L(R)$, that does not contain the iterated shuffle operator.*

Proof. Let L be a finite language with $L = \mathrm{perm}(L)$. Clearly, there is a regular expression R_L, with $L(R_L) = L$, that uses only the catenation and union operations. As L is perm-closed, the α-SHUF expression R obtained from R_L by replacing all catenation by shuffle operators satisfies $L(R) = \mathrm{perm}(L(R_L)) = L$ by Lemma 3 and does not contain the iterated shuffle operator. Conversely, let R be an α-SHUF expression that does not contain the iterated shuffle operator. By combining Theorem 3 with Corollary 1, we know that $L(R)$ is perm-closed. It is rather straightforward that $L(R)$ is also finite.   □

Let us now see an example for the class $\mathcal{JFA}$.

Example 1. The finite machine $M = (\{s, r, t, f\}, \{a, b\}, R, s, \{f\})$ with $R = \{sa \rightarrow r, sb \rightarrow f, ra \rightarrow t, rb \rightarrow r, ta \rightarrow f, tb \rightarrow s, fa \rightarrow r, fb \rightarrow s\}$ accepts (in terms of traditional regular expressions) $L = L_{\mathrm{FA}}(M)$ with $L = L\big(((ab^*ab)^*((ab^*aa)+b)(ab^*aa)^*((ab^*ab)+b))^*(ab^*ab)^*((ab^*aa)+b)(ab^*aa)^*\big).$

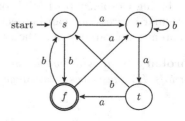

The same M accepts (in terms of α-SHUF expressions) $L = L_{\mathrm{JFA}}(M)$ with
$L = L((((a \sqcup b^{\sqcup,*} \sqcup a \sqcup b)^{\sqcup,*} \sqcup ((a \sqcup b^{\sqcup,*} \sqcup a \sqcup a) + b) \sqcup (a \sqcup b^{\sqcup,*} \sqcup a \sqcup a)^{\sqcup,*} \sqcup$
$((a \sqcup b^{\sqcup,*} \sqcup a \sqcup b) + b))^{\sqcup,*} \sqcup (a \sqcup b^{\sqcup,*} \sqcup a \sqcup b)^{\sqcup,*} \sqcup ((a \sqcup b^{\sqcup,*} \sqcup a \sqcup a) + b) \sqcup$
$(a \sqcup b^{\sqcup,*} \sqcup a \sqcup a)^{\sqcup,*})$.

4   Representations and Normal Forms

Our desired representation theorem can be stated as follows.

Theorem 4. *Let $L \in \mathfrak{JFA}$. Then there exists a number $n \geq 1$ and finite sets M_i, N_i for $1 \leq i \leq n$, so that the following representation is valid.*

$$L = \bigcup_{i=1}^{n} \mathrm{perm}(M_i) \sqcup (\mathrm{perm}(N_i))^{\sqcup,*} \tag{1}$$

We will prove this representation theorem on the level of α-SHUF expressions, so that we actually get a normal form theorem for these. A central tool in the proofs of this normal form theorem is the following notion that corresponds to the well-known star-height of regular expressions.

Definition 6. *We can inductively associate the (shuffle iteration) height h to any α-SHUF expression S as follows.*

- *If S is a base case, then $h(S) = 0$.*
- *If $S = (S_1 + S_2)$ or $S = (S_1 \sqcup S_2)$, then $h(S) = \max\{h(S_1), h(S_2)\}$.*
- *If $S = S_1^{\sqcup,*}$, then $h(S) = h(S_1) + 1$.*

The shuffle iteration height of a $\mathfrak{JFA}$-language L is then the smallest shuffle iteration height of any α-SHUF expression S describing L.

Let us mention the following interesting consequence obtained by combining Theorem 4 with Theorem 3, Lemma 3 and Theorem 1.

Corollary 4. *$L \in \mathfrak{JFA}$ if and only if there is a regular language R of star height at most one such that $L = \mathrm{perm}(R)$.*

Immediately from the Definition 6, we obtain from Proposition 4:

Corollary 5. *A language is finite and perm-closed if and only if it can be described by some α-SHUF expression of shuffle iteration height zero.*

Combining Corollary 5 with Theorem 1 and the well-known fact that finiteness of regular expressions can be decided, we immediately obtain the following, as Theorem 4 guarantees that the height of $\mathfrak{JFA}$ languages is zero or one:

Corollary 6. *It is decidable, given some JFA and some integer k, whether or not this JFA describes a language of shuffle iteration height at most k.*

Notice that we have formulated, in this corollary, the shuffle analogue of the famous star height problem, which has been a major open problem for regular languages [6]. Recall that Eggan's Theorem [3] relates the star height of a regular language to its so-called cycle rank, which formalizes loop-nesting in NFA's. Again, the characterization theorems that we derived allow us to conclude that, in short, for any $L \in \mathcal{JFA}$ there exists some finite machine M of cycle rank at most one such that $L_{\mathrm{JFA}}(M) = L$.

Corollary 5 means that, in order to show Theorem 4, it is sufficient (and in a sense stronger) to prove the following normal form theorem for α-SHUF expressions. The proof resembles the one given by Jantzen [8] for a different variant of shuffle expressions, but we keep it here, as it shows several technicalities with these notions.

Theorem 5. *For any α-SHUF expression R, an equivalent α-SHUF expression S with $h(S) = 1$ can be constructed that is the union of n α-SHUF expressions $S_1, \ldots, S_n$ such that $S_i = F_i \amalg G_i^{\amalg,*}$, where $h(F_i) = h(G_i) = 0$, $1 \le i \le n$. Moreover, we can assume that $F_i = \bigcup_{j=1}^{n(i)} u_j$ and $G_i = \bigcup_{j=1}^{m(i)} v_j$, where all u_j and v_j are α-SHUF expressions with $\amalg$ as their only operators.*

Proof. We show the claim by induction on the height of R. If $h(R) = 0$, then $S = R \amalg \emptyset^{\amalg,*}$ is an equivalent expression in the desired normal form. Let $h > 0$. Assume now that the result is true for all α-SHUF expressions of height less than h and consider some α-SHUF expression R with $h(R) = h$. By repeatedly applying the distributive law, we can obtain an equivalent α-SHUF expression R' that is of the following form:

$$R' = \bigcup_{j=1}^{m} \coprod_{k=1}^{k(j)} S_{j,k} \, ,$$

where each expression $S_{j,k}$ contains only the operators shuffle and iterated shuffle. In a first step, by applying the commutative law of the shuffle, we can order the $S_{j,k}$ such that, slightly abusing notation, $S_{j,1}, \ldots, S_{j,b(j)}$ are base cases, and $S_{j,b(j)+1}, \ldots, S_{j,k(j)}$ are of the form $S_{j,i} = (T_{j,i})^{\amalg,*}$. To simplify the further discussions, we can assume that none of the base cases $S_{j,1}, \ldots, S_{j,b(j)}$ is $\emptyset$, as this would mean that the language $L(\coprod_{k=1}^{k(j)} S_{j,k})$ is empty, and we can omit this part immediately from the union. In the next step, we form $F_j' := \coprod_{k=1}^{b(j)} S_{j,k}$. Notice that, by Corollary 5, each F_j' represents a finite perm-closed set. Moreover, we define α-SHUF expressions G_j' of iteration height less than h as follows. If $b(j) = k(j)$, then $G_j' := \emptyset$. Otherwise, $G_j' := \bigcup_{i=b(j)+1}^{k(j)} T_{j,i}$. By using Rule 4 from Proposition 1, one can see that

$$R'' := \bigcup_{j=1}^{m} F_j' \amalg (G_j')^{\amalg,*}$$

is equivalent to R'. As all G'_j have iteration height less than h, we can apply the induction hypothesis to them and replace G'_j by equivalent expressions

$$\bigcup_{i=1}^{n(j)} F_{j,i} \shuffle G_{j,i}^{\shuffle,*},$$

where each $F_{j,i}$ and each $G_{j,i}$ are α-SHUF expressions of height zero. Rule 4 now yields the following equivalent expression:

$$R''' := \bigcup_{j=1}^{m} F'_j \shuffle \biguplus_{i=1}^{n(j)} \left(F_{j,i} \shuffle G_{j,i}^{\shuffle,*}\right)^{\shuffle,*}$$

Now, we can apply Rule 6 to avoid nesting of the iterated shuffle. Hence, the following expression is again equivalent:

$$R^{iv} := \bigcup_{j=1}^{m} F'_j \shuffle \biguplus_{i=1}^{n(j)} \left(F_{j,i} \shuffle (F_{j,i} \cup G_{j,i})^{\shuffle,*} \cup \{\varepsilon\}\right)$$

Finally, setting $F_{j,I} := F'_j \shuffle \bigsqcup\bigsqcup_{i \in I} F'_{j,i}$ and $G_{j,I} := \bigcup_{i \in I}(F_{j,i} \cup G_{j,i})$ for $I \subseteq I(j) := \{1, \ldots, n(j)\}$, with $F_{j,\emptyset} = F'_j$ and $G_{j,\emptyset} = \emptyset$, and observing that also these α-SHUF expressions are of height zero, we define

$$S := \bigcup_{j=1}^{m} \bigcup_{I \subseteq I(j)} F_{j,I} \shuffle G_{j,I}^{\shuffle,*}.$$

By the commutative and distributive laws and by Rule 4, S is equivalent to R^{iv} and satisfies all the properties of the theorem, possibly apart from the last sentence, which can be enforced by exhaustively applying the distributive law.   □

Unfortunately, the construction of Theorem 5 could blow up the size of the resulting expression exponentially. This does not harm the statement of the theorem, and also Theorem 4 follows immediately. For algorithmic purposes, this is indeed a drawback, because this also means that the running time of an algorithm (derived from the proof of Theorem 5) would be exponential in the length of the input expression.

Therefore, we establish the following weaker normal form result that can be, however, obtained in time that can be described within the framework of parameterized complexity [2]. In this framework, certain parts of the input are singled out as so-called parameters. In our case, it will be the number of iterated shuffle operator occurrences, as well as the shuffle iteration height of the expression. We will then present an algorithm whose only exponential-time dependencies is on these two parts of the input. In other words, if both are fixed (or if we consider only expressions with a certain upper bound on these parameters as inputs), we obtain a polynomial-time transformation algorithm.

This is an interesting fact in itself, as it also raises the descriptional complexity question if the blow-up formally described below is indeed necessary. We are not aware of any work that can be considered as "parameterized descriptional complexity", which might be therefore an interesting (new) subject on its own, motivated by the construction below.

Let us first describe the idea and some of the details of the construction that we have in mind here. As we are aiming at obtaining some equivalent α-SHUF expression of shuffle iteration height at most one, we can assume that the expression that we start with has a height of at least two. When we want to measure the size of an α-SHUF expression E, we simply count the number of all occurrences of operators in the expression, and we denote this by $s(E)$. Clearly, if we consider E as a word over Σ (plus operator symbols and parentheses), then the length of E is bounded by a linear function in $s(E)$. First of all, observe that each iterated shuffle operator occurrence in some α-SHUF expression E can be viewed as the outermost operator of a subexpression F of E that is of a certain shuffle iteration height $h(F)$. For the sake of convenience, we can hence associate a shuffle iteration height also to occurrences of shuffle operators. Let $ISO(E)$ collect all iterated shuffle operator occurrences of expression E and $ISO_h(E)$ those of shuffle iteration height h. Hence,

$$ISO(E) = \bigcup_{h=1}^{h(E)} ISO_h(E).$$

If E is an α-SHUF expresssion over the alphabet Σ, then let $\Sigma_1, \ldots, \Sigma_{h(E)}$ be fresh alphabets containing new letters, with $|\Sigma_i| = |ISO_i(E)|$ and hence a natural bijection $\psi_i : ISO_i(E) \to \Sigma_i$. Now, consider the α-SHUF expresssion E' obtained from E by replacing, for $h = 1, \ldots, h(E) - 2$, the subexpression whose outermost operator is some iterated shuffle occurrence $j \in ISO_h(E)$ by the letter $\psi_h(j)$, for all occurrences in $ISO_h(E)$. As by our assumption $h(E) \geq 2$, $h(E') = 2$. So, $ISO_{h(E)}(E) = ISO_2(E')$ and $ISO_{h(E)-1}(E) = ISO_1(E')$. In the following, we are considering all $2^{|ISO_1(E')|}$ many subsets of $ISO_1(E')$. We will convert accordingly derived expressions into equivalent ones of star height one. Proceeding inductively, we can finally show:

Theorem 6. (*) *For any α-SHUF expression R, an equivalent α-SHUF expression S with $h(S) \leq 1$ can be constructed in time $O^*(2^{|ISO(R)|}2^{h(R)})$; the resulting expression could be as big as this.*

Notice that the O^*-notation suppresses polynomial factors, which is a very suitable notation in the area of Parameterized Complexity. This shows that the transformation of R into normal form is in FPT, with parameter $|ISO(R)|$.

5 Comparing $\mathcal{JFA}$ and $\mathcal{REG}$

By the results of Meduna and Zemek, we know that $\mathcal{JFA}$ and $\mathcal{REG}$ are two incomparable families of languages. Above, we already derived several characterizations of $\mathcal{JFA} \cap \mathcal{FIN} \subset \mathcal{REG}$. Let us first explicitly write up a characterization of $\mathcal{JFA} \cap \mathcal{REG}$ that can be easily deduced from our previous results.

Proposition 5. $L \in \mathcal{JFA} \cap \mathcal{REG}$ iff $L \in \mathcal{REG}$ and L is perm-closed.

We mention this, as the class $\mathcal{JFA} \cap \mathcal{REG}$ can be also characterized as follows according to Ehrenfeucht, Haussler and Rozenberg [4]. Namely, they describe this class of (what they call) commutative regular languages as finite unions of periodic languages. We are not giving a definition of this notion here, but rather state an immediate consequence of their characterization in our terminology.

Theorem 7. Let $L \subseteq \Sigma^*$. Then, $L \in \mathcal{JFA} \cap \mathcal{REG}$ if and only if there exists a number $n \geq 1$, words w_i and finite sets N_i for $1 \leq i \leq n$, where each N_i is given as $\bigcup_{a \in \Sigma_i} a^{n_i(a)}$ for some $\Sigma_i \subseteq \Sigma$ and some $n_i : \Sigma_i \to \mathbb{N}$, so that the following representation is valid.

$$L = \bigcup_{i=1}^{n} \operatorname{perm}(w_i) \sqcup (\operatorname{perm}(N_i))^{\sqcup,*}$$

Let us finally mention that yet another characterization of $\mathcal{JFA} \cap \mathcal{REG}$ was derived in [14, Theorem 3].

6   Complexity of Parsing

For a fixed JFA M, we can decide, for a given word w, whether $w \in L(M) \subseteq \Sigma^*$ in the following way. We scan over w and construct its Parikh mapping $\pi_\Sigma(w)$. Then we simulate a computation of M on w by nondeterministically choosing in every state the transition labelled by some symbol and decrementing the corresponding component of $\pi_\Sigma(w)$. If we reach an accepting state with all components of $\pi_\Sigma(w)$ being 0, then we conclude $w \in L(M)$. In this procedure, we only have to store the Parikh mapping, which only requires logarithmic space; thus, this shows $\mathcal{JFA} \subseteq \mathsf{NL} \subseteq \mathsf{P}^1$.

These considerations show that the *fixed* word problem can be solved in polynomial time. In the following, we look at the *universal* word problem for (generalized) jumping finite automata, which is to decide for a given (general) finite machine M with input alphabet Σ and a word $w \in \Sigma^*$, whether or not $w \in L_{\mathrm{JFA}}(M)$. The study of this problem was explicitly suggested in [15], where only the mere decidability status was resolved.

We first show that the universal word problem for jumping finite automata can be solved in polynomial time, provided that the alphabet is fixed.

Theorem 8. *For any fixed alphabet, the universal word problem for jumping finite automata is polynomial-time solvable.*

Proof. Let $M = (Q, \Sigma, R, s, F)$ be a finite machine over $\Sigma = \{a_1, a_2, \ldots, a_k\}$ and let $w \in \Sigma^*$. We define a directed graph $\mathcal{G}_M = (V_M, E_M)$, where V_M contains all elements $(p, (\ell_1, \ell_2, \ldots, \ell_k))$ with $p \in Q$ and, for every i, $1 \leq i \leq k$, $0 \leq \ell_i \leq |w|_{a_i}$,

---

[1] We wish to point out that this also follows from results in [1], where containment in NP is shown for a superclass of $\mathcal{JFA}$.

and $E_M \subseteq V_M \times V_M$ contains all pairs $((p, (\ell_1, \ell_2, \ldots, \ell_k)), (p', (\ell'_1, \ell'_2, \ldots, \ell'_k)))$ such that there is a rule $pa_i \to p' \in R$, $\ell'_i = \ell_i - 1$ and, for every j, $1 \leq j \leq k$, with $i \neq j$, $\ell'_j = \ell_j$. We note that $|\mathcal{G}_M| \leq (|w|^k|Q|)^2$ and that $\mathcal{G}_M$ can be constructed in time $\mathcal{O}(|\mathcal{G}_M|)$. The graph $\mathcal{G}_M$ corresponds to the computation of M on input w: a vertex is a configuration consisting of the current state and the Parikh mapping of the remaining input and there is an edge between two configurations if it is possible to reach one from the other by the application of a rule. Hence, $w \in L_{\mathrm{JFA}}(M)$ if and only if there exists a path in $\mathcal{G}_M$ from $(s, \pi_\Sigma(w))$ to some vertex $(q, (0, 0, \ldots, 0))$ with $q \in F$. This property can be decided in time $\mathcal{O}(|\mathcal{G}_M|)$.   □

The decision procedure of Theorem 8 is only polynomial if the alphabet size is a constant, which for most real-world applications is the case. From a theoretical point of view, it would nevertheless be interesting to know whether a polynomial time procedure for unbounded alphabets is possible.

Next, we show that if the JFA-language is given as an α-SHUF expression in the normal form of Theorem 5, then the universal word problem can be solved in polynomial time also for unbounded alphabets.

Theorem 9. *The universal word problem is polynomial-time solvable for α-SHUF expressions in normal form.*

Proof. Let $R = \bigcup_{i=1}^n R_i \shuffle (R'_i)^{\shuffle,*}$ be the α-SHUF expression in the normal form of Theorem 5. Moreover, we define $M_i = L(\widehat{R_i})$ and $N_i = L(\widehat{R'_i})$, where $\widehat{R_i}$ and $\widehat{R'_i}$ are the regular expressions obtained from R_i and R'_i by replacing every shuffle operation by a catenation operation. We can convert each of the n parts of the union into linear equations as follows. Let $w \in \Sigma^*$ be the input word. Then, $w \in L(R_i \shuffle (R'_i)^{\shuffle,*})$ if and only if there is a non-negative integer solution of one of the linear equations

$$\pi_\Sigma(w) = \pi_\Sigma(u) + \sum_{v \in N_i} x_v \pi_\Sigma(v),$$

where $u \in M_i$. Since the expressions R_i, R'_i are unions of shuffles of single symbols, $\sum_{u \in M_i} |u|$ and $\sum_{v \in N_i} |v|$ are linear in $|R_i|$ and $|R'_i|$, respectively. Thus, each of the equations is of polynomial size in terms of the size of R. Each of these linear equations can be analyzed by Gaussian elimination in polynomial time. Altogether, this proves the claim.   □

If the input finite machine is allowed to be a *general* finite machine, then the complexity of the universal word problem increases considerably, i. e., it becomes NP-complete even for general finite machines accepting *finite* language over *binary* alphabets.

Theorem 10. *The universal word problem is NP-complete for generalized jumping finite automata (even for finite languages over binary alphabets).*

We can simulate a given generalized jumping finite automaton on a word by guessing where to jump and which rules to apply. Since the number of guesses is cleary bounded by the length of the input word, this shows that the universal word problem is in NP.

It remains to prove the NP-hardness of this problem, which can be done by a reduction from the following problem.

EXACT BLOCK COVER (EBC)

Instance: Words $u_1, u_2, \ldots, u_k$ and v over some alphabet Σ.
Question: Does there exist a permutation $\pi : \{1, 2, \ldots, k\} \to \{1, 2, \ldots, k\}$ such that $v = u_{\pi(1)} u_{\pi(2)} \cdots u_{\pi(k)}$?

By EBC$_2$, we denote the restricted version of EBC, where Σ is a fixed binary alphabet. It has recently been shown in [12] that EBC$_2$ is NP-complete.

Let $u_1, u_2, \ldots, u_k, v \in \Sigma^*$ be an instance of EBC$_2$, where $\Sigma = \{a, b\}$. For the sake of convenience, we define $u_i = s_{i,1} s_{i,2} \ldots s_{i,\ell_i}$, $s_{i,j} \in \Sigma$, $1 \leq i \leq k$, $1 \leq j \leq \ell_i$, and $v = t_1 t_2 \ldots t_m$, $t_j \in \Sigma$, $1 \leq j \leq m$. Furthermore, for every j, $1 \leq j \leq 2m$, we define the jth *separator* $\star_j = a b^{j+m} a$. For every i, $1 \leq i \leq k$, u_i is transformed into $A_i = \{\star_j s_{i,1} \star_{j+1} s_{i,2} \ldots \star_{j+\ell_i-1} s_{i,\ell_i} : 1 \leq j \leq m\}$ and v is transformed into $\widehat{v} = \star_1 t_1 \star_2 t_2 \ldots \star_m t_m$. We note that, for every j, $1 \leq j \leq m$, there is exactly one unique occurrence of the jth separator in $\widehat{v}$ and all these occurrences of separators are non-overlapping. Finally, we define a general finite machine $M = (Q, \Sigma, R, q_0, F)$ by $Q = \{q_0, q_1, q_2, \ldots, q_k\}$, $R = \bigcup_{i=1}^{k} \{q_{i-1} w \to q_i : w \in A_i\}$ and $F = \{q_k\}$. This reduction is obviously polynomial.

We give a proof sketch for the correctness of this reduction. To this end, let $(u_1, u_2, \ldots, u_k, v)$ be a positive instance of EBC$_2$. Then $v = u_{\pi(1)} \cdots u_{\pi(k)}$ for some permutation $\pi : \{1, 2, \ldots, k\} \to \{1, 2, \ldots, k\}$. If we insert the jth separator after the jth symbol of $u_{\pi(1)} \cdots u_{\pi(k)}$, then we obtain $\widehat{v} = w_{\pi(1)} w_{\pi(2)} \cdots w_{\pi(k)}$ with $w_{\pi(i)} \in A_{\pi(i)}$, $1 \leq i \leq k$; thus, $\widehat{v} \in L_{\mathrm{JFA}}(M)$, which yields the following.

Lemma 4. *(*) If $(u_1, u_2, \ldots, u_k, v) \in$ EBC$_2$, then $\widehat{v} \in L_{JFA}(M)$.*

If, on the other hand, $\widehat{v} \in L_{\mathrm{JFA}}(M)$, then $v = u_{\pi(1)} u_{\pi(2)} \cdots u_{\pi(k)}$ can only be concluded if M never erases a factor that does not correspond to an original factor of $\widehat{v}$ (or, equivalently, if M never erases a factor that contains consecutive symbols that do not correspond to consecutive symbols of $\widehat{v}$). This property is enforced by the separator words; thus, we can conlude the following.

Lemma 5. *(*) If $\widehat{v} \in L_{JFA}(M)$, then $(u_1, u_2, \ldots, u_k, v) \in$ EBC$_2$.*

Theorems 8 and 10 point out that the difference between finite machines and general finite machines is crucial if we interpret them as jumping finite automata. In contrast to this, the universal word problem for (classical) finite automata on the one hand and (classical) general finite automata on the other is very similar in terms of complexity, i. e., in both cases it can be solved in polynomial time.

References

1. Crespi-Reghizzi, S., San Pietro, P.: Commutative languages and their composition by consensual methods. In: Ésik, V., Fülöp, Z. (eds.) Proceedings 14th International Conference on Automata and Formal Languages (AFL), vol. 151 of EPTCS, pp. 216–230 (2014)
2. Downey, R.G., Fellows, M.R.: Fundamentals of Parameterized Complexity. Texts in Computer Science. Springer, Heidelberg (2013)
3. Eggan, L.C.: Transition graphs and the star-height of regular events. Mich. Math. J. **10**(4), 385–397 (1963)
4. Ehrenfeucht, A., Haussler, D., Rozenberg, G.: On regularity of context-free languages. Theor. Comput. Sci. **27**, 311–332 (1983)
5. Esparza, J., Ganty, P., Kiefer, S., Luttenberger, M.: Parikh's theorem: A simple and direct automaton construction. Inf. Process. Lett. **111**(12), 614–619 (2011)
6. Hashiguchi, K.: Algorithms for determining relative star height and star height. Inf. Comput. **78**(2), 124–169 (1988)
7. Höpner, M., Opp, M.: About three equations classes of languages built up by shuffle operations. In: Mazurkiewicz, A.W. (ed.) MFCS 1976. LNCS, vol. 45, pp. 337–344. Springer, Heidelberg (1976)
8. Jantzen, M.: Eigenschaften von Petrinetzsprachen. Technical report IFI-HH-B-64, Fachbereich Informatik, Universität Hamburg, Germany (1979)
9. Jantzen, M.: The power of synchronizing operations on strings. Theor. Comput. Sci. **14**, 127–154 (1981)
10. Jantzen, M.: Extending regular expressions with iterated shuffle. Theor. Comput. Sci. **38**, 223–247 (1985)
11. Jedrzejowicz, J., Szepietowski, A.: Shuffle languages are in P. Theor. Comput. Sci. **250**(1–2), 31–53 (2001)
12. Jiang, H., Su, B., Xiao, M., Xu, Y., Zhong, F., Zhu, B.: On the exact block cover problem. In: Gu, Q., Hell, P., Yang, B. (eds.) AAIM 2014. LNCS, vol. 8546, pp. 13–22. Springer, Heidelberg (2014)
13. Klíma, O., Polák, L.: On biautomata. RAIRO Informatique théorique et Appl. Theor. Inf. Appl. **46**, 573–592 (2012)
14. Latteux, M., Rozenberg, G.: Commutative one-counter languages are regular. J. Comput. Sys. Sci. **1**, 54–57 (1984)
15. Meduna, A., Zemek, P.: Jumping finite automata. Int. J. Found. Comput. Sci. **23**(7), 1555–1578 (2012)
16. Meduna, A., Zemek, P.: Chapter 17: Jumping finite automata. In: Meduna, A., Zemek, P. (eds.) Regulated Grammars and Automata, pp. 567–585. Springer, New York (2014)
17. Otto, F.: Restarting automata. In: Ésik, Z., Martín-Vide, C., Mitrana, V. (eds.) FCT 1995, vol. 965, pp. 269–303. Springer, Heidelberg (2006)
18. Parikh, R.J.: On context-free languages. J. ACM **13**(4), 570–581 (1966)

Run-Length Encoded Nondeterministic KMP and Suffix Automata

Emanuele Giaquinta[(✉)]

Department of Computer Science, Aalto University, Espoo, Finland
emanuele.giaquinta@aalto.fi

Abstract. We present a novel bit-parallel representation, based on the run-length encoding, of the nondeterministic KMP and suffix automata for a string P with at least two distinct symbols. Our encoding requires $O((\sigma + m)\lceil \rho/w \rceil)$ space and allows one to simulate the automata on a string in time $O(\lceil \rho/w \rceil)$ per transition, where σ is the alphabet size, m is the length of P, ρ is the length of the run-length encoding of P and w is the machine word size in bits. The input string can be given in either unencoded or run-length encoded form. Finally, we present practical variants of the Shift-And and BNDM algorithms based on this encoding.

1 Introduction

The string matching problem consists in finding all the occurrences of a string P of length m in a string T of length n, both over a finite alphabet Σ of size σ. The matching can be either exact or approximate, according to some metric which measures the closeness of a match. The finite automata for the languages $\Sigma^* P$ (prefix automaton) and $\mathit{Suff}(P)$ (suffix automaton), where $\mathit{Suff}(P)$ is the set of suffixes of P, are the main building blocks of very efficient algorithms for the exact and approximate string matching problem. Two fundamental algorithms for the exact problem, based on the deterministic version of these automata, are the KMP and BDM algorithms, which run in $O(n)$ and $O(nm)$ worst-case time, respectively, using $O(m)$ space [6,14]. In the average case, the BDM algorithm achieves the optimal $\mathcal{O}(n \log_\sigma(m)/m)$ time bound. The nondeterministic version of the prefix and suffix automata can be simulated using an encoding, known as bit-parallelism, based on bit-vectors and word-level parallelism [16]. The variants of the KMP algorithm based on the nondeterministic prefix automaton, known as SHIFT-OR and SHIFT-AND, run in $O(n\lceil m/w \rceil)$ worst-case time and use $O(\sigma\lceil m/w \rceil)$ space, where w is the machine word size in bits [2,18]. Similarly, the variant of the BDM algorithm based on the nondeterministic suffix automaton, known as BNDM, runs in $\mathcal{O}(nm\lceil m/w \rceil)$ worst-case time and uses $O(\sigma\lceil m/w \rceil)$ space [15]. In the average case, the BNDM algorithm runs in $\mathcal{O}(n \log_\sigma(m)/w)$ time, which is suboptimal for patterns whose length is greater than w. There also exist practical variants of BNDM [7,17], and a variant of SHIFT-OR which achieves $\mathcal{O}(n \log_\sigma(m)/w)$ time in the average case [10]. As for the approximate string matching problem, there are also various algorithms based on the nondeterministic prefix and suffix automata [3,9,11,12,15,18].

© Springer International Publishing Switzerland 2015
F. Drewes (Ed.): CIAA 2015, LNCS 9223, pp. 102–113, 2015.
DOI: 10.1007/978-3-319-22360-5_9

In general, the bit-parallel algorithms are suboptimal if compared to their "deterministic" counterparts in the case $m > w$, because of the additional $\lceil m/w \rceil$ term in the time complexity. A way to overcome this problem is to use a filtering method, namely, searching for the prefix of P of length w and verifying each occurrence with a naive algorithm. Assuming uniformly random strings, the average time complexity of SHIFT-AND and BNDM with this method is $O(n)$ and $O(n \log_\sigma w/w)$, respectively. Recently, a few approaches were proposed to improve the case of long patterns. In 2010 Durian et al. presented three variants of BNDM tuned for the case of long patterns, two of which are optimal in the average case [8]. In the same year, Cantone et al. presented a different encoding of the prefix and suffix automata, based on word-level parallelism and on a particular factorization on strings [5]. The general approach is to devise, given a factorization f on strings, a bit-parallel encoding of the automata based on f such that one transition can be performed in $O(\lceil |f(P)|/w \rceil)$ time instead of $O(\lceil m/w \rceil)$, at the price of more space. The gain is two-fold: i) if $|f(P)| < m$, then the overhead of the simulation is reduced. In particular, there is no overhead if $|f(P)| \leq w$, which is preferable if $|f(P)| < m$; ii) if we use the filtering method, we can search for the longest substring P' of P such that $|f(P')| \leq w$. This yields $O(n \log_\sigma |P'|/|P'|)$ average time for BNDM, which is preferable if $|P'| > w$. The factorization introduced by Cantone et al. is such that $\lceil m/\sigma \rceil \leq |f(P)| \leq m$ and their encoding requires $O(\sigma^2 \lceil |f(P)|/w \rceil)$ space.

In this paper we present a novel encoding of the prefix and suffix automata, based on this approach, where $f(P)$ is the run-length encoding of P, provided that P has at least two distinct symbols. The run-length encoding of a string is a simple encoding where each maximal consecutive sequence of the same symbol is encoded as a pair consisting of the symbol plus the length of the sequence. Our encoding requires $O((\sigma + m) \lceil \rho/w \rceil)$ space and allows one to simulate the automata in $O(\lceil \rho/w \rceil)$ time per transition, where ρ is the length of the run-length encoding of P. While the present algorithm uses the run-length encoding, the input string can be given in either unencoded or run-length encoded form. Finally, we present practical variants of the SHIFT-AND and BNDM algorithms based on this encoding.

2   Notions and Basic Definitions

Let Σ be a finite alphabet of symbols and let Σ^* be the set of all strings over Σ. The empty string ε is a string of length 0. Given a string S, we denote by $|S|$ the length of S and by $S[i]$ the i-th symbol of S, for $0 \leq i < |S|$. The concatenation of two strings S and $\bar{S}$ is denoted by $S\bar{S}$. Given two strings S and $\bar{S}$, S is a substring of $\bar{S}$ if there are indices $0 \leq i, j < |S|$ such that $\bar{S} = S[i]...S[j]$. If $i = 0$ ($j = |S|-1$) then $\bar{S}$ is a prefix (suffix) of S. The set $Suff(S)$ is the set of all suffixes of S. We denote by $S[i..j]$ the substring $S[i]...S[j]$ of S. For $i > j$ $S[i..j] = \varepsilon$. We denote by S^k the concatenation of k strings S's, for $S \in \Sigma^*$ and $k \geq 1$. The string S^R is the reverse of the string S, i.e., $S^R = S[|S|-1]S[|S|-2]...S[0]$. A pattern with character classes is a sequence $C_1 C_2 ... C_m$ where $C_i \subseteq \Sigma$. Given a string S, we write $C_1 C_2 ... C_m = S$ if $|S| = m$ and $S[i-1] \in C_i$, for $1 \leq i \leq m$.

Given a string $P \in \Sigma^*$ of length m, we denote by $\mathcal{A}(P) = (Q, \Sigma, \delta, q_0, F)$ the nondeterministic finite automaton (NFA) for the language $\Sigma^* P$ of all strings in Σ^* whose suffix of length m is P, where:

- $Q = \{q_0, q_1, \ldots, q_m\}$ (q_0 is the initial state)
- the transition function $\delta : Q \times \Sigma \longrightarrow \mathscr{P}(Q)$ is defined by:

$$\delta(q_i, c) =_{Def} \begin{cases} \{q_0, q_1\} & \text{if } i = 0 \text{ and } c = P[0] \\ \{q_0\} & \text{if } i = 0 \text{ and } c \neq P[0] \\ \{q_{i+1}\} & \text{if } 1 \leq i < m \text{ and } c = P[i] \\ \emptyset & \text{otherwise} \end{cases}$$

- $F = \{q_m\}$ (F is the set of final states).

Similarly, we denote by $\mathcal{S}(P) = (Q, \Sigma, \delta, I, F)$ the nondeterministic suffix automaton with ε-transitions for the language $Suff(P)$ of the suffixes of P, where:

- $Q = \{I, q_0, q_1, \ldots, q_m\}$ (I is the initial state)
- the transition function $\delta : Q \times (\Sigma \cup \{\varepsilon\}) \longrightarrow \mathscr{P}(Q)$ is defined by:

$$\delta(q, c) =_{Def} \begin{cases} \{q_{i+1}\} & \text{if } q = q_i \text{ and } c = P[i] \ (0 \leq i < m) \\ Q & \text{if } q = I \text{ and } c = \varepsilon \\ \emptyset & \text{otherwise} \end{cases}$$

- $F = \{q_m\}$ (F is the set of final states).

We use the notation q_I to indicate the initial state of the automaton, i.e., q_I is q_0 for $\mathcal{A}(P)$ and I for $\mathcal{S}(P)$. The valid configurations $\delta^*(q_I, S)$ which are reachable by the automata $\mathcal{A}(P)$ and $\mathcal{S}(P)$ on input $S \in \Sigma^*$ are defined recursively as follows:

$$\delta^*(q_I, S) =_{Def} \begin{cases} E(q_I) & \text{if } S = \varepsilon, \\ \bigcup_{q' \in \delta^*(q_I, S')} \delta(q', c) & \text{if } S = S'c, \text{ for some } c \in \Sigma \text{ and } S' \in \Sigma^*. \end{cases}$$

where $E(q_I)$ denotes the ε-closure of q_I.

Given a string P, a run of P is a maximal substring of P containing exactly one distinct symbol. The run-length encoding (RLE) of a string P, denoted by $\text{RLE}(P)$, is a sequence of pairs (runs) $\langle (c_0, l_0), (c_1, l_1), \ldots, (c_{\rho-1}, l_{\rho-1}) \rangle$ such that $c_i \in \Sigma$, $l_i \geq 1$, $c_i \neq c_{i+1}$ for $0 \leq i < \rho$, and $P = c_0^{l_0} c_1^{l_1} \ldots c_{\rho-1}^{l_{\rho-1}}$. The starting and ending position in P of the run (c_i, l_i) are $\alpha_P(i) = \sum_{j=0}^{i-1} l_j$ and $\beta_P(i) = \sum_{j=0}^{i} l_j - 1$, for $i = 0, \ldots, \rho - 1$. We also put $\alpha_P(\rho) = |P|$. The length of the run (c_i, l_i) is denoted by $\ell_P(i)$.

Finally, we recall the notation of some bitwise infix operators on computer words, namely the bitwise **and** "&", the bitwise **or** "|", the **left shift** "$\ll$" operator (which shifts to the left its first argument by a number of bits equal to its second argument), and the unary bitwise **not** operator "$\sim$".

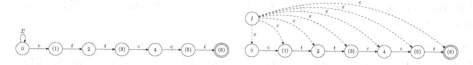

Fig. 1. (a) The automata $\mathcal{A}(P)$ and $\mathcal{S}(P)$ for the pattern $P = cttcct$. The state labels corresponding to the starting positions of the runs of $\mathrm{RLE}(P)$ are in parentheses.

3 The Shift-And and BNDM Algorithms

In this section we briefly describe the SHIFT-AND and BNDM algorithms. Given a pattern P of length m and a text T of length n, the SHIFT-AND and BNDM algorithms find all the occurrences of P in T. The SHIFT-AND algorithm works by simulating the $\mathcal{A}(P)$ automaton on T and reporting all the positions j in T such that the final state of $\mathcal{A}(P)$ is active in the corresponding configuration $\delta^*(q_I, T[0..j])$. Instead, the BNDM algorithm works by sliding a window of length m along T. For a given window ending at position j, the algorithm simulates the automaton $\mathcal{S}(P^R)$ on $(T[j - m + 1 .. j])^R$. Based on the simulation, the algorithm computes the length k and k' of the longest suffix of $T[j - m + 1 .. j]$ which is a prefix and a proper prefix, respectively, of P (i.e., a suffix of P^R). If $k = m$ then $T[j - m + 1 .. j] = P$ and the algorithm reports an occurrence of P at position j. The window is then shifted by $m - k'$ positions to the right, so as to align it with the longest proper prefix of P found. The automata are simulated using an encoding based on bit-vectors and word-level parallelism. The algorithms run in $O(n\lceil m/w\rceil)$ and $O(nm\lceil m/w\rceil)$ time, respectively, using $O(\sigma\lceil m/w\rceil)$ space, where w is the word size in bits. The automata and the associated encoding can also be extended to the case of a pattern with character classes.

4 RLE-Based Encoding of the Nondeterministic KMP and Suffix Automata

Given a string P of length m defined over an alphabet Σ of size σ, let $\mathrm{RLE}(P) = \langle (c_0, l_0), (c_1, l_1,), \dots, (c_{\rho-1}, l_{\rho-1}) \rangle$ be the run-length encoding of P. In the following, we describe how to simulate the $\mathcal{A}(P)$ and $\mathcal{S}(P)$ automata, using word-level parallelism, on a string S of length n in $O(\lceil \rho/w\rceil)$ time per transition and $O((m + \sigma)\lceil \rho/w\rceil)$ space. We recall that the simulation of the automaton $\mathcal{A}(P)$ on a string S detects all the prefixes of S whose suffix of length m is P. Similarly, the simulation of the automaton $\mathcal{S}(P)$ detects all the prefixes of S which are suffixes of P.

Let $\mathcal{I}(S) = \{\alpha_S(i) \mid 0 \leq i \leq |\mathrm{RLE}(S)|\}$ be the set of starting positions of the runs of S, for a given string S. Note that $0 \in \mathcal{I}(S)$. Given a string S, we denote by $D_j = \delta^*(q_I, S[0 .. j - 1])$ the configuration of $\mathcal{A}(P)$ or $\mathcal{S}(P)$ after reading $S[0 .. j - 1]$, for any $0 \leq j \leq |S|$. We start with the following Lemma:

Lemma 1. *Let* $j \in \mathcal{I}(S)$. *Then, for any* $q_i \in D_j$ *such that* $i \notin \mathcal{I}(P)$, *we have* $\delta(q_i, S[j]) = \emptyset$.

Proof. Let $q_i \in D_j$ with $i \notin \mathcal{I}(P)$. By definition of q_i and by $i \notin \mathcal{I}(P)$ it follows that $S[j-1] = P[i-1]$ and $P[i-1] = P[i]$, respectively. Moreover, by $j \in \mathcal{I}(S)$, we have $S[j] \neq S[j-1]$. Suppose that $\delta(q_i, S[j]) \neq \emptyset$, which implies $S[j] = P[i]$. Then we have $S[j] = P[i] = P[i-1] = S[j-1]$, which yields a contradiction. □

This Lemma states that, for any $j \in \mathcal{I}(S)$, any state $q_i \in D_j$ with $i \notin \mathcal{I}(P)$ is dead, as no transition is possible from it on $S[j]$. Figure 1 shows the automata $\mathcal{A}(P)$ and $\mathcal{S}(P)$ for $P = cttcct$; the state labels corresponding to indexes in $\mathcal{I}(P)$ are in parentheses. In this case $\mathcal{I}(P) = \{0, 1, 3, 5, 6\}$ and therefore states q_2 and q_4 are dead in any configuration D_j with $j \in \mathcal{I}(S)$.

We assume that P has at least two distinct symbols, i.e., $\rho \geq 2$. The following Lemma shows that, under this assumption, there can be at most one prefix of S in the language of $\mathcal{A}(P)$ or of $\mathcal{S}(P)$ ending in a position between $\alpha_S(i)$ and $\beta_S(i)$ in S, for any $1 \leq i \leq |\text{RLE}(S)|$ (note that $i \geq 1$ implies that the corresponding prefix of S in the language has at least two distinct symbols).

Lemma 2. *Let* $i \in \{1, \ldots, |\text{RLE}(S)| - 1\}$. *If* $\rho \geq 2$, *there exists at most one* j *in the interval* $[\alpha_S(i), \beta_S(i)]$ *such that* $q_m \in \delta^*(q_I, S[0 .. j])$.

Proof. The claim follows by observing that for any two strings S_1 and S_2 in the language with at least two distinct symbols we have $\ell_{S_1}(|\text{RLE}(S_1)| - 1) = \ell_{S_2}(|\text{RLE}(S_2)| - 1)$.

Specifically, the only prefix in the language, if any, corresponds to index $\alpha_S(i) + l_{\rho-1} - 1$. For $i = 0$ it is easy to see that: i) in the case of the prefix automaton, since $\rho \geq 2$, $S[0 .. j]$ is not in the language for $j \in [\alpha_S(0), \beta_S(0)]$; ii) in the case of the suffix automaton, if $S[0] = P[m-1]$ then $S[0 .. j]$ is in the language for $0 \leq j < \min(\ell_S(0), l_{\rho-1})$, and is not otherwise. Hence, in the case of the suffix automaton, we can detect all the prefixes of S in the language with one distinct symbol by comparing the first run of S with the last run of P.

By definition of D_j and by Lemma 1, we have

$$
\begin{aligned}
D_{\alpha_S(j+1)} &= \delta^*(q_I, S[0 .. \beta_S(j)]) \\
&= \bigcup_{q \in D_{\alpha_S(j)}} \delta^*(q, S[\alpha_S(j) .. \beta_S(j)]) \\
&= \bigcup_{q \in D_{\alpha_S(j)} \cap \{q_i \,|\, i \in \mathcal{I}(P)\}} \delta^*(q, S[\alpha_S(j)]^{\ell_S(j)})
\end{aligned}
$$

for any position $\alpha_S(j + 1)$. The idea is to compute the configurations D_j, restricted to the states with index in $\mathcal{I}(P)$, corresponding to positions $j \in \mathcal{I}(S)$ only by reading S run-wise. Observe that it is not possible to detect the single prefix of S in the language, if any, ending at a position between $\alpha_S(j-1)$ and $\beta_S(j-1)$ using $D_{\alpha_S(j)}$, because $q_m \notin D_{\alpha_S(j)}$ if the prefix does not end at position $\beta_S(j-1)$, or equivalently if $\ell_S(j-1) > l_{\rho-1}$.

Let $\bar{D}_j$ be the set such that $i \in \bar{D}_j$ iff $q_{\alpha_P(i)} \in D_{\alpha_S(j)}$, for $1 \leq i < \rho$, and $\rho \in \bar{D}_j$ iff $j \geq 2$ and $q_m \in D_{\alpha_S(j-1)+1} \cup \ldots \cup D_{\alpha_S(j)}$. The set $\bar{D}_j$ is the encoding of the configuration of $\mathcal{A}(P)$ or $\mathcal{S}(P)$ after reading $S[0 \mathinner{.\,.} \beta_S(j-1)]$, for $0 \leq j \leq |\mathrm{RLE}(S)|$, such that ρ is present iff a prefix of S in the interval $[\alpha_S(j-1), \beta_S(j-1)]$ with at least two distinct symbols is in the language. Observe that, if $\rho \in \bar{D}_j$, then $\rho \notin \bar{D}_{j+1}$, since $S[\alpha_S(j)] \neq S[\alpha_S(j+1)]$.

The following example shows the configurations $\bar{D}_j$ of $\mathcal{A}(P)$ on S, for $P = cttcct$ and $S = cttccttcct$, and the starting positions of the runs of P and S:

$P = cttcct$	
$S = cttccttcct$	
$\bar{D}_1 = \{1\}$	$\bar{D}_2 = \{2\}$
$\bar{D}_3 = \{1,3\}$	$\bar{D}_4 = \{2,4\}$
$\bar{D}_5 = \{1,3\}$	$\bar{D}_6 = \{4\}$

i	0	1	2	3	4
$\alpha_P(i)$	0	1	3	5	6

i	0	1	2	3	4	5	6
$\alpha_S(i)$	0	1	3	5	7	9	10

Note that q_0 is not represented and that $\bar{D}_0$ is equal to $\emptyset$ and $\{1, \ldots, \rho\}$ for $\mathcal{A}(P)$ and $\mathcal{S}(P)$, respectively. We now describe how to compute the configurations $\bar{D}_j$.

4.1 Computation of $\bar{D}_j$

Let $P_a = P[\alpha_P(0)]P[\alpha_P(1)] \ldots P[\alpha_P(\rho-1)]$ and $P_b = \ell_P(0)\ell_P(1) \ldots \ell_P(\rho-1)$ be the strings corresponding to the concatenation of the symbols and lengths of the runs in the run-length encoding of P, respectively. For example, if $P = cttcct$ we have $P_a = ctct$ and $P_b = 1221$. Let S_a and S_b be defined analogously. Observe that the strings S_a and S_b can be computed on the fly in constant space from S. It is well known that the string matching problem on $\mathrm{RLE}(S)$ can be reduced to that of searching for P_a in S_a and for P_b in S_b [1]. Indeed, we have the following Lemma:

Lemma 3. *Let $i \in \{1, \ldots, |\mathrm{RLE}(S)| - 1\}$. We have $P = S[j - m + 1 \mathinner{.\,.} j]$, for some $j \in [\alpha_S(i), \beta_S(i)]$, iff the following conditions hold:*

1. $P_a = S_a[i - \rho + 1 \mathinner{.\,.} i]$;
2. $P_b[1 \mathinner{.\,.} \rho - 2] = S_b[i - \rho + 2 \mathinner{.\,.} i - 1]$;
3. $P_b[\rho - 1] \leq S_b[i]$;
4. $P_b[0] \leq S_b[i - \rho + 1]$.

Suppose now that we want to determine whether a prefix of S matches a suffix of P with at least two distinct symbols. In this case, we have the following Lemma:

Lemma 4. *Let $i \in \{1, \ldots, |\mathrm{RLE}(S)| - 1\}$ and $i' \in \{0, \ldots, \rho - 2\}$. We have $P[j' \mathinner{.\,.} m - 1] = S[0 \mathinner{.\,.} j]$, for some $j \in [\alpha_S(i), \beta_S(i)]$ and $j' \in [\alpha_P(i'), \beta_P(i')]$, iff the following conditions hold:*

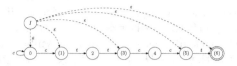

Fig. 2. (a) The automaton $S_r(P)$ for the pattern $P = cttcct$.

1. $P_a[i' \mathbin{..} \rho - 1] = S_a[0 \mathbin{..} i]$;
2. $P_b[i' + 1 \mathbin{..} \rho - 2] = S_b[1 \mathbin{..} i - 1]$;
3. $P_b[\rho - 1] \leq S_b[i]$;
4. $S_b[0] \leq P_b[i']$.

Let $S_{b'}$ be the string of length n such that $S_{b'}[i] = \min(\ell_S(i), m + 1)$. The string $S_{b'}$ corresponds to the concatenation of the run lengths of $\mathrm{RLE}(S)$ such that lengths longer than m are replaced with $m + 1$. The idea is to perform the computation of $\bar{D}_j$ by means of a joint simulation of $\mathcal{A}(P_a)$ on S_a and of a modified version of $\mathcal{A}(P_b)$ on $S_{b'}$, and analogously for $\mathcal{S}(P)$.

Consider the automata $\mathcal{A}(P_a)$ and $\mathcal{A}(P_b)$. Observe that both automata have $\rho + 1$ states and are defined over the alphabet Σ and $\{1, 2, \ldots, m\}$, respectively. Let $\mathcal{A}'(P_b)$ be the automaton $\mathcal{A}(P_b)$ modified by augmenting the transition function as follows:

- $\delta(q_{\rho-1}, l) = \{q_\rho\}$, for $P_b[\rho - 1] + 1 \leq l \leq m + 1$;
- $\delta(q_0, l) = \{q_0, q_1\}$, for $P_b[0] + 1 \leq l \leq m + 1$.

Similarly, Let $\mathcal{S}'(P_b)$ be the automaton $\mathcal{S}(P_b)$ modified by augmenting the transition function as follows:

- $\delta(q_{\rho-1}, l) = \{q_\rho\}$, for $P_b[\rho - 1] + 1 \leq l \leq m + 1$;
- $\delta(I, l) = \{q_i \mid i \leq \rho - 1 \wedge l \leq P_b[i]\}$, for $1 \leq l \leq m$.

The first and second changes to $\mathcal{A}(P_b)$ ($\mathcal{S}(P_b)$) correspond to handling conditions 3 and 4 of Lemma 3 (4), respectively. In other words, by means of character classes, we turn condition 3 of Lemmas 3 and 4 into the equivalent condition $S_{b'}[i] \in \{P_b[\rho - 1], P_b[\rho - 1] + 1, \ldots, m + 1\}$, condition 4 of Lemma 3 in $S_{b'}[i - \rho + 1] \in \{P_b[0], P_b[0] + 1, \ldots, m + 1\}$ and condition 4 of Lemma 4 in $S_{b'}[0] \in \{1, 2, \ldots, P_b[i']\}$.

Let $\bar{D}_j^a = \{1 \leq i \leq \rho \mid q_i^a \in \delta^*(q_I^a, S_a[0 \mathbin{..} j])\}$ be the configuration of $\mathcal{A}(P_a)$ or $\mathcal{S}(P_a)$ after reading $S_a[0 \mathbin{..} j]$ and analogously for $\bar{D}_j^b$. It is not hard to verify that

$$\bar{D}_j = \bar{D}_j^a \cap \bar{D}_j^b \tag{1}$$

and, therefore, the computation of the configurations $\bar{D}_j$ can be reduced to the simulation of $\mathcal{A}(P_a)$ on S_a and of $\mathcal{A}'(P_b)$ on $S_{b'}$, and analogously for $\mathcal{S}(P)$.

Let $P_{b'}$ be the pattern with character classes $\{\ell_P(0), \ell_P(0) + 1, \ldots, m + 1\}\ell_P(1) \ldots \ell_P(\rho - 2)\{\ell_P(\rho - 1), \ell_P(\rho - 1) + 1, \ldots, m + 1\}$. Observe that the automaton $\mathcal{A}'(P_b)$ corresponds to $\mathcal{A}(P_{b'})$. Instead, if we replace $\mathcal{S}'(P_b)$ with $\mathcal{S}(P_{b'})$, we obtain a simulation of the automaton $S_r(P)$ whose language is the subset of $Suff(P)$ $\{P[i..m-1] \mid i \geq 1 \wedge P[i] \neq P[i-1]\}$ plus $P[0]^*P$ and whose transition function is defined as follows:

$$\delta(q,c) =_{Def} \begin{cases} \{q_0, q_1\} & \text{if } q = q_0 \text{ and } c = P[0] \\ \{q_{i+1}\} & \text{if } q = q_i \text{ and } c = P[i] \ (0 \leq i < m) \\ \{q_i \mid i = 0 \vee P[i] \neq P[i-1]\} & \text{if } q = I \text{ and } c = \varepsilon \\ \emptyset & \text{otherwise} \end{cases}$$

Fig. 2 shows the automaton $\mathcal{S}_r(P)$ for $P = cttcct$.

We now describe the encoding of $\mathcal{A}(P)$ and $\mathcal{S}(P)$. Let

$$B_1(c) = \{1 \leq i \leq \rho \mid c = P_a[i-1]\},$$

for any $c \in \Sigma$. The set $B_1(c)$ includes the indices of all the runs whose symbol is equal to c. It is well known that for $\mathcal{A}(P_a)$ we have

$$\bar{D}^a_{j+1} = \{i+1 \mid i \in \bar{D}^a_j \cup \{0\}\} \cap B_1(S_a[j])$$

for $j \geq 0$, while for $\mathcal{S}(P_a)$ we have

$$\begin{aligned} \bar{D}^a_1 &= B_1(S_a[0]) \\ \bar{D}^a_{j+1} &= \{i+1 \mid i \in \bar{D}^a_j\} \cap B_1(S_a[j]) \end{aligned}$$

for $j \geq 1$. Consider now the automata $\mathcal{A}'(P_b)$ and $\mathcal{S}'(P_b)$, and let $B_2(l) = \{1 \leq i \leq \rho \mid l = P_b[i-1]\}$. The set $B_2(l)$ includes the indices of all the runs whose length is equal to l. Note that $B_2(l) = \emptyset$, for any $l > m$; thus, we can define B_2 up to $m+1$ and map any integer greater than m onto $m+1$. By using B_2 in place of B_1 we simulate the automaton $\mathcal{A}(P_b)$. To account for the first and second change in $\mathcal{A}'(P_b)$ we add ρ to $B_2(l)$, for $P_b[\rho-1]+1 \leq l \leq m+1$, and 1 to $B_2(l)$, for $P_b[0]+1 \leq l \leq m+1$, respectively. For example, for $P = cttcct$ we have:

$$\begin{aligned} B_1(c) &= \{1,3\} \quad B_2(1) = \{1,4\} \\ B_1(t) &= \{2,4\} \quad B_2(2) = \{1,2,3,4\} \end{aligned}$$

and $B_2(l) = \{1,4\}$, for $3 \leq l \leq 7$. Concerning $\mathcal{S}'(P_b)$, while the first change is the same as for $\mathcal{A}'(P_b)$, the second change is different and affects the first transition only. To account for the first change we add ρ to $B_2(l)$, for $P_b[\rho-1]+1 \leq l \leq m+1$. Instead, to account for the second change, we define the set

$$B_3(l) = \{1 \leq i \leq \rho-1 \mid l \leq P_b[i-1]\}$$

for $1 \leq l \leq m+1$ and use it to compute $\bar{D}^b_1$. The set $B_3(l)$ includes the indices of all the runs, except the last, whose length is not less than l.

We now show how to implement Eq. 1 efficiently using word-level parallelism. We represent the configurations $\bar{D}$ and the sets B as bit-vectors of ρ bits, denoted by D and B, respectively. Based on the encodings described above, Eq. 1 for $\mathcal{A}(P)$ can be written as

$$\bar{D}_{j+1} = \{i+1 \mid i \in \bar{D}_j \cup \{0\}\} \cap B_1(S_a[j]) \cap B_2(S_{b'}[j]),$$

which corresponds to the following bitwise operations

$$\mathsf{D}_{j+1} = ((\mathsf{D}_j \ll 1) \mid 0^{\rho-1}1) \ \& \ \mathsf{B}_1(S_a[j]) \ \& \ \mathsf{B}_2(S_{b'}[j]).$$

Similarly, in the case of $\mathcal{S}(P)$ we have

$$\begin{aligned}
\bar{D}_1 &= B_1(S_a[0]) \cap B_3(S_{b'}[0]),\\
\bar{D}_{j+1} &= \{i+1 \mid i \in \bar{D}_j\} \cap B_1(S_a[j]) \cap B_2(S_{b'}[j]),
\end{aligned}$$

which corresponds to the following bitwise operations

$$\begin{aligned}
\mathsf{D}_1 &= \mathsf{B}_1(S_a[0]) \ \& \ \mathsf{B}_3(S_{b'}[0]),\\
\mathsf{D}_{j+1} &= (\mathsf{D}_j \ll 1) \ \& \ \mathsf{B}_1(S_a[j]) \ \& \ \mathsf{B}_2(S_{b'}[j]).
\end{aligned}$$

We now analyze the complexity of the described encodings. The computation of a single configuration $\bar{D}_j$ requires $O(\lceil \rho/w \rceil)$ time. The total time complexity of the simulation is thus $O(|S|\lceil \rho/w \rceil)$, as the total number of configurations is $|\mathrm{RLE}(S)| \leq |S|$. The bit-vectors B can be preprocessed in $O(m + (\sigma + m)\lceil \rho/w \rceil)$ time and require $O((\sigma + m)\lceil \rho/w \rceil)$ space. The string P or S can be given in either unencoded or run-length encoded form. In the former case its run-length encoding does not need to be stored. It can be computed on the fly in $O(m)$ or $O(|S|)$ time, using constant space, during the preprocessing or searching phase.

5   The Variants of Shift-And and BNDM

The variants of the SHIFT-AND and BNDM algorithms based on the encoding described in the previous section run in $O(n\lceil \rho/w \rceil)$ and $O(nm\lceil \rho/w \rceil)$ time, respectively, using $O((\sigma + m)\lceil \rho/w \rceil)$ space. The encoding of the suffix automaton is however not ideal in practice, due to the different first transition. We now describe a variant of BNDM, based on Lemma 3, where the first transition of the automaton is equal to the subsequent ones. The idea is to search for P_a in S_a and $P_{b'}$ in $S_{b'}$ with the BNDM algorithm maintaining a single window in both strings. For a given window of length ρ ending at position j in S_a and in $S_{b'}$, we compute the largest $k' \leq \rho-1$ and $k \leq \rho$ such that $P_a[0 \mathinner{..} k'-1] = S_a[j-k'+1 \mathinner{..} j]$ and $P_{b'}[0 \mathinner{..} k'-1] = S_{b'}[j - k' + 1 \mathinner{..} j]$, and analogously for k.

Let $\bar{k}'$ be the length of the longest suffix of $S[\alpha_S(j - \rho + 1) \mathinner{..} \beta_S(j)]$ which is a proper prefix of P and such that $P[\bar{k}' - 1] \neq P[\bar{k}']$. It is not hard to see that $k' = |\mathrm{RLE}(P[0 \mathinner{..} \bar{k}' - 1])|$. Observe that, if $P[\bar{k}' - 1] = P[\bar{k}']$, then the window starting at position $\beta_S(j) - \bar{k}' + 1$ does not contain an occurrence of P, because $P[\bar{k}' - 1] = S[\beta_S(j)]$ and $S[\beta_S(j)] \neq S[\beta_S(j) + 1]$. Therefore, $\rho - k'$ is a safe shift and we can thus slide the window in S_a and $S_{b'}$ by $\rho - k'$. Concerning k, either $k < \rho$ and $k = k'$ or $S[i - m + 1 \mathinner{..} i] = P$, where $i = \alpha_S(j) + \ell_P(\rho - 1) - 1$. The lengths k' and k can be found by computing the intersection $\bar{D}_i^a \cap \bar{D}_i^{b'}$, for $j - \rho + 1 \leq i \leq j$, of the configurations of the automata $\mathcal{S}(P_a)$ and $\mathcal{S}(P_{b'})$, which is equivalent to simulating the automaton $\mathcal{S}_r(P^R)$ on $(S[\alpha_S(j - \rho + 1) \mathinner{..} \beta_S(j)])^R$. The pseudocode of the variants of the SHIFT-AND and BNDM algorithms based on the run-length encoding is shown in Fig. 3.

```
RL-PREPROCESS(P)
   1. ρ ← |RLE(P)|
   2. for c ∈ Σ do B₁[c] ← 0^ρ
   3. for i ← 1 to |P| + 1 do B₂[i] ← 0^ρ
   4. i ← 0
   5. for (c, l) ∈ RLE(P) do
   6.      H ← 0^{ρ−1}1 ≪ i
   7.      B₁[c] ← B₁[c] | H
   8.      if i = 0 or i = ρ − 1 then
   9.           ℓ = l
  10.           for j ← l to |P| + 1 do
  11.                B₂[j] ← B₂[j] | H
  12.      else B₂[l] ← B₂[l] | H
  13.      i ← i + 1
  14. return(B₁, B₂, ρ, ℓ)
```

```
RL-SHIFT-AND(P, T)
  1. (B₁, B₂, ρ, ℓ) ← RL-PREPROCESS(P)
  2. D ← 0^ρ
  3. j ← 0
  4. for (c, l) ∈ RLE(T) do
  5.      D ← ((D ≪ 1) | 0^{ρ−1}1) & B₁[c]
  6.      D ← D & B₂[min(l, |P| + 1)]
  7.      if D & 10^{ρ−1} ≠ 0^ρ then
  8.           Output(j + ℓ)
  9.      j ← j + l
```

```
RL-BNDM(P, T)
  1. (B₁, B₂, ρ, ℓ) ← RL-PREPROCESS(Pᴿ)
  2. s ← |P| − 1
  3. while s < |T| do
  4.      D ← 1^ρ
  5.      b ← s − |P| + 1
  6.      while s + 1 < |T| and T[s] = T[s + 1] do
  7.           s ← s + 1
  8.      j ← 0, k ← 1
  9.      for (c, l) ∈ RLE(T[b .. s]ᴿ) do
 10.           D ← D & B₁[c]
 11.           D ← D & B₂[min(l, |P| + 1)]
 12.           if D & 10^{ρ−1} ≠ 0^ρ then
 13.                if (j + ℓ ≥ |P|) then
 14.                     Output(s − j − ℓ)
 15.                else k ← j + ℓ
 16.           D ← D ≪ 1
 17.           j ← j + l
 18.      s ← s + |P| − k
```

Fig. 3. The variants of Shift-And and BNDM based on the run-length encoding.

6   Comparison with the 1-Factorization Encoding

Consider the greedy 1-factorization proposed by Cantone et al., that is defined as follows:

Definition 1. *The greedy 1-factorization of a string P is the sequence $\langle u_1, u_2, \ldots, u_k \rangle$ of nonempty substrings of P such that:*

(a) $P = u_1 u_2 \ldots u_k$;
(b) u_j is the longest prefix of $P[i .. |P| - 1]$ that contains at most one *occurrence of any of the symbols in the alphabet Σ, where $i = |u_1 u_2 \ldots u_{j-1}|$, for $j = 1, \ldots, k$.*

Let P be a string of length m and let Σ_P be the set of symbols occurring in P. The size k of the greedy 1-factorization of P satisfies the condition $\left\lceil \frac{m}{|\Sigma_P|} \right\rceil \leq k \leq m$. Instead the size ρ of the run-length encoding of P satisfies the condition

$|\Sigma_P| \leq \rho \leq m$. If $|\Sigma_P| = m$, as in the case $P = acg$, we have $k = 1$ and $\rho = m$. Instead, if $|\Sigma_P| = 1$, as in the case $P = aaa$, we have $k = m$ and $\rho = 1$. In other words, the best case, with respect to size, for the greedy 1-factorization is the worst case for the run-length encoding, and *vice versa*. Assume now that P is a uniformly random string and let X and X' be the random variables corresponding to the length of a factor in the 1-factorization and run-length encoding, respectively. It is easy to verify that $Pr[X \geq i] = \frac{(\sigma-1)!}{(\sigma-i)!\sigma^{i-1}}$ and $Pr[X' \geq i] = \frac{1}{\sigma^{i-1}}$. Then, we have $E[X] = Q(\sigma) = \Theta(\sqrt{\sigma})$ by Knuth's analysis of the $Q(n)$ function [13, 1.2.11.3], and $E[X'] = \sigma/(\sigma - 1)$. Therefore, in the case of uniformly random strings, k is smaller than ρ on average, as we have $k = \Theta(n/\sqrt{\sigma})$ and $\rho = \Theta(n)$. In many common domains of real strings, such as DNA, protein and natural language sequences, k is also smaller than ρ on average. For this reason, we do not provide experimental results, even though we have implemented the algorithms and experimentally verified their correctness: in the general case the proposed algorithms are not preferable. However, they can be useful in specific application domains where the run-length encoding is effective, such as the ones of bi-level images and of time series analysis using clipping [4].

One potential way to exploit the run-length encoding in the general case could be to combine it with the 1-factorization, by computing the 1-factorization of the string induced by the run-length encoding (over the alphabet of distinct pairs in the sequence). The challenge here is to devise a combination of the algorithms based on the two encodings which still is simple enough to be practical and fast.

7   Conclusions

In this paper we have shown that the nondeterministic KMP and suffix automata of a string P can be simulated, using an encoding based on word-level parallelism, in $O(\lceil \rho/w \rceil)$ time per transition, where ρ is the length of the run-length encoding of P. We have also presented practical variants of the SHIFT-AND and BNDM algorithms based on this encoding. An open problem is whether there exist other factorizations which can be used to obtain similar efficient encodings.

Acknowledgments. The author thanks Jorma Tarhio, Djamal Belazzougui and the anonymous reviewers for helpful comments.

References

1. Amir, A., Landau, G.M., Vishkin, U.: Efficient pattern matching with scaling. J. Algorithms **13**(1), 2–32 (1992)
2. Baeza-Yates, R.A., Gonnet, G.H.: A new approach to text searching. Commun. ACM **35**(10), 74–82 (1992)
3. Baeza-Yates, R.A., Navarro, G.: Faster approximate string matching. Algorithmica **23**(2), 127–158 (1999)

4. Bagnall, A.J., Ratanamahatana, C.A., Keogh, E.J., Lonardi, S., Janacek, G.J.: A bit level representation for time series data mining with shape based similarity. Data Min. Knowl. Discov. **13**(1), 11–40 (2006)
5. Cantone, D., Faro, S., Giaquinta, E.: A compact representation of nondeterministic (suffix) automata for the bit-parallel approach. Inf. Comput. **213**, 3–12 (2012)
6. Crochemore, M., Rytter, W.: Text Algorithms. Oxford University Press, New York (1994)
7. Durian, B., Holub, J., Peltola, H., Tarhio, J.: Improving practical exact string matching. Inf. Process. Lett. **110**(4), 148–152 (2010)
8. Durian, B., Peltola, H., Salmela, L., Tarhio, J.: Bit-parallel search algorithms for long patterns. In: Festa, P. (ed.) SEA 2010. LNCS, vol. 6049, pp. 129–140. Springer, Heidelberg (2010)
9. Fredriksson, K., Giaquinta, E.: On a compact encoding of the swap automaton. Inf. Process. Lett. **114**(7), 392–396 (2014)
10. Fredriksson, K., Grabowski, S.: Average-optimal string matching. J. Discrete Algorithms **7**(4), 579–594 (2009)
11. Hyyrö, H.: Improving the bit-parallel NFA of Baeza-Yates and Navarro for approximate string matching. Inf. Process. Lett. **108**(5), 313–319 (2008)
12. Hyyrö, H., Navarro, G.: Bit-parallel witnesses and their applications to approximate string matching. Algorithmica **41**(3), 203–231 (2005)
13. Knuth, D.E.: The Art of Computer Programming, Volume I: Fundamental Algorithms, 2nd edn. Addison-Wesley, Boston (1973)
14. Knuth, D.E., Pratt, V.R., Morris Jr., J.H.: Fast pattern matching in strings. SIAM J. Comput. **6**(2), 323–350 (1977)
15. Navarro, G., Raffinot, M.: Fast and flexible string matching by combining bit-parallelism and suffix automata. ACM J. Exp. Algorithmics **5**, 4 (2000)
16. Navarro, G., Raffinot, M.: Flexible Pattern Matching in Strings - Practical on-Line Search Algorithms for Texts and Biological Sequences. Cambridge University Press, Cambridge (2002)
17. Peltola, H., Tarhio, J.: Alternative algorithms for bit-parallel string matching. In: Nascimento, M.A., de Moura, E.S., Oliveira, A.L. (eds.) SPIRE 2003. LNCS, vol. 2857, pp. 80–93. Springer, Heidelberg (2003)
18. Wu, S., Manber, U.: Fast text searching allowing errors. Commun. ACM **35**(10), 83–91 (1992)

More on Deterministic and Nondeterministic Finite Cover Automata

Extended Abstract

Hermann Gruber[1], Markus Holzer[2](✉), and Sebastian Jakobi[2]

[1] Knowledgepark AG, Leonrodstr. 68, 80636 Munich, Germany
hermann.gruber@knowledgepark-ag.de
[2] Institut für Informatik, Universität Giessen, Arndtstr. 2,
35392 Giessen, Germany
{holzer,sebastian.jakobi}@informatik.uni-giessen.de

Abstract. Finite languages are an important sub-regular language family, which were intensively studied during the last two decades in particular from a descriptional complexity perspective. An important contribution to the theory of finite languages are the deterministic and the recently introduced nondeterministic finite *cover* automata (DFCAs and NFCAs, respectively) as an alternative representation of finite languages by ordinary finite automata. We compare these two types of cover automata from a descriptional complexity point of view, showing that these devices have a lot in common with ordinary finite automata. In particular, we study how to adapt lower bound techniques for nondeterministic finite automata to NFCAs such as, e.g., the biclique edge cover technique, solving an open problem from the literature. Moreover, the trade-off of conversions between DFCAs and NFCAs as well as between finite cover automata and ordinary finite automata are investigated. Finally, we present some results on the average size of finite cover automata.

1 Introduction

If one tries to describe formal objects such as, e.g., Boolean functions, graphs, trees, languages, as compact as possible we are faced with the question, which representation to use. This quest for compact representations of formal objects dates back to the early beginnings of theoretical computer science. For instance, one can prove by a *simple* counting argument that most Boolean functions have exponential circuit complexity [27]. For other representations of Boolean functions than circuits, such as formulas, ordered binary decision diagrams, etc. a similar result applies. This incompressibility is inherent in almost all possible representations of formal objects.

Part of the work was done while the first author was at Institut für Informatik, Ludwig-Maximilians-Universität München, Oettingenstraße 67, 80538 München, Germany and the second author was at Institut für Informatik, Technische Universität München, Boltzmannstraße 3, 85748 Garching bei München, Germany.

© Springer International Publishing Switzerland 2015
F. Drewes (Ed.): CIAA 2015, LNCS 9223, pp. 114–126, 2015.
DOI: 10.1007/978-3-319-22360-5_10

When considering formal languages, automata are the preferred choice of representation. In particular, for regular languages and subfamilies one may use deterministic (DFAs) or nondeterministic finite automata (NFAs) or variants thereof to describe these languages. It is well known that these two formalisms are equivalent. The obvious way to obtain a DFA form a given NFA is by applying the *subset* or *power-set construction* [24]. This construction allows to show an upper bound of 2^n states in the DFA obtained from an n-state NFA, and this bound is known to be tight. For finite languages a slightly smaller bound on the determinization problem is given in [25]. Here the tight bound depends on the alphabet size k and reads as $\Theta(k^{\frac{n}{1+\log_2 k}})$. Thus, for a two-letter input alphabet $\Theta(2^{\frac{n}{2}})$ states are sufficient and necessary in the worst case for a DFA to accept a language specified by an n-state NFA. There are a lot of other results known for finite automata accepting finite languages such as, e.g., the maximal number of states of the minimal DFA accepting a subset of Σ^ℓ or $\Sigma^{\leq \ell}$ [5,12], or the average case size of DFAs and NFAs w.r.t. the number of states and transitions accepting a subset of Σ^ℓ or $\Sigma^{\leq \ell}$ [17].

Since regular languages and finite automata are widely used in applications, and most of them use actually finite languages only, it is worth considering further representations for finite languages that may be more compact, but still bare nice handling in applications. Such a representation is based on finite automata and is known as finite cover automata. The idea is quite simple, namely a finite cover automaton A of a finite language $L \subseteq \Sigma^*$ is a finite automaton that accepts all words in L and possibly other words that are longer than any word in L. Formally, this reads as $L = L(A) \cap \Sigma^{\leq \ell}$, where ℓ is the length of the longest word(s) in L; then we say that A *covers* the finite language L. Originally deterministic finite cover automata (DFCAs) were introduced in [10], where an efficient minimization algorithm for these devices was given. Further results on important aspects of DFCAs can be found in, e.g., [8–11,21]. Recently, DFCAs were generalized to nondeterministic finite cover automata (NFCAs) in [4] and it was shown that they can even give a more compact representation of finite languages than both NFAs and DFCAs. To our knowledge this was the first systematic study on this subject, although it has been suggested already earlier in a survey paper on cover automata [28].

We further develop the theory of finite cover automata in this paper. At first we introduce the necessary definitions in the next section. Then we briefly recall what is known on lower bound techniques for both types of finite cover automata. In particular, we first reconsider the fooling set techniques known for nondeterministic finite automata (NFAs) and secondly we show how to alter the biclique edge cover technique from [16] to make it applicable for NFCAs, too. This positively answers a question stated in [4], whether the biclique edge cover technique can be used at all to prove lower bounds for NFCAs. As a byproduct we develop a lower bound method for E-equivalent NFAs. This concept was recently introduced in [19]. Two languages are E-equivalent if their symmetric difference lies in the so called *error language* E. Thus, E-equivalence is a generalization of ordinary equivalence and also of cover-automata. In particular, setting $E = \Sigma^{>\ell}$, thus not taking care of words that are too long, we are back to covering languages and cover automata. Section 4 is devoted to conversions between finite automata and finite cover automata. First we provide a large family of languages where cover state complexity meets ordinary state complexity (up to one state for deterministic devices). Hence, for the conversions from finite automata to finite cover automata not much state savings are possible. For the opposite direction we show that an n-state finite cover automaton for a language of order ℓ can be converted to an equivalent finite automaton with about $n \cdot \ell$ states; the exact bounds are shown to be tight for all n and ℓ. In particular, this shows that roughly speaking the number of states of a finite cover automaton is *at least* an ℓ-th fraction of the state size of the equivalent finite automaton. Then we take a closer look on determinizing NFCAs by the well known power-set construction. We show that here the state blow-up heavily depends on the order ℓ of the finite language represented by the NFCA. When the order is large enough, we get a tight exponential blow-up of 2^n, just as in the case of ordinary finite automata. We give a range of conditions that imply sub-exponential, polynomial, and even linear determinization blow-ups. These results are presented in Sect. 5. In the penultimate section, we perform average case comparisons of the descriptional complexity of finite cover automata. For ordinary finite automata this was already done in, e.g., [17], where it was shown that almost all DFAs accepting finite languages of order ℓ over a binary input alphabet have state complexity $\Theta(2^\ell/\ell)$, while NFAs are shown to perform better, namely the nondeterministic state complexity is in $\Theta(\sqrt{2^\ell})$. Interestingly, in both cases the aforementioned bounds are asymptotically like in the worst case. For finite cover automata exactly the same picture as for ordinary finite automata emerges. Finally, we summarize our results in the conclusions section and state some open problems for future research. Due to space limitations all proofs are omitted.

2   Preliminaries

We recall some definitions on finite automata as contained in [18]. A *nondeterministic finite automaton* (NFA) is a quintuple $A = (Q, \Sigma, \delta, q_0, F)$, where Q is the finite set of *states*, Σ is the finite set of *input symbols*, $q_0 \in Q$ is the *initial state*, $F \subseteq Q$ is the set of *accepting states*, and $\delta \colon Q \times \Sigma \to 2^Q$ is the *transition function*. The *language accepted* by the NFA A is defined as

$$L(A) = \{\, w \in \Sigma^* \mid \delta(q_0, w) \cap F \neq \emptyset \,\},$$

where the transition function is recursively extended to $\delta \colon Q \times \Sigma^* \to 2^Q$. An NFA is *deterministic* (DFA), if and only if $|\delta(q, a)| = 1$, for every $q \in Q$ and $a \in \Sigma$. In this case we simply write $\delta(q, a) = p$ instead of $\delta(q, a) = \{p\}$, assuming that the transition function $\delta \colon Q \times \Sigma \to Q$ is a *total* mapping. Two automata A and B are *equivalent* if they accept the same language, that is, $L(A) = L(B)$. An NFA (DFA, respectively) A is *minimal* if any equivalent NFA (DFA, respectively) needs at least as many states as A. It is a well known fact that minimal DFAs are unique up to isomorphism, while minimal NFAs are *not* necessarily unique in general. Let $\mathsf{nsc}(L)$ ($\mathsf{sc}(L)$, respectively) refer to the number of states a minimal NFA (DFA, respectively) needs to accept the language L. By definition and the seminal result in [24] we have $\mathsf{nsc}(L) \leq \mathsf{sc}(L) \leq 2^{\mathsf{nsc}(L)}$, if L is a language accepted by a finite automaton. Proving lower bounds for $\mathsf{nsc}(L)$ can be done by applying, e.g., the extended fooling set technique, which reads as follows [1]:

Theorem 1. *Let $L \subseteq \Sigma^*$ be a regular language and suppose there exists a set of pairs $S = \{ (x_i, y_i) \mid 1 \leq i \leq n \}$ such that (i) $x_i y_i \in L$, for $1 \leq i \leq n$ and (ii) $i \neq j$ implies $x_i y_j \notin L$ or $x_j y_i \notin L$, for $1 \leq i, j \leq n$. Then any nondeterministic finite automaton for L has at least n states, i.e., $n \leq \mathsf{nsc}(L)$. Here S is called an* extended fooling set *for L.*

A non-empty finite language $L \subseteq \Sigma^*$ is said to be of *order* ℓ, if ℓ is the length of the longest word(s) in the set L, i.e., $L \subseteq \Sigma^{\leq \ell}$, where $\Sigma^{\leq \ell}$ refers to the set $\{ w \in \Sigma^* \mid |w| \leq \ell \}$, where $|w|$ denotes the length of the word w. In particular, the length of the empty word λ is zero. A *deterministic finite cover automaton* (DFCA) for a language $L \subseteq \Sigma^*$ of order ℓ is a DFA A such that $L(A) \cap \Sigma^{\leq \ell} = L$; these devices were introduced in [10]. This definition naturally carries over to NFAs, hence leading to *nondeterministic finite cover automata* (NFCA), which were recently introduced in [4]. Two cover automata A and B are *equivalent* if they cover the same finite language $L \subseteq \Sigma^*$, that is, $L(A) \cap \Sigma^{\leq \ell} = L(B) \cap \Sigma^{\leq \ell}$, where ℓ is the order of L. A DFCA (NFCA, respectively) A for a finite language L is *minimal* if any equivalent automaton of same type needs at least as many states as A. Let $\mathsf{ncsc}(L)$ ($\mathsf{csc}(L)$, respectively) refer to the number of states a minimal NFCA (DFCA, respectively) needs to accept the finite language L. By definition we have $\mathsf{ncsc}(L) \leq \mathsf{csc}(L)$, if L is a finite language. Moreover, since any cover automaton can be at most as large as an ordinary finite automaton of the same type for a finite language L, we have $\mathsf{csc}(L) \leq \mathsf{sc}(L)$ as well as $\mathsf{ncsc}(L) \leq \mathsf{nsc}(L)$. A useful tool for the study of minimal DFCAs is the notion the similarity relation, which plays a similar role as the Myhill-Nerode relation[1] in case of DFAs. For a finite language $L \subseteq \Sigma^*$ of order ℓ the similarity relation $\approx_L$ on words is defined as follows: for $u, v \in \Sigma^*$ let $u \approx_L v$ if and only if we have $uw \in L \iff vw \in L$, for all $w \in \Sigma^*$, whenever $|uw| \leq \ell$ and $|vw| \leq \ell$. Observe, that $\approx_L$ is not an equivalence relation in general. The relation $\approx_L$ can also be defined for states of a DFCA $A = (Q, \Sigma, \delta, q_0, F)$. Two states p and q are *similar*, denoted by $p \approx_L q$, if $\delta(p, w) \in F \iff \delta(q, w) \in F$ holds for all $w \in \Sigma^{\leq \ell - m}$, with $m = \max(lev_A(p), lev_A(q))$—here $lev_A(p) = \min\{ |u| \mid \delta(q_0, u) = p \}$. If $p \not\approx_L q$ then p and q are *dissimilar*. It is known [10] that a DFCA is minimal if all its states are pairwise dissimilar.

3 Lower Bound Techniques for Cover Automata

The problem to estimate the necessary number of states of a minimal NFA accepting a given regular language is complicated. Several authors have introduced methods for proving lower bounds. The most widely used lower bound

---

[1] For a language $L \subseteq \Sigma^*$ define the Myhill-Nerode relation $\equiv_L$ on words as follows: for $u, v \in \Sigma^*$ let $u \equiv_L v$ if and only if $uw \in L \iff vw \in L$, for all $w \in \Sigma^*$.

techniques for NFAs are the so-called *fooling set* techniques—the fooling set technique [14] and the extended fooling set method [1]. Recently, in [4] both fooling set methods were adapted to work for NFCAs as well. Here we first reconsider the fooling set techniques and then show how to modify yet another lower bound method, the biclique edge cover technique of [16], to work with NFCAs. Whether this latter technique can be generalized to NFCAs was stated as an open problem in [4].

In [4] it was argued that there is no doubt that any fooling set type technique used to prove a lower bound for NFCAs must explicitly consider the order of the language under consideration. In this vein, both fooling set techniques were adapted. In fact, we show that the original fooling set technique of [14] (not the extended version of [1]) already gives a lower bound for NFCAs without modifying the technique to explicitly deal with the order of the language under consideration.

Theorem 2. *Let $L \subseteq \Sigma^*$ be a finite language and suppose there exists a set of pairs $S = \{ (x_i, y_i) \mid 1 \leq i \leq n \}$ such that (i) $x_i y_i \in L$, for $1 \leq i \leq n$, and (ii) $x_i y_j \notin L$, for $1 \leq i, j \leq n$, and $i \neq j$. Then any nondeterministic finite cover automaton for L has at least n states, i.e., $n \leq \mathsf{ncsc}(L)$. Here S is called a fooling set for L.*   □

In contrast the more powerful extended fooling set technique presented in [1] does not work as a lower bound technique for NFCAs as the following example illustrates, and therefore the modification of this technique presented in [4] is the right generalization.

Example 3. Consider the unary finite language $L = \{a\}^{\leq \ell}$, for $\ell \geq 1$. Clearly, this language can be covered by an NFCA with a single state. However, the set $S = \{ (a^i, a^{\ell-i}) \mid 0 \leq i \leq \ell \}$ is an extended fooling set for L, proving a lower bound of $\ell + 1$ on the nondeterministic state complexity of L.   □

In the remainder of this subsection we turn our attention to the biclique edge cover technique from [16]. A central role in this technique plays the notion of the *bipartite dimension* $\dim(G)$ of a bipartite graph G, which is the minimum number of bicliques in G needed to cover all edges of G. The following example shows that this technique cannot be applied to NFCAs without any modification.

Example 4. Let $\ell \geq 1$ and consider the finite language $L = \{a\}^{\leq \ell}$. Clearly the single-state DFA accepting for the language $\{a\}^*$ is a cover automaton for L, hence we have $\mathsf{ncsc}(L) = 1$. However, the bipartite dimension of the graph $G = (X, Y, E)$, with $X = Y = L$ and $E = \{ (x, y) \in X \times Y \mid xy \in L \}$, is $\ell + 1 > 1$. This can be seen as follows. Notice that $(a^i, a^j) \in E$ if and only if $i + j \leq \ell$. In particular, for $0 \leq i \leq \ell$, the edge $e_i = (a^i, a^{\ell-i})$ belongs to E. Therefore, every such e_i has to be covered by some biclique $H_i = (X_i, Y_i, E_i)$ with $a^i \in X_i$, $a^{\ell-i} \in Y_i$, and $E_i = X_i \times Y_i$. Now we see that distinct edges e_i and e_j must be covered by distinct bicliques, that is, $H_i \neq H_j$, for $1 \leq i, j \leq \ell$, with $i \neq j$: if $H_i = H_j$ then we have $a^i, a^j \in X_i$ and $a^{\ell-i}, a^{\ell-j} \in Y_i$, and since H_i is a biclique, its set of edges E_i contains both $(a^i, a^{\ell-j})$ and $(a^j, a^{\ell-i})$. But since $i \neq j$, either

$i + \ell - j > \ell$ or $j + \ell - i > \ell$, which means that one of the two edges does not belong to E—a contradiction to $H_0, H_1, \ldots H_\ell$ being a biclique edge cover. This shows that the bipartite dimension of G is at least $\ell + 1$. Equality is witnessed by the bicliques $H_i = (X_i, Y_i, E_i)$ with $X_i = \{a^i\}$, $Y_i = \{a\}^{\leq \ell - i}$, and $E_i = X_i \times Y_i$, for $0 \leq i \leq \ell$.   □

In the following we want to generalize the biclique edge cover technique so that it can also be used to prove lower bounds for the size of NFCAs. In fact, we present a generalization that can be used even for the more general notion of E-equivalent automata, which was recently introduced in [19]. In order to avoid confusion with the set of edges of a graph, we use here the term D-equivalence instead of E-equivalence. Let $D \subseteq \Sigma^*$ be some language, the so called *error language*. Two languages L and L' over the alphabet Σ are called *D-equivalent* if they differ only on elements from the error language D, that is, if

$$(L \setminus L') \cup (L' \setminus L) \subseteq D.$$

In this case we write $L \sim_D L'$. Similarly, two automata A and B are D-equivalent, if $L(A) \sim_D L(B)$. The connection between D-equivalence and cover automata is as follows. Assume $L \subseteq \Sigma^{\leq \ell}$ is some finite language of order ℓ. Then a language $L' \subseteq \Sigma^*$ is a cover language for L if and only if $L \sim_D L'$, for the error language $D = \Sigma^{>\ell}$. In other words, any two cover languages L' and L'' for a finite language of order ℓ are D-equivalent, for $D = \Sigma^{>\ell}$.

We now come to our generalization of the biclique edge cover technique. In the original technique we have to find bicliques $H_i = (X_i, Y_i, E_i)$ with $1 \leq i \leq k$, for some k, of a bipartite graph $G = (X, Y, E)$, such that $E = \bigcup_{i=1}^k E_i$. In our generalization, we use two sets of edges in the bipartite graph G, namely a set $\underline{E}$ of edges that *must* be covered, and a set $\overline{E}$, with $\underline{E} \subseteq \overline{E}$, of edges that *may* be covered by bicliques. We use the notation $G = (X, Y, \underline{E}, \overline{E})$ to denote such a bipartite graph. Now an $(\underline{E}, \overline{E})$-*approximation* of G is a collection of bicliques $H_i = (X_i, Y_i, E_i)$ of G, with $1 \leq i \leq k$ for some k, such that

$$\underline{E} \subseteq \bigcup_{i=1}^k E_i \subseteq \overline{E}.$$

The $(\underline{E}, \overline{E})$-*dimension* of G, denoted by $\dim^*(G)$, is defined as the minimal number of bicliques that constitute an $(\underline{E}, \overline{E})$-*approximation* of G.

Now we are ready to present our lower bound technique for D-equivalent automata. Notice that the sets $\underline{E}$ and $\overline{E}$ of edges of graph G in the following theorem depend on the given language L and error set D by definition.

Theorem 5. *Let L and D be languages over some alphabet Σ. Moreover, let $X, Y \subseteq \Sigma^*$ and $G = (X, Y, \underline{E}, \overline{E})$, with $\underline{E} = \{(x,y) \in X \times Y \mid xy \in L \setminus D\}$ and $\overline{E} = \{(x,y) \in X \times Y \mid xy \in L \cup D\}$. Then the number of states of any nondeterministic finite automaton A, with $L(A) \sim_D L$, is at least $\dim^*(G)$.*   □

Notice that Theorem 5 yields the original biclique edge cover technique when choosing the error language $D = \emptyset$, that is, when considering the special case of classical language equivalence. Moreover, with the error language $D = \Sigma^{>\ell}$ we obtain the following technique for proving lower bounds on the state complexity of nondeterministic cover automata for finite languages of order ℓ.

Corollary 6. *Let $L \subseteq \Sigma^*$ be some finite language of order ℓ. Moreover, let $X, Y \subseteq \Sigma^*$ and $G = (X, Y, \underline{E}, \overline{E})$, with $\underline{E} = \{ (x, y) \in X \times Y \mid xy \in L \}$ and $\overline{E} = \{ (x, y) \in X \times Y \mid xy \in L \cup \Sigma^{>\ell}, \}$. Then the number of states of any nondeterministic finite cover automaton for L is at least $\dim^*(G)$, that is, $\dim^*(G) \leq \mathsf{ncsc}(L)$.*   □

4   Conversions Between Finite Automata and Cover Automata

In this section we compare the descriptional complexity of finite automata and cover automata, by studying the cost of conversions between these models. We consider nondeterministic as well as deterministic automata.

4.1   From Finite Automata to Cover Automata

Clearly, a finite automaton for a finite language L is also a cover automaton for that language. So the bounds $\mathsf{ncsc}(L) \leq \mathsf{nsc}(L)$ and $\mathsf{csc}(L) \leq \mathsf{sc}(L)$ are obvious. However, the question is whether these bounds are tight in the following sense: does there exist, for every integer $n \geq 1$, a regular language L_n that is accepted by a DFA (NFA, respectively) with n states such that the minimal DFCA (NFCA, respectively) needs n states, too? The next result answers this question in the affirmative for nondeterministic automata, while for deterministic devices the bound is off by one.

Theorem 7. *If L is a finite language with all words having the same length ℓ, then $\mathsf{ncsc}(L) = \mathsf{nsc}(L)$ and $\mathsf{csc}(L) = \mathsf{sc}(L) - 1$.*   □

From Theorem 7 and the obvious upper bound $\mathsf{ncsc}(L) \leq \mathsf{nsc}(L)$ we obtain the following result. In fact, Theorem 7 provides the lower bound already by *unary* witness languages.

Corollary 8. *Let $n \geq 1$ and L be a finite language accepted by a nondeterministic finite automaton with n states. Then n states are sufficient and necessary in the worst case for a nondeterministic finite cover automaton to accept L. This bound is tight already for a unary alphabet.*   □

Next we want to close the gap between the lower and upper bound for the conversion from DFAs to DFCAs.

Theorem 9. *Let L be a finite language accepted by a deterministic finite automaton with n states. If $n = 1$ or $n \geq 4$ then n states are sufficient and necessary in the worst case for a deterministic finite cover automaton to accept L. These bounds are tight already for binary alphabets. If $n \in \{2, 3\}$, or if $n \geq 2$ and L is a unary language, then $n - 1$ states are sufficient and necessary in the worst case.*   □

We also note that the conversion from NFAs to DFCAs was investigated already in [6]. They present binary languages L_n that can be accepted by an n-state NFA, while $2^{n-t} - 2^{t-2} + 2^t - 1$ states are necessary, with $t = \lfloor \frac{n}{2} \rfloor$, for a deterministic finite cover automaton to accept L_n. Then they generalize their examples to larger alphabets. The lower bound is known to be tight if n is even, but the tight bound for odd n remains to be determined.

4.2   From Cover Automata to Finite Automata

In the previous subsection we have seen that there are finite languages where the description size cannot be reduced when changing the descriptional model from finite automata to cover automata. In this section we now consider the inverse conversion: given a cover automaton for a finite language, how large can a minimal finite automaton for that language become? In this setting we will see that the number of states of a cover automaton alone is not a fair size measure. In fact, we propose that a reasonable size measure for cover automata must also take the cover length into account: for every integer $\ell \geq 0$ the finite language $\{a\}^{\leq \ell}$ can be covered by a single-state cover automaton, but a NFA for this language has at least $\ell + 1$ states. Therefore, if we start with a cover automaton with n states that describes a finite language of order ℓ, then the number of states of an equivalent finite automaton should be a function in n and ℓ.

Since the language L described by a cover automaton A with cover length ℓ satisfies $L = L(A) \cap \Sigma^{\leq \ell}$, a finite automaton for L can be obtained by applying a cross product construction on A and an automaton for $\Sigma^{\leq \ell}$. The states of the constructed automaton are pairs (q, i), where q is a state of A, and i is a counter for the word length. This yields upper the upper bounds $\mathsf{nsc}(L) \leq \mathsf{ncsc}(L) \cdot (\ell + 1)$ and $\mathsf{sc}(L) \leq \mathsf{csc}(L) \cdot (\ell + 2)$ for finite languages L of order ℓ. In the upcoming lemma we show that these bounds can be slightly reduced. In the following we do not consider languages of order $\ell = 0$, because the only such language is $\{\lambda\}$, which is accepted by a single-state NFA and a two-state DFA. Moreover, the case where $\mathsf{ncsc}(L) = 1$ is also omitted—here it is easy to see that the upper bounds $\mathsf{nsc}(L) \leq \ell + 1$ and $\mathsf{sc}(L) \leq \ell + 2$ apply, and optimality is witnessed by the language $L = \Sigma^{\leq \ell}$.

Lemma 10. *Let $n \geq 2$ and A be an n-state nondeterministic cover automaton for a finite language L of order $\ell \geq 1$. Then one can construct a nondeterministic finite automaton for L that has at most $n \cdot (\ell - 1) + 2$ states. If A is deterministic, then one can construct a deterministic finite automaton for L with $n \cdot (\ell - 1) + 3$ states.* $\square$

Next we show that the constructions from Lemma 10 cannot be improved in general by providing a matching lower bounds. Observe that the following lemma even provides a lower bound for the conversion from *deterministic* cover automata to *nondeterministic* finite automata.

Lemma 11. *For every integers $n \geq 2$ and $\ell \geq 1$ there exists a finite language L of order ℓ that is described by a deterministic n-state cover automaton, such that* *any nondeterministic finite automaton for L needs* $n \cdot (\ell - 1) + 2$ *states, and any deterministic finite automaton for L needs* $n \cdot (\ell - 1) + 3$ *states.*   □

From Lemmata 10 and 11 we obtain the following result.

Theorem 12. *Let L be a finite language of order* $\ell \geq 1$ *that is described by a nondeterministic cover automaton A with* $n \geq 2$ *states. Then* $n \cdot (\ell - 1) + 2$ *states are sufficient and necessary in the worst case for a nondeterministic finite automaton to accept L. Moreover, if A is a deterministic cover automaton for L, then* $n \cdot (\ell - 1) + 3$ *states are sufficient and necessary in the worst case for a deterministic finite automaton to accept L.*   □

The proof for the lower bound from Lemma 11 uses $2n - 2$ alphabet symbols. In fact, one can also show that the bounds $\mathsf{nsc}(L) \leq \mathsf{ncsc}(L) \cdot (\ell - 1) + 2$ and $\mathsf{sc}(L) \leq \mathsf{csc}(L) \cdot (\ell - 1) + 3$ for the conversions from cover automata to finite automata are *not* tight for languages over an alphabet of constant size. For the deterministic case, this is easy to see: assuming a k-letter alphabet Σ, at most k different states of the form $(q, 1)$ are reachable from the initial state $(q_0, 0)$ in the DFA constructed from a DFCA as shown in the proof of Lemma 10.

Although this argumentation does not hold for nondeterministic automata, where every state of the given NFCA could be reachable in one step from the initial state, the number of states of an equivalent minimal NFA still depends on the number of alphabet symbols: when using the construction from Lemma 10 to obtain an NFA A' for the language $L \subseteq \Sigma^{\leq \ell}$, the automaton A' has a distinguished "last" accepting state $(\bullet, \ell)$, which has no outgoing transitions. This state is only reachable from states of the form $(q, \ell - 1)$, and from such states no other state is reachable. Assume that two such states $(p, \ell - 1)$ and $(q, \ell - 1)$ go to state $(\bullet, \ell)$ on the same set of input letters. If additionally p and q are of same acceptance value, then clearly they can be merged into a single state. Since a k-letter alphabet Σ has $2^k - 1$ non-empty subsets, the number of accepting states of the form $(q, \ell - 1)$ can always be reduced to $2^k - 1$, and similarly for the non-accepting states. So in total there are at most $2 \cdot (2^k - 1)$ states of the form $(q, \ell - 1)$, which may be large compared to k, but it is still a constant.

5   Determinization of Finite Cover Automata

In this section we continue our descriptional complexity studies of cover automata: we investigate the cost of determinization, that is, the conversion from a nondeterministic to a deterministic cover automaton. A classical result in the theory of finite automata is that every n-state NFA can be converted by the so-called power-set construction to an equivalent DFA with at most 2^n states [24]. Moreover, it is known that this bound is tight in the sense that for every $n \geq 1$ there exists a language accepted by a minimal n-state NFA, and for which the minimal DFA needs exactly 2^n states [23]. Now the question is to which extent these results carry over to cover automata. Clearly, since the power-set construction for finite automata preserves the accepted language, it can be used to convert an NFCA into an equivalent DFCA. Thus, the following is immediate.

Lemma 13. *Let L be a finite language described by a nondeterministic cover automaton with $n \geq 1$ states. Then one can construct a deterministic cover automaton for L that has at most 2^n states.* □

Our next goal is to prove a matching lower bound of 2^n states for the determinization of n-state NFCAs. The next fact we present is useful to show that a number of worst case results known for the state complexity of deterministic finite automata carry over to the setting of cover automata.

Theorem 14. *Assume L is a regular language over Σ with $\mathsf{sc}(L) = n$, and let $L' = L \cap \Sigma^{\leq n+2^n}$. Then $\mathsf{csc}(L') = n$.* □

Theorem 14 implies that if the order of the language is large compared to the size of the NFA, then determinization of cover automata is as expensive as for usual finite automata. In particular, classical examples for finite automata [23] show that the full blow-up from n states to 2^n states may be necessary for converting an NFCA into an equivalent DFCA. Together with Lemma 13 we obtain the following result.

Corollary 15. *Let L be a finite language that is described by a nondeterministic cover automaton with $n \geq 1$ states. Then 2^n states are sufficient and necessary in the worst case for a deterministic cover automaton to accept L.* □

A natural question is now whether the full blow-up can be reached if the order of the described language is small compared to the number of states in the given NFCA. First, recall that every finite language L of order ℓ over a k-letter alphabet satisfies $\mathsf{sc}(L) \leq (1 + \mathrm{o}(1))\frac{k^{\ell+2}}{d_k \ell}$ with $d_k = (k-1)^2 \log k$; see [5]. This shows that the full blow-up cannot be reached if ℓ is too small compared to n. From that result and the fact that $\mathsf{csc}(L) \leq \mathsf{sc}(L)$, the following bounds for the size of a deterministic cover automaton can be derived. In fact, since the proof of the next result only uses the above bound on $\mathsf{sc}(L)$, the statements also hold for the determinization of finite automata.

Theorem 16. *Let L be a finite language of order ℓ over a k-letter alphabet Σ and assume L is described by a nondeterministic finite cover automaton with n states.*

1. *If $(\ell + 2) \cdot \log k - \log \ell + 1 < n$, then $\mathsf{csc}(L) < 2^n$, for large enough n.*
2. *if $\ell \in \mathrm{o}(n)$, then $\mathsf{csc}(L) \in 2^{\mathrm{o}(n)}$,*
3. *if $\ell \in \mathrm{O}(\log n)$, then $\mathsf{csc}(L) \in n^{\mathrm{O}(1)}$,*
4. *if $(\ell + 2) \cdot \log k - \log \ell + 1 < \log n$, then $\mathsf{csc}(L) < n$, for large enough n.* □

The fourth statement in the above theorem is of particular practical relevance: in this case, the given n-state NFCA is not minimal, and determinization followed by minimization yields a smaller cover automaton. In contrast to languages of order less than n, where the blow-up of 2^n states cannot be achieved, there are quite natural examples reaching the full blow-up already for order linear in the number of states of the NFCA. The example used in the following proof is essentially due to [22, Lemma 2]:

Theorem 17. *Let* $L_n = \left(a + (a \cdot b^*)^{n-1} \cdot a\right)^* \cap \Sigma^{\leq 5n-2}$. *Then* L_k *can be covered by an n-state nondeterministic cover automaton, but the smallest deterministic cover automaton for* L_n *has at least* 2^n *states.*   □

6   Average Size Comparisons of Finite Cover Automata

This section is devoted to the average case state complexity of DFCAs and NFCAs, when choosing a finite language of a certain "size" ℓ uniformly at random from all finite languages of that particular size. Here size means that all words of the language are either of the same length ℓ, or of length at most ℓ. This model was used in [17] to compare the number of states or transitions of ordinary finite automata on average. There it is shown that almost all DFAs accepting finite languages over a binary input alphabet have state complexity $\Theta(2^\ell/\ell)$, while NFAs are shown to perform better, namely the nondeterministic state complexity is in $\Theta(\sqrt{2^\ell})$. Interestingly, in both cases the aforementioned bounds are asymptotically like in the worst case. As we will see, a similar situation emerges for finite cover automata as well. The first theorem gives us the expected number of states a DFCA has on average, if we assume that all finite languages from $\mathfrak{P}(\Sigma^{\leq \ell})$, that is, the power-set of $\Sigma^{\leq \ell}$, are equiprobable.

Theorem 18. *Let* Σ *be an alphabet of size* k *and* $c_k = (k-1)\log k$. *Then* $\mathbb{E}[csc(L)] \geq (1 - o(1))\frac{k^\ell}{c_k \ell}$, *if* L *is a language drawn uniformly at random from the power-set of* $\Sigma^{\leq \ell}$.   □

Regarding an upper bound, it is known from [5] that $sc(L) \leq (1 + o(1))\frac{k^{\ell+2}}{d_k \ell}$, as ℓ tends to infinity, with $d_k = (k-1)^2 \log_2 k$, for languages $L \subseteq \Sigma^{\leq \ell}$ and alphabet size k. This generalized a previous result of [12]. Recall that the size of a minimum DFA for a finite language is an upper bound for the size of a minimum DFCA; and the state complexity in the worst case is of course an upper bound for the average state complexity. So the above average case result is tight up to a factor of at most $(1 + o(1))\frac{k^2}{(k-1)}$. Next we turn our attention to the average state complexity of NFCAs.

Theorem 19. *Let* Σ *be an alphabet of size* k. *Then for large enough* ℓ *we have* $\mathbb{E}[ncsc(L)] > k^{\frac{\ell}{2}-1}$, *if* L *is a language drawn uniformly at random from the power-set of* $\Sigma^{\leq \ell}$.   □

A worst case upper bound for the nondeterministic state complexity of subsets of $\Sigma^{\leq \ell}$ is given in [17] for binary alphabets. Generalizing this result to cover automata and larger alphabets, the bound reads as follows:

Theorem 20. *Let* Σ *be an alphabet of size* k. *Then* $ncsc(L) \leq nsc(L) < \frac{3}{k-1}\sqrt{k^\ell}$, *if* L *is any subset of* $\Sigma^{\leq \ell}$, *i.e.,* $L \subseteq \Sigma^{\leq \ell}$.   □

7   Conclusions

We completed the picture of lower bound techniques for nondeterministic finite cover automata, and solved the problems left open in [4]. Then we determined the precise best-case and worst-case bounds for conversions between DFCAs and DFAs, as well as between NFCAs and NFAs. In [6], almost tight bounds for the conversion between NFAs and DFCAs were given. Determining the precise bound in this case remains an open problem.

When the length ℓ of the longest word is much smaller than the number n of states in a minimal cover automaton, then the succinctness gain offered by finite cover automata over finite automata is very modest, even in the best case. We note that this is the case in the area of natural language processing: in [20] they construct a minimum 29317-state DFA accepting 81142 English words. Of course, almost all common English words have $\ell < 20$. Similarly, in [26] they construct an NFA accepting roughly 230000 Greek words.

Finally, a recent experimental study [7] showed that for binary finite languages, the expected reduction in the number of states provided by DFCAs is negligible. Our analysis of the average case provides a theoretical underpinning of their observations. One may study further random models of finite languages, e.g., a Bernoulli-type model [17], and one based on the sum of word lengths [2].

References

1. Birget, J.C.: Intersection and union of regular languages and state complexity. Inform. Process. Lett. **43**, 185–190 (1992)
2. Bassino, F., Giambruno, L., Nicaud, C.: The average state complexity of rational operations on finite languages. Internat. J. Found. Comput. Sci. **21**(4), 495–516 (2010)
3. Brzozowski, J.A.: Canonical Regular Expressions and Minimal State Graphs for Definite Events. Mathematical Theory of Automata, MRI Symposia Series. Polytechnic Press, New York (1962)
4. Câmpeanu, C.: Non-deterministic finite cover automata. Sci. Ann. Comput. Sci. **25**(1), 3–28 (2015)
5. Câmpeanu, C., Ho, W.H.: The maximum state complexity for finite languages. J. Autom. Lang. Comb. **9**(2–3), 189–202 (2004)
6. Câmpeanu, C., Kari, L., Păun, A.: Results on Transforming NFA into DFCA. Fundam. Inform. **64**(1–4), 53–63 (2005)
7. Câmpeanu, C., Moreira, N., Reis, R.: Expected compression ratio for DFCA: experimental average case analysis. Universidade do Porto (2011), Technical report DCC-2011-07
8. Câmpeanu, C., Păun, A., Smith, J.R.: Tight bounds for the state complexity of deterministic cover automata. In: Leung, H., Pighizzini, G. (eds.) Proceedings of the 8th Workshop on Descriptional Complexity of Formal Systems, pp. 223–231, Las Cruces (2006), Computer Science Technical report NMSU-CS-2006-001
9. Câmpeanu, C., Păun, A., Yu, S.: An efficient algorithm for constructing minimal cover automata for finite languages. Internat. J. Found. Comput. Sci. **13**(1), 83–97 (2002)

10. Câmpeanu, C., Sântean, N., Yu, S.: Minimal cover-automata for finite languages. Theoret. Comput. Sci. **267**(1–2), 3–16 (2001)
11. Champarnaud, J.M., Guingne, F., Hansel, G.: Similarity relations and cover automata. RAIRO-Informatique théorique et Appl. Theor. Inform. Appl. **39**(1), 115–123 (2005)
12. Champarnaud, J.M., Pin, J.E.: A maxmin problem on finite automata. Discrete Appl. Math. **23**, 91–96 (1989)
13. Domaratzki, M., Kisman, D., Shallit, J.: On the number of distinct languages accepted by finite automata with n states. J. Autom. Lang. Comb. **7**(4), 469–486 (2002)
14. Glaister, I., Shallit, J.: A lower bound technique for the size of nondeterministic finite automata. Inform. Process. Lett. **59**, 75–77 (1996)
15. Gramlich, G., Schnitger, G.: Minimizing nfa's and regular expressions. J. Comput. System Sci. **73**(6), 908–923 (2007)
16. Gruber, H., Holzer, M.: Finding lower bounds for nondeterministic state complexity is hard. In: Ibarra, O.H., Dang, Z. (eds.) DLT 2006. LNCS, vol. 4036, pp. 363–374. Springer, Heidelberg (2006)
17. Gruber, H., Holzer, M.: Results on the average state and transition complexity of finite automata accepting finite languages. Theoret. Comput. Sci. **387**(2), 155–166 (2007)
18. Harrison, M.A.: Introduction to Formal Language Theory. Addison-Wesley, Newyork (1978)
19. Holzer, M., Jakobi, S.: From equivalence to almost-equivalence, and beyond: minimizing automata with errors. Internat. J. Found. Comput. Sci. **24**(7), 1083–1134 (2013)
20. Lucchesi, C.L., Kowaltowski, T.: Applications of finite automata representing large vocabularies. Softw. Pract. Exper. **23**(1), 15–30 (1993)
21. Körner, H.: A time and space efficient algorithm for minimizing cover automata for finite languages. Internat. J. Found. Comput. Sci. **14**(6), 1071–1086 (2003)
22. Leung, H.: Separating exponentially ambiguous finite automata from polynomially ambiguous finite automata. SIAM J. Comput. **27**(4), 1073–1082 (1998)
23. Lupanov, O.B.: Über den Vergleich zweier Typen endlicher Quellen. Probleme der Kybernetik **6**, 328–335 (1966)
24. Rabin, M.O., Scott, D.: Finite automata and their decision problems. IBM J. Res. Dev. **3**, 114–125 (1959)
25. Salomaa, K., Yu, S.: NFA to DFA transformation for finite language over arbitrary alphabets. J. Autom. Lang. Comb. **2**(3), 177–186 (1997)
26. Sgarbas, K.N., Fakotakis, N.D., Kokkinakis, G.K.: Incremental construction of compact acyclic NFAs. In: 39th Annual Meeting of the Association for Computational Linguistics, pp. 482–489. Association for Computational Linguistics (2001)
27. Shannon, C.E.: The synthesis of two-terminal switching circuits. Bell Sys. Tech. J. **28**(1), 59–98 (1949)
28. Yu, S.: Cover automata for finite language. Bull. EATCS **92**, 65–74 (2007)

On the Number of Synchronizing Colorings of Digraphs

Vladimir V. Gusev[1,2(✉)] and Marek Szykuła[3(✉)]

[1] ICTEAM Institute, Université catholique de Louvain,
Louvain-la-Neuve, Belgium
vl.gusev@gmail.com
[2] Institute of Mathematics and Computer Science, Ural Federal University,
Ekaterinburg, Russia
[3] Institute of Computer Science, University of Wrocław,
Wrocław, Poland
msz@cs.uni.wroc.pl

Abstract. We deal with k-out-regular directed multigraphs with loops (called simply *digraphs*). The edges of such a digraph can be colored by elements of some fixed k-element set in such a way that outgoing edges of every vertex have different colors. Such a coloring corresponds naturally to an automaton. The road coloring theorem states that every primitive digraph has a synchronizing coloring.

In the present paper we study how many synchronizing colorings can exist for a digraph with n vertices. We performed an extensive experimental investigation of digraphs with small number of vertices. This was done by using our dedicated algorithm exhaustively enumerating all small digraphs. We also present a series of digraphs whose fraction of synchronizing colorings is equal to $1 - 1/k^d$, for every $d \geq 1$ and the number of vertices large enough.

On the basis of our results we state several conjectures and open problems. In particular, we conjecture that $1 - 1/k$ is the smallest possible fraction of synchronizing colorings, except for a single exceptional example on 6 vertices for $k = 2$.

1 Introduction

Throughout the paper we deal with directed multigraphs $\mathcal{G} = \langle V, E \rangle$ of a fixed out-degree k with loops, where V is a finite set of n *vertices* and E is a finite multiset of *edges*. For each $v \in V$ there are exactly k outgoing edges $(v, w) \in E$.

V.V. Gusev—Supported by the Communauté française de Belgique - Actions de Recherche Concertées, by the Belgian Programme on Interuniversity Attraction Poles, by the Russian foundation for basic research (grant 13-01-00852), Ministry of Education and Science of the Russian Federation (project no. 1.1999.2014/K), Presidential Program for Young Researchers (grant MK-3160.2014.1) and the Competitiveness Program of Ural Federal University.
M. Szykuła—Supported in part by Polish NCN grant DEC-2013/09/N/ST6/01194.

© Springer International Publishing Switzerland 2015
F. Drewes (Ed.): CIAA 2015, LNCS 9223, pp. 127–139, 2015.
DOI: 10.1007/978-3-319-22360-5_11

These are simply called *digraphs* throughout the paper. For every vertex $v \in V$ of a digraph $\mathcal{G}$, the k outgoing edges can be colored differently by one of the k colors from a finite set Σ, giving raise to a *deterministic finite (semi)automaton* $\mathscr{A} = \langle V, \Sigma, \delta \rangle$ with the set of states V, the alphabet Σ, and the transition function δ, where $\delta(v, a) = w$ whenever an edge $(v, w) \in E$ was colored by a. Every such automaton $\mathscr{A}$ is called a *coloring* of the digraph $\mathcal{G}$. Thus we identify automata with colorings of their underlying digraphs. We extend the transition function $\delta \colon V \times \Sigma \to V$ to $\delta \colon 2^V \times \Sigma^* \to 2^V$ on subsets and words in the natural way. For $\delta(S, w)$, where $S \subseteq V$ and $w \in \Sigma^*$, we also write shortly Sw.

Automaton $\mathscr{A}$ is called *synchronizing* if there exist a word w and a state p such that for every state $q \in Q$ we have $\delta(q, w) = p$. Such a word w is called *reset* (or *synchronizing*) word for $\mathscr{A}$. The length of the shortest synchronizing word of $\mathscr{A}$ is called *reset threshold* and is denoted by $\mathrm{rt}(\mathscr{A})$. Recent surveys of the theory of synchronizing automata may be found in [11,20].

Call a digraph *primitive* (or *aperiodic*) if it is strongly connected and the gcd of all its cycles is equal to 1. It is easy to show that an underlying digraph of a synchronizing automaton is primitive. In 1977 Adler, Goodwyn and Weiss conjectured [1] that every primitive digraph has a synchronizing coloring. This conjecture became widely known as the *road coloring* problem. It was arguably one of the most important conjectures in automata theory until it was finally proved by Trahtman in 2007 [18]. One of the goals of the present paper is to find the right quantitative formulation of the road coloring theorem.

Another part of our motivation comes from the algorithmic issues related to the road coloring problem. How to find a synchronizing coloring of a given digraph? A non-trivial algorithm working in time $O(kn^2)$ is known for this task [3]. On the other hand, M.-P. Béal suggested during her talk at CANT 2012 that a random sampling of colorings in a search for a synchronizing one may lead to a simple and practically effective algorithm for the problem. Since one can check whether a coloring is synchronizing in $O(kn^2)$ time, it remains to show that a random coloring is synchronizing with high probability. In our research we were partially motivated by this observation.

There are other computational problems related to the synchronizing colorings of digraphs, such as deciding existence of a synchronizing coloring for a fixed reset word [21], or for a fixed reset threshold [16]. Also, several open problems concerning synchronizing automata and the road coloring problem have been stated by M.V. Volkov [19].

For a given k-out-regular digraph $\mathcal{G}$ with n vertices, the *synchronizing ratio* is the number of synchronizing colorings to the number $(k!)^n$ of all possible colorings. Note that we distinguish edges of $\mathcal{G}$, i.e. two colorings are the same if all edges have the same color. Therefore, there is always exactly $(k!)^n$ different colorings of $\mathcal{G}$. A digraph $\mathcal{G}$ is *totally synchronizing* if its synchronizing ratio of $\mathcal{G}$ is equal to 1.

In this paper we perform an experimental and theoretical study on the synchronizing ratio of digraphs. Our main contributions are as follows:

1. We developed an efficient algorithm for enumerating and checking synchronizing ratios of nonisomorphic digraphs.

2. Using the algorithm, we performed extensive experiments revealing various phenomena concerning the synchronizing ratio. These provide evidence to state several conjectures and form a basis for further investigation.
3. We found out that for small n and k there are no primitive strongly connected digraphs with synchronizing ratio less than $1 - 1/k$, except for a single particular example for $n = 6$ and $k = 2$.
4. We constructed digraphs with synchronizing ratio $1 - 1/k^d$, for every $d \geq 1$ and $n \geq 3d$. This shows that there are many examples with different synchronizing ratio in the range $[1 - 1/k, 1]$.

2 General Statements

A strongly connected component S of a digraph $\mathcal{G} = \langle V, E \rangle$ is called a *sink component* if there are no edges going from S to $V \setminus S$. It is *reachable* if for any vertex $v \in V$ there is a directed path from v to a vertex in S.

Proposition 1. *If a digraph $\mathcal{G}$ has a synchronizing coloring then it has a unique reachable sink component S. Furthermore, the synchronizing ratio of $\mathcal{G}$ is equal to the synchronizing ratio of the digraph induced by S.*

Proof. The proof of the first statement belongs to folklore. It is not hard to see that an arbitrary coloring $\mathscr{A}$ of digraph $\mathcal{G}$ is synchronizing if and only if the subautomaton $\mathscr{A}'$ induced by the sink component S is synchronizing. Therefore, the set of all colorings of $\mathcal{G}$ can be divided into groups of equal size, each group containing the colorings with the same induced subcoloring of S. Since colorings from each group are altogether synchronizing or non-synchronizing, we obtain that the synchronizing ratio of $\mathcal{G}$ is equal to the synchronizing ratio of the digraph induced by S. $\qquad\square$

Since a one-vertex digraph is totally synchronizing we have the following corollary:

Corollary 1. *A digraph with a sink state is either totally synchronizing or none of its colorings is synchronizing.*

Due to Proposition 1 the study of synchronizing ratios and totally synchronizing digraphs can be reduced to the case of strongly connected digraphs.

Surprisingly, the underlying digraphs of several automata presented in the literature appear to be totally synchronizing. One important example of such a digraph is well known to the community, see for example [19]. Recall that the Černý automaton $\mathscr{C}_n$ ([6]) can be defined as $\langle \{0, \ldots, n-1\}, \{a, b\}, \delta \rangle$, where $\delta(i, a) = i + 1$ for $i < n - 1$, $\delta(n - 1, a) = 0$, $\delta(n - 1, b) = 0$, and $\delta(i, b) = i$ for $i < n - 1$. The proof of the following folklore result has not yet appeared in the literature.

Proposition 2. *The underlying digraph of $\mathscr{C}_n$ is totally synchronizing.*

Proof. Let $\mathscr{C}'_n$ be an arbitrary coloring of the underlying digraph of $\mathscr{C}_n$. It is well known that an automaton is synchronizing if and only if every pair of states i, j is synchronizing, i.e. there is a word w such that $iw = jw$ (see [6], or [20, Proposition 2.1]).

We will show that any pair of states (i, j) of $\mathscr{C}'_n$ satisfy this condition. Let $d(i, j)$ be the length of the shortest path from i to j. We will proceed by induction on $d(i, j)$. Consider a pair (i, j); without loss of generality we may assume that $d(i, j) \leq d(j, i)$.

If $j = n - 1$ then let y be the letter on the edge from i to $i + 1$. We apply y so $(iy, jy) = (i + 1, 0)$, and $d(i, j) = d(i + 1, 0)$.

Consider the case $j \neq n - 1$. Let x be the letter on the loop on the state j. If $i < n - 1$ then let y be the letter on the edge from i to $i + 1$; otherwise let $y = x$. If $x = y$ then $d(ix, jx) = d(iy, j) < d(i, j)$. Otherwise we apply the letter y, and in the same manner consider the pair $(iy, jy) = (i + 1, j + 1)$. Note that $d(i, j) = d(i + 1, j + 1)$. Following in this way, after at most $n - 1 - i$ steps, we will reach a pair $(n - 1, k)$. For $(n - 1, k)$ we choose $y = x$, we obtain $d((n - 1)x, kx) = d(0, k) < d(n - 1, k)$.   □

Underlying digraphs of many other automata that appeared in literature are also totally synchronizing. In a similar fashion one can show that the underlying digraphs of the series of slowly synchronizing automata (see [2,13]) are totally synchronizing. Also, almost all presented examples of automata with two cycle lengths have this property [9].

For the sake of completeness we mention the following notions from related topics. A word w is called *totally synchronizing* if w is a reset word for any coloring of totally synchronizing digraph $\mathcal{G}$. See [5] for an analysis of totally synchronizing digraphs and words in some special classes of digraphs. A word over an alphabet Σ is called *n-synchronizing* if it is a reset word for all synchronizing automata with $n + 1$ states over the alphabet Σ. See [7] for the introduction to the topic.

3   Experimental Investigation of Digraphs

We performed a series of experiments to reveal some properties of the synchronizing ratio of digraphs. These include both exhaustive enumeration of small digraphs and larger random digraphs. We are interested mostly in primitive strongly connected digraphs (cf. Proposition 1). In the case of exhaustive enumeration we checked the synchronizing ratio of all nonisomorphic k-out-regular digraphs with a given n vertices.

3.1   Algorithms

To check as many cases as possible and obtain a large data set, we needed to design and implement our algorithms carefully. This is especially important during the exhaustive search, since the number of digraphs grows very fast with

n and k. Here we briefly describe our algorithms, skipping numerous technical improvements and tricks in the implementation. Some of our ideas are based on [12], where the Černý conjecture was verified by an exhaustive enumeration for all binary automata up to $n \le 11$ states.

To determine the synchronizing ratio of a digraph, we can just enumerate all its colorings and count the synchronizing ones. Checking whether a coloring (automaton) is synchronizing can be easily done in $O(kn^2)$ time [6,8]. Note that in many cases, some colorings give rise to the same particular automaton (e.g. if there are two or more parallel edges (v, w) then we can permute its colors obtaining the same automaton). Also, every coloring has $k!$ equivalent colorings obtained only by permuting the colors. Using these facts we could greatly reduce the total number of really checked colorings for synchronization.

Checking whether a digraph is strongly connected and the gcd of its cycles is 1 can be effectively done in $O(kn)$ time basing on the algorithms from [10, 17] respectively.

Now, computing the synchronizing ratios of a set of random digraphs follows easily, and we can proceed this in parallel on a grid. However, in an exhaustive enumeration, the number of digraphs grows very fast in terms of n and k (see Table 2), and the main problem was to deal with it.

Algorithm 1. Exhaustive checking of digraphs.

Require: n – the number of vertices (states)
Require: k – the out-degree (size of the alphabet)
1: $G_n \leftarrow$ the set of all simple graphs with n vertices.
2: **for all** simple graphs $\mathcal{G} \in G_n$ **do**   ▷ In parallel
3:     $CanSet \leftarrow$ EMPTYSET
4:     **for all** digraphs $\mathcal{D}_{n,k}$ with underlying graph $\mathcal{G}$ **do** ▷ Orient and multiply the edges of $\mathcal{G}$, and add loops
5:         **if** $\mathcal{D}_{n,k}$ is primitive **then**
6:             $\mathcal{R}_{n,k} \leftarrow$ the canonical representation of $\mathcal{D}_{n,k}$.
7:             **if** $\mathcal{R}_{n,k} \notin CanSet$ **then**
8:                 $CanSet$.INSERT($\mathcal{R}_{n,k}$)
9:                 Count synchronizing colorings of $\mathcal{R}_{n,k}$.
10:             **end if**
11:         **end if**
12:     **end for**
13: **end for**

Our algorithm for exhaustive checking of digraphs is summarized in Algorithm 1. First, in line 1, we generate all nonisomorphic simple graphs with n vertices. A *simple graph* is a graph with undirected edges joining two distinct vertices. This can be done effectively by the algorithm from [14], implemented in package **nauty**. Now, we can process each such a simple graph in parallel. In line 4, for every simple graph $\mathcal{G}$ we orient and multiply its edges so that there are at most k outgoing edges for each vertex. Then we interpret the missing edges as loops. Clearly, an isomorphic copy of every digraph can be obtained in this way from its underlying simple graph, and the digraphs obtained from two nonisomorphic simple graphs are also nonisomorphic. We can, however, obtain isomorphic digraphs from the same simple graph. In line 5 we skip non-strongly connected and non-primitive digraphs. In line 6 we compute the *canonical representation* of a generated digraph $\mathcal{D}_{n,k}$; this is the lexicographically minimal representation among all digraphs isomorphic to $\mathcal{D}_{n,k}$ (cf. [14]). To skip isomorphic copies obtained from the same simple graph, in line 3 we introduce the set *CanSet* of canonical representations of generated digraphs. Then in line 7, we check if an isomorphic copy of the digraph $\mathcal{D}_{n,k}$ was already generated; if not, in line 8 we insert it to the set. The set *CanSet* can be effectively implemented as a radix trie, allowing to perform both membership test and insertion in linear time, and providing some compression (which is also important in view of the number of generated digraphs). Finally, we can count synchronizing colorings of the generated digraph (line 9).

3.2   Experimental Results from Exhaustive Enumeration

The algorithms in C++ and compiled with GCC 4.8.1. The computations were performed in parallel on a small grid consisted of computers with 8 processors Quad-Core AMD Opteron(tm) 8350 (2 GHz) and 64GB of RAM.

We were able to check all 2-out-regular digraphs with up to 10 vertices, 3-out-regular up to 7 states, 4-out-regular up to 5 states, and 5-out-regular up to 4 states. In the case of $k = 2$-out-regular digraphs with $n = 10$ states, the total processor time was more than 60 days (about 1 day of parallelized computation). The case of $k = 3$ and $n = 7$ took even more, about 72 days; the total number of colorings was $\sim 7 \times 10^{14}$, but, thanks to optimization, we required to check only $\sim 10^{13}$ automata.

The results concerning synchronizing ratios are summarized in Table 1. In Table 2 we present the exact number of strongly connected aperiodic digraphs, and totally synchronizing digraphs. We observe that the fraction of totally synchronizing digraphs within the class of strongly connected aperiodic digraphs is growing.

In Table 3, for $k = 2$ and $n = 8$, we present the number of nonisomorphic digraphs with particular numbers of synchronizing colorings. Interestingly, there are several graphs in the distribution, and the gaps grow for smaller number of synchronizing colorings. This picture is similar for the other values of n and k that we checked. The number of gaps seems to grow with n and k.

3.3   Experiments on Random Digraphs

To deal also with larger digraphs, we performed additional experiments with random digraphs. We used the uniform model of a random digraph, that is, for every outgoing edge from a vertex v we choose the destination vertex uniformly at random and independently from the other choices.

Table 1. The minimum, average, and standard deviation of the number of synchronizing colorings of all strongly connected aperiodic k-out-regular digraphs with n vertices

k	n	Min	Min ratio	Avg	Avg ratio	Std dev
2	2	2	0.5	3	0.750	1.000
2	3	4	0.5	6.833	0.854	1.280
2	4	8	0.5	14.640	0.915	2.243
2	5	16	0.5	30.987	0.968	2.146
2	6	30	0.469	63.139	0.986	2.381
2	7	64	0.5	127.365	0.995	2.033
2	8	128	0.5	255.483	0.998	1.866
2	9	256	0.5	511.563	0.999	1.617
2	10	512	0.5	1,023.607	≈ 1.000	1.468
3	2	24	0.667	31.2	0.867	5.879
3	3	144	0.667	208.800	0.967	14.163
3	4	864	0.667	1,284.987	0.991	36.346
3	5	5,184	0.667	7,765.775	0.999	50.091
3	6	31,104	0.667	46,643.953	≈ 1.000	78.679
3	7	186,624	0.667	279,921.191	≈ 1.000	108.167
4	2	432	0.75	533.333	0.926	61.738
4	3	10,368	0.75	13,704.874	0.991	367.767
4	4	248,832	0.75	331,421.072	0.999	2,233.171
4	5	5,971,968	0.75	7,961,941.49	≈ 1.000	7,104.373
5	2	11,520	0.75	13,782.857	0.957	1,048.941
5	3	1,382,400	0.75	1,723,468.312	0.997	720,951.433
5	4	165,888,000	0.75	207,324,196.845	≈ 1.000	412,162.118

For $k = 2$ we checked $1000,000$ digraphs for every $n = 4, \ldots, 15$, and $100,000$ for $n = 16, \ldots, 27$. Since the number of possible colorings grow very fast with k, for $k = 3$ we tested $n = 4, \ldots, 12$, and for $k = 4$ we tested only $n = 4, \ldots, 8$. We additionally checked the same numbers of random of digraphs in the class of strongly connected aperiodic digraphs, within the same range of n and k.

The results from random tests for larger n show the same patterns as observed in those from exhaustive search. Figure 1 shows the fraction of totally synchronizing digraphs in random samples of strongly connected aperiodic digraphs. The picture is very similar in the class of all digraphs.

4   Digraphs with Specific Synchronizing Ratios

In this section we present different examples of digraphs with particular values of the synchronizing ratio. According to our computational experiments the smallest possible value of the synchronizing ratio among all digraphs is equal to $\frac{30}{64}$.

Table 2. The number of nonisomorphic strongly connected aperiodic digraphs, the number of totally synchronizing digraphs, and their fraction

k	n	S.c. aperiodic	Totally synchronizing	Fraction
2	2	2	1	0.500
2	3	12	6	0.500
2	4	100	66	0.660
2	5	1220	890	0.729
2	6	19,064	14,973	0.785
2	7	361,157	296,303	0.82
2	8	8,001,589	6,754,895	0.844
2	9	202,635,930	174,246,295	0.860
2	10	5,765,318,112	5,026,305,042	0.872
3	2	5	3	0.600
3	3	85	63	0.741
3	4	3,148	2,672	0.849
3	5	199,489	182,326	0.914
3	6	19,059,581	18,006,297	0.945
3	7	2,537,475,117	2,443,850,969	0.963
4	2	9	6	0.666
4	3	357	302	0.846
4	4	39,680	36,762	0.926
4	5	9,089,413	8,779,342	0.966
5	2	14	10	0.714
5	3	1,102	990	0.898
5	4	304,082	291,530	0.959

Table 3. The number of nonisomorphic digraphs with the given number of synchronizing colorings for $k = 2$ and $n = 8$

# synch. col.	128	130	...	158	160	162	...	174	176	178	180	182	184	186	188	190	192
# digraphs	72	0	...	0	24	0	...	0	1	0	0	0	5	0	0	1	813

# synch. col.	194	196	198	200	202	204	206	208	210	212	214	216	218	220	222	224	226
# digraphs	0	1	1	12	1	1	6	202	0	2	1	134	4	22	14	4,022	60

# synch. col.	228	230	232	234	236	238	240	242
# digraphs	73	170	852	179	1,226	610	21,933	699

# synch. col.	244	246	248	250	252	254	256
# digraphs	4,523	3,171	44,230	27,438	310,400	825,791	6,754,895

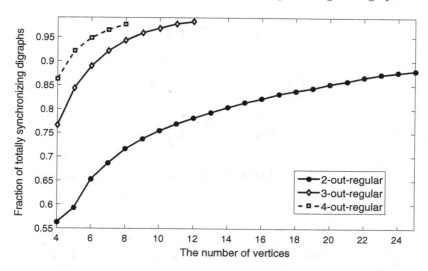

Fig. 1. The fraction of totally synchronizing digraphs in the class of strongly connected and aperiodic digraphs

This value is achieved by the digraph $\mathcal{G}_{30}$ (Fig. 2). By direct computation one can verify that only 30 colorings of $\mathcal{G}_{30}$ are synchronizing.

Proposition 3. *There is a 2-out-regular digraph with 6 states and the synchronizing ratio* $\frac{30}{64}$.

The exceptional example $\mathcal{G}_{30}$ seems to be unique. We did not find any other digraph with this particular value of the synchronizing ratio. Furthermore, according to our computational experiments the smallest value of the synchronizing ratio among all other k-out-regular digraphs seems to be equal to $\frac{k-1}{k}$. There are many examples that reach this bound, and in the following theorem we construct a series of digraphs with this property.

Theorem 1. *For every $n > 3$ there is a k-out-regular digraph with n vertices and the synchronizing ratio* $\frac{k-1}{k}$.

Proof. We will define the digraph $\mathcal{G}_{n,k}$ as follows; see Fig. 3. The set of vertices V is $\{0, 1, \ldots n - 1\}$. The edges $(0, 1)$ and $(1, 2)$ are of multiplicity 1, the edges $(1, 1)$ and $(0, 2)$ are of multiplicity $k - 1$, and the edges $(2, 3), (3, 4), \ldots, (i, i + 1), \ldots, (n - 1, 0)$ are of multiplicity k.

Consider now an arbitrary coloring of the digraph $\mathcal{G}_{n,k}$. Let x be the letter on the edge $(0, 1)$ and y be the letter on the edge $(1, 2)$. If $x = y$ then every letter acts as a permutation, and so the automaton is not synchronizing. If $x \neq y$ then x^{n-1} is a reset word for the given coloring. Hence, a coloring is synchronizing if and only if $x = y$. Therefore, the synchronizing ratio of $\mathcal{G}_{n,k}$ is equal to $\frac{k-1}{k}$. $\square$

We generalize the above result to obtain digraphs with different values of the synchronizing ratio.

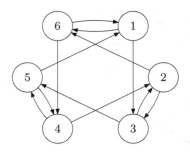

Fig. 2. Digraph $\mathcal{G}_{30}$.

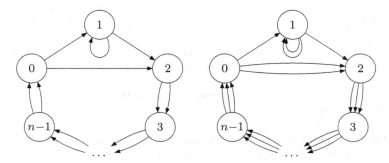

Fig. 3. The digraphs $\mathcal{G}_{n,2}$ and $\mathcal{G}_{n,3}$.

Theorem 2. *For every integers $d \geq 1$ and $n \geq 3d$ there is a k-out-regular digraph with n vertices and the synchronizing ratio $1 - \frac{1}{k^d}$.*

Proof. We will define a digraph $\mathcal{H}_{n,k}^d$ as follows; see Fig. 4. The set of vertices V is $\{0, 1, \ldots n-1\}$. There are edges $(i, i+1)$ of multiplicity k for every $2d \leq i \leq n-1$. There are edges $(i, i+1)$ of multiplicity $k-1$ for every $0 \leq i < 2d$. For every $0 \leq i < d$ the vertex $2i+1$ has a loop. The remaining edges of multiplicity 1 are of the form $(2i, 2i+2)$ for every $0 \leq i < d$.

Consider now an arbitrary coloring $\mathscr{H}_{n,k}^d$ of the digraph $\mathcal{H}_{n,k}^d$. Let x_i be the letter of the edge $(2i, 2i+2)$ and y_i be the letter of the loop $2i+1$, for $0 \leq i < d$. We will show that the automaton $\mathscr{H}_{n,k}^d$ is synchronizing if and only if $x_i = y_i$ for every i. It will immediately imply that the synchronizing ratio of $\mathcal{H}_{n,k}^d$ is equal to $1 - \frac{1}{k^d}$.

If $x_i = y_i$ for every i then every letter acts as a permutation; thus the automaton is not synchronizing. Assume now that there is ℓ such $x_\ell \neq y_\ell$, and let ℓ be the smallest integer with this property. In order to show that $\mathscr{H}_{n,k}^d$ is synchronizing it remains to prove that any pair can be synchronized. Let p and q be a pair of states. Since the automaton is strongly connected, there is a word u mapping p to $2d$. Let $q' = \delta(q, u)$. Let v be a shortest word mapping q' to a state in $D = \{i \in V \mid i \geq 2d\} \cup \{0\}$. Note that $|v| \leq d$. Now we have $\delta(q, uv) \in D$, and also $\delta(p, uv) = \delta(2d, v) \in D$, because $n \geq 3d$. By the fact that ℓ is the

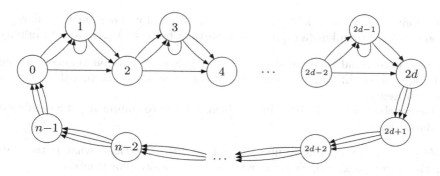

Fig. 4. The digraph $\mathcal{H}^d_{n,3}$.

smallest integer with the property $x_\ell \neq y_\ell$ we obtain $\delta(s', y^n_\ell) = \delta(t', y^n_\ell)$, which concludes the proof.   □

Remark 1. There exist many other digraphs with synchronizing ratio $1 - \frac{1}{k^d}$. Note that we can replace the k-path $(2d, \ldots, n-1, 0)$ of $\mathcal{H}^d_{n,k}$ with any acyclic multigraph such that: any vertex is reachable from the vertex $2d$; from any vertex we can reach the vertex 0; and every path from $2d$ to 0 is of the same length $\geq 3d$.

5   Conclusions and Open Problems

In this section we summarize all the conjectures and open problems. All of the conjectures are supported by experimental evidence.

Conjecture 1. The minimum value of the synchronizing ratio among all k-out-regular digraphs with n vertices is equal to $\frac{k-1}{k}$, except for the case $k = 2$ and $n = 6$ when it is equal to $\frac{30}{64}$.

The conjecture was verified for small values of n and k (see Table 1). It implies that a uniformly random coloring of a primitive strongly connected digraph is synchronizing with probability at least $1/2$, and hence it would justify the algorithm finding synchronizing coloring randomly.

To state the next conjecture, let say that a *gap* in the distribution of the number of synchronizing colorings is a maximal interval of integers divisible by $k!$, such that there are no digraphs with the number of synchronizing colorings in this interval. Thus the conjecture above states that for every k and $n \neq 6$ large enough there is the gap $[k!, \frac{k-1}{k}(k!)^n - k!]$.

Conjecture 2. For every k and $g \geq 1$, there is an n large enough such that there are at least g gaps in the distribution of the number of synchronizing colorings of k-out-regular digraphs with n vertices.

The following conjecture can be stated either in the class of strongly connected and aperiodic digraphs, or in the class of all digraphs.

Conjecture 3. For every $k \geq 2$, the fraction of totally synchronizing digraphs among all k-out-regular digraphs with n vertices tends to 1 as n goes to infinity.

A recent non-trivial theorem states that a random automaton is synchronizing with high probability [4,15]. Conjecture 3 can be seen as a further development of this statement.

We conclude with the following problem related to computing the number of synchronizing colorings.

Problem 1. Given a k-out-regular digraph $\mathcal{G}$ with n vertices, what is the computational complexity of checking whether $\mathcal{G}$ is totally synchronizing.

Acknowledgment. The authors want to thank Mikhail Volkov for his significant contributions to the theory of synchronizing automata on the occasion of his 60th birthday.

References

1. Adler, R.L., Goodwyn, L.W., Weiss, B.: Equivalence of topological Markov shifts. Israel J. Math. **27**(1), 49–63 (1977)
2. Ananichev, D.S., Volkov, M.V., Gusev, V.V.: Primitive digraphs with large exponents and slowly synchronizing automata. J. Math. Sci. **192**(3), 263–278 (2013)
3. Béal, M.-P., Perrin, D.: A quadratic algorithm for road coloring. Discrete Appl. Math. **169**, 15–29 (2014)
4. Berlinkov, M.V.: On the probability of being synchronizable (2013). http://arxiv.org/abs/1304.5774
5. Cardoso, Â.: The Černý Conjecture and Other Synchronization Problems. Ph.D. thesis, University of Porto, Portugal (2014). http://hdl.handle.net/10216/73496
6. Černý, J.: Poznámka k homogénnym eksperimentom s konečnými automatami. Matematicko-fyzikálny Časopis Slovenskej Akadémie Vied **14**(3), 208–216 (1964). In Slovak
7. Cherubini, A.: Synchronizing and collapsing words. Milan J. Math. **75**(1), 305–321 (2007)
8. Eppstein, D.: Reset sequences for monotonic automata. SIAM J. Comput. **19**, 500–510 (1990)
9. Gusev, V.V., Pribavkina, E.V.: Reset thresholds of automata with two cycle lengths. In: Holzer, M., Kutrib, M. (eds.) CIAA 2014. LNCS, vol. 8587, pp. 200–210. Springer, Heidelberg (2014)
10. Jarvis, J.P., Shier, D.R.: Applied Mathematical Modeling: A Multidisciplinary Approach. Graph-theoretic analysis of finite Markov chains. CRC Press, Boca Raton (1996)
11. Kari, J., Volkov, M.V.: Černý's conjecture and the road coloring problem. In: Handbook of Automata. European Science Foundation, to appear
12. Kisielewicz, A., Szykuła, M.: Generating small automata and the Černý conjecture. In: Konstantinidis, S. (ed.) CIAA 2013. LNCS, vol. 7982, pp. 340–348. Springer, Heidelberg (2013)
13. Kisielewicz, A., Szykuła, M.: Synchronizing Automata with Large Reset Lengths (2014). http://arxiv.org/abs/1404.3311

14. McKay, B.D., Piperno, A.: Practical graph isomorphism II. J. Symbolic Comput. **60**, 94–112 (2014)
15. Nicaud, C.: Fast synchronization of random automata (2014). http://arxiv.org/abs/1404.6962
16. Roman, A.: P–NP threshold for synchronizing road coloring. In: Dediu, A.-H., Martín-Vide, C. (eds.) LATA 2012. LNCS, vol. 7183, pp. 480–489. Springer, Heidelberg (2012)
17. Tarjan, R.: Depth-first search and linear graph algorithms. SIAM J. Comput. **1**(2), 146–160 (1972)
18. Trahtman, A.N.: The road coloring problem. Isr. J. Math. **172**(1), 51–60 (2009)
19. Volkov, M.V.: Open problems on synchronizing automata. Workshop "Around the Černý conjecture", Wrocław (2008). http://csseminar.kadm.usu.ru/SLIDES/WroclawABCD2008/volkov_abcd_problems.pdf
20. Volkov, M.V.: Synchronizing automata and the černý conjecture. In: Martín-Vide, C., Otto, F., Fernau, H. (eds.) LATA 2008. LNCS, vol. 5196, pp. 11–27. Springer, Heidelberg (2008)
21. Vorel, V., Roman, A.: Complexity of road coloring with prescribed reset words. In: Dediu, A.-H., Formenti, E., Martín-Vide, C., Truthe, B. (eds.) LATA 2015. LNCS, vol. 8977, pp. 161–172. Springer, Heidelberg (2015)

On the Uniform Random Generation of Non Deterministic Automata Up to Isomorphism

Pierre-Cyrille Héam[✉] and Jean-Luc Joly

FEMTO-ST, CNRS UMR 6174, Université de Franche-Comté,
INRIA 16 Route de Gray, 25030 Besançon Cedex, France
{pheam,jean-luc.joly}@femto-st.fr

Abstract. In this paper we address the problem of the uniform random generation of non deterministic automata (NFA) up to isomorphism. First, we show how to use a Monte-Carlo approach to uniformly sample a NFA. Secondly, we show how to use the Metropolis-Hastings Algorithm to uniformly generate NFAs up to isomorphism. Using labeling techniques, we show that in practice it is possible to move into the modified Markov Chain efficiently, allowing the random generation of NFAs up to isomorphism with dozens of states. This general approach is also applied to several interesting subclasses of NFAs (up to isomorphism), such as NFAs having a unique initial states and a bounded output degree. Finally, we prove that for these interesting subclasses of NFAs, moving into the Metropolis Markov chain can be done in polynomial time. Promising experimental results constitute a practical contribution.

1 Introduction

Finite automata play a central role in the field of formal language theory and are intensively used to address algorithmic problems from model-checking to text processing. Many automata based algorithms have been developed and are still being developed, proposing new approaches and heuristics, even for basic problems like the inclusion problem[1]. Evaluating new algorithms is a challenging problem that cannot be addressed only by the theoretical computation of the worst case complexity. Several other complementary techniques can be used to measure the efficiency of an algorithm: average complexity, generic case complexity, benchmarking, evaluation on hard instances, evaluations on random instances. The first two approaches are hard theoretical problems, particularly for algorithms using heuristics and optimizations. Benchmarks, as well as known hard instances, are not always available. Nevertheless, in practice, random generation of inputs is a good way to estimate the efficiency of an algorithm. Designing uniform random generator for classes of finite automata is a challenging problem that has been addressed mostly for deterministic automata [CP05, BN07, AMR07, CN12] -the interested reader is referred to [Nic14] for a recent survey. However, the problem of uniform random generation of non deterministic automata (NFAs) is more complex, particularly for a

---

[1] see http://www.languageinclusion.org/.

© Springer International Publishing Switzerland 2015
F. Drewes (Ed.): CIAA 2015, LNCS 9223, pp. 140–152, 2015.
DOI: 10.1007/978-3-319-22360-5_12

random generation up to isomorphism: the size of the automorphism group of a n-state non deterministic automata may vary from 1 to $n!$. For most applications, the complexity of the algorithm is related to the structure of the automata, not to the names of the states: randomly generated NFAs, regardless of the number of isomorphic automata, may therefore lead to an over representation of some isomorphism classes of automata. Moreover, as discussed in the conclusion of [Nic14], the random generation of non deterministic automata has to be done on particular subclasses of automata in order to obtain a better sampler for the evaluation of algorithm (since most of the NFAs, for the uniform distribution, will accept all words).

In this paper we address the problem of the uniform generation of some classes of non deterministic automata (up to isomorphism) by using Monte-Carlo techniques. We propose this approach for the class of n-state non deterministic automata as well as for (a priori) more interesting sub-classes. Determining the most interesting subclasses of NFAs for testing practical applications is not the purpose of this paper. We would like to point out that Monte Carlo approaches are very flexible and can be applied quite easily for many classes of NFAs. More precisely:

1. We propose in Sect. 2 several ergodic Markov Chains whose stationary distributions are respectively uniform on the set of n-state NFAs, n-state NFAs with a fixed maximal output degree and n-state NFAs with a fixed maximal output degree for each letter. In addition, these tools can be adapted for these three classes including automata with a fixed single initial state. Moving into these Markov chains can be done in time polynomial in n.
2. The main idea of this paper is exposed in Sect. 3.1, where we show how to modify these Markov Chains using the Metropolis-Hastings Algorithm in order to obtain stationary distributions that are uniform for the given classes of automata but up to isomorphism. Moving into these new Markov chains requires computing the sizes of the automorphism group of the occurring NFAs.
3. The main contributions of this paper are given in Sect. 3. We show in Sect. 3.2 that, for the classes with a bounded output degree, moving into the modified Markov chains can be done in polynomial time. In Sect. 3.4 we explain how to use labeling techniques to do it efficiently in practice for all NFAs. Promising experiments are described in Sect. 3.5.

The random generation of non deterministic automata is explored in [TV05] using random graph techniques (without considering the obtained distribution relative to automata or to the isomorphism classes). In [CHPZ02], the random generation of NFAs is performed using bitstream generation. In [Nic09, NPR10] NFAs are obtained by the random generation of a regular expression and by transforming it into an equivalent automaton using Glushkov Algorithm. The use of Markov chains based techniques to randomly generate finite automata was introduced in [CF11, CF12] for acyclic automata.

1.1   Theoretical Background on NFA

For a general reference on finite automata see [HU79]. In this paper Σ is a fixed finite alphabet of cardinal $|\Sigma| \geq 2$, and m is an integer satisfying $m \geq 2$.

A *non-deterministic automaton* (NFA) on Σ is a tuple (Q, Δ, I, F) where Q is a finite set of *states*, Σ is a finite alphabet, $\Delta \subset Q \times \Sigma \times Q$ is the set of transitions, $F \subset Q$ is the set of final states and $I \subseteq Q$ is the set of initial states. For any state p and any letter a, we denote by $p \cdot a$ the set of states q such that $(p, a, q) \in \Delta$. The set of transitions Δ is *deterministic* if for every pair (p, a) in $Q \times \Sigma$ there is at most one $q \in Q$ such that $(p, a, q) \in \Delta$. Two NFAs are depicted on Fig. 1. A NFA is *complete* if for every pair (p, a) in $Q \times \Sigma$ there is at least one $q \in Q$ such that $(p, a, q) \in \Delta$. A *path* in a NFA is a sequence of transitions $(p_0, a_0, q_0)(p_1, a_1, q_1) \ldots (p_k, a_k, q_k)$ such that $q_i = p_{i+1}$. The word $a_0 \ldots a_k$ is the *label* of the path and k its *length*. If $p_0 \in I$ and $q_k \in F$ the path is successful. A word is *accepted* by a NFA if it's the label of a successful path. A NFA is *accessible* (resp. *co-accessible*) if for every state q there exists a path from an initial state to q (resp. if for every state q there exists a path from q to a final state). A NFA is *trim* if it is both accessible and co-accessible A *deterministic automaton* is a NFA where $|I| = 1$ and whose set of transitions is deterministic.

Let $\mathfrak{A}(n)$ be the class of finite automata whose set of states is $\{0, \ldots, n-1\}$. Let $\mathfrak{N}(n)$ be the subclass of $\mathfrak{A}(n)$ of trim finite automata. Let $\mathfrak{N}_m(n)$ be the class of finite automata in $\mathfrak{N}(n)$ such that, for each state p, there is at most m pairs (a, q) such that (p, a, q) is a transition. Let $\mathfrak{N}'_m(n)$ be the class of finite automata in $\mathfrak{N}_m(n)$ such that, for each state p and each letter a, there is at most m states q such that (p, a, q) is a transition. For any class $\mathfrak{X}$ of finite automata, we denote by $\mathfrak{X}^\bullet$ the subclass of $\mathfrak{X}$ of automata whose set of initial states is reduced to $\{1\}$. One has

$$\mathfrak{N}_m(n) \subseteq \mathfrak{N}'_m(n) \subseteq \mathfrak{N}(n) \subseteq \mathfrak{A}(n).$$

Two NFAs are *isomorphic* if there exists a bijection between their sets of states preserving the sets initial states, final states and transitions. More precisely, let $\mathcal{A} = (Q, \Sigma, \Delta, I, F)$ and let φ be a bijection from Q into a finite set $\varphi(Q)$. We denote by $\varphi(\mathcal{A})$ the automaton $(\varphi(Q), \Sigma, \Delta', \varphi(I), \varphi(F))$, with $\Delta' = \{(\varphi(p), a, \varphi(q)) \mid (p, a, q) \in \Delta\}$. Two automata $\mathcal{A}_1$ and $\mathcal{A}_2$ are isomorphic if there exists a bijection φ such that $\varphi(\mathcal{A}_1) = \mathcal{A}_2$.

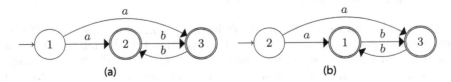

Fig. 1. Two isomorphic automata

Two isomorphic NFAs have the same number of states and are equal, up to the states names. The relation *is isomorphic to* is an equivalence relation.

For instance, the two automata depicted on Fig. 1 are isomorphic, with $\varphi(1) = 2$, $\varphi(2) = 1$ and $\varphi(3) = 3$. An *automorphism* for a NFA is an isomorphism between this NFA and itself. Given a NFA $\mathcal{A} = (Q, \Sigma, \Delta, I, F)$, the set of automorphisms of $\mathcal{A}$ is a finite group denoted $\mathrm{Aut}(\mathcal{A})$. For $Q' \subseteq Q$, $\mathrm{Aut}_{Q'}(\mathcal{A})$ denotes the subset of $\mathrm{Aut}(\mathcal{A})$ of automorphisms ϕ fixing each element of Q': for each $q \in Q'$, $\phi(q) = q$. Particularly $\mathrm{Aut}_\emptyset(\mathcal{A}) = \mathrm{Aut}(\mathcal{A})$, and $\mathrm{Aut}_Q(\mathcal{A})$ is reduce to the identity. For instance, the automorphism group of the automaton depicted on Fig. 1(a) has two elements, the identity and the isomorphism switching 2 and 3.

The size of the automorphism group of a non deterministic n-state automaton may vary from 1 to $n!$. For instance, any deterministic trim automaton whose states are all final has an automorphism group reduce to the identity. The non deterministic n-state automaton with no transition and where all states are both initial and final has for automorphism group the symmetric group.

The isomorphism problem consists in deciding whether two finite automata are isomorphic. It is investigated for deterministic automata in [Boo78]. It is naturally closed to the same problem for directed graph and the following result [Luk82] will be useful in this paper.

Theorem 1. *Let m be a fixed positive integer. The isomorphism problem for directed graphs with degree bounded by m is polynomial.*

1.2 Theoretical Background on Markov Chains

For a general reference on Markov Chains see [DLW08]. Basic probability notions will not be defined in this paper. The reader is referred for instance to [MU05].

Let Ω be a finite set. A *Markov chain* on Ω is a sequence $X_0, \ldots, X_t, \ldots$ of random variables on Ω such that $\mathbb{P}(X_{t+1} = x_{t+1} \mid X_t = x_t) = \mathbb{P}(X_{t+1} = x_{t+1} \mid X_t = x_t, \ldots, X_i = x_i, \ldots, X_0 = x_0)$, for all $x_i \in \Omega$. A Markov chain is defined by its *transition matrix* M, which is a function from $\Omega \times \Omega$ into $[0, 1]$ satisfying $M(x, y) = \mathbb{P}(X_{t+1} = y \mid X_t = x)$. The underlying graph of a Markov chain is the graph whose set of vertices is Ω and there is an edge from x to y if $M(x, y) \neq 0$. A Markov chain is *irreducible* if its underlying graph is strongly connected. It is *aperiodic* if for all node x, the gcd of the lengths of all cycles visiting x is 1. Particularly, if for each x, $M(x, x) \neq 0$, the Markov chain is aperiodic. A Markov chain is *ergodic* if it is irreducible and aperiodic. A Markov chain is symmetric if $M(x, y) = M(y, x)$ for all $x, y \in \Omega$. A distribution π on Ω is a stationary distribution for the Markov Chain if $\pi M = \pi$. It is known that an ergodic Markov chain has a unique stationary distribution [DLW08, Chapter 1]. Moreover, if the chain is symmetric, this distribution is the uniform distribution on Ω.

Given an ergodic Markov chain $X_0, \ldots, X_t, \ldots$ with stationary distribution π, it is known that, whatever is the value of X_0, the distribution of X_t converges to π when $t \to +\infty$: $\max \|M^t(x, \cdot) - \pi\|_{TV} \underset{t \to +\infty}{\to} 0$, where $\|\,\|_{TV}$ designates the total variation distance between two distributions [DLW08, Chapter 4]. This leads to the Monte-Carlo technique to randomly generate elements of Ω according to the distribution π by choosing arbitrarily X_0, computing $X_1, X_2, \ldots$, and returning

X_t for t large enough. The convergence rate is known to be exponential, but computing the constants is a very difficult problem: choosing the step t to stop is a challenging question depending both on how close to π we want to be and on the convergence rate of $M^t(x, \cdot)$ to π. For this purpose, the ε-*mixing time* of an ergodic Markov chain of matrix M and stationary distribution π is defined by $t_{\text{mix}}(\varepsilon) = \min\{t \mid \max_{x \in \Omega} \|P_t(x, \cdot) - \pi\|_{TV} \leq \varepsilon\}$. Computing mixing time bounds is a central question on Markov Chains.

The Metropolis-Hasting Algorithm is based on the Monte-Carlo technique and aims at modifying the transition matrix of the Markov chain in order to obtain a particular stationary distribution [DLW08, Chapter 3]. Suppose that M is an ergodic symmetric transition matrix of a symmetric Markov chain on Ω and ν is a distribution on Ω. The transition matrix P_ν for ν is defined by:

$$P_\nu(x, y) = \begin{cases} \min\left\{1, \frac{\nu(y)}{\nu(x)}\right\} M(x, y) & \text{if } x \neq y, \\ 1 - \sum_{z \neq x} \min\left\{1, \frac{\nu(z)}{\nu(x)}\right\} M(x, z) & \text{if } x = y. \end{cases}$$

The chain defined by P_ν is called *the Metropolis Chain for* ν. It is known [DLW08, Chapter 3] that it is an ergodic Markov chain whose stationary distribution is ν.

2   Random Generation of Non Deterministic Automata Using Markov Chain

In this section, we propose families of symmetric ergodic Markov chains on $\mathfrak{A}(n)$, $\mathfrak{N}(n)$, $\mathfrak{N}_m(n)$ and $\mathfrak{N}'_m(n)$, as well as on the respective corresponding doted classes of NFAs.

Let $\mathcal{A} = (Q, \Sigma, \Delta, I, F)$ be a finite automaton. For any q in Q and any (p, a, q) in $Q \times \Sigma \times Q$, the automata $\mathsf{Ch}_{\text{init}}(\mathcal{A}, q)$, $\mathsf{Ch}_{\text{final}}(\mathcal{A}, q)$ and $\mathsf{Ch}_{\text{trans.}}(\mathcal{A}, (p, a, q))$ are defined as follows:

- If $q \in I$, then $\mathsf{Ch}_{\text{init}}(\mathcal{A}, q) = (Q, \Sigma, \Delta, I \setminus \{q\}, F)$ and $\mathsf{Ch}_{\text{init}}(\mathcal{A}, q) = (Q, \Sigma, \Delta, I \cup \{q\}, F)$ otherwise.
- If $q \in F$, then $\mathsf{Ch}_{\text{final}}(\mathcal{A}, q) = (Q, \Sigma, \Delta, I, F \setminus \{q\})$, and $\mathsf{Ch}_{\text{final}}(\mathcal{A}, q) = (Q, \Sigma, \Delta, I, F \cup \{q\})$ otherwise.
- If $(p, a, q) \in \Delta$, then $\mathsf{Ch}_{\text{trans.}}(\mathcal{A}, (p, a, q)) = (Q, \Sigma, \Delta \setminus \{(p, a, q)\}, I, F)$, and $\mathsf{Ch}_{\text{trans.}}(\mathcal{A}, (p, a, q)) = (Q, \Sigma, \Delta \cup \{(p, a, q)\}, I, F)$ otherwise.

Let ρ_1, ρ_2, ρ_3 be three real numbers satisfying $0 \leq \rho_i \leq 1$ and $\rho_1 + \rho_2 + \rho_3 \leq 1$. Let $\mathfrak{X}$ be a class of automata whose set of states is Q. We define the transition matrix $S^{\mathfrak{X}}_{\rho_1, \rho_2, \rho_3}(x, y)$ on $\mathfrak{X}$ by:

- If there exists q such that $y = \mathsf{Ch}_{\text{init}}(x, q)$, then $S^{\mathfrak{X}}_{\rho_1, \rho_2, \rho_3}(x, y) = \frac{\rho_1}{|Q|}$.
- If there exists q such that $y = \mathsf{Ch}_{\text{final}}(x, q)$, then $S^{\mathfrak{X}}_{\rho_1, \rho_2, \rho_3}(x, y) = \frac{\rho_2}{|Q|}$.
- If there exists $(p, a, q) \in Q \times \Sigma \times Q$ such that $y = \mathsf{Ch}_{\text{trans.}}(x, q)$, then $S^{\mathfrak{X}}_{\rho_1, \rho_2, \rho_3}(x, y) = \frac{\rho_3}{|\Sigma| . |Q|^2}$.
- If y is different of x and has not one of the above forms, $S^{\mathfrak{X}}_{\rho_1, \rho_2, \rho_3}(x, y) = 0$.
- $S^{\mathfrak{X}}_{\rho_1, \rho_2, \rho_3}(x, x) = 1 - \sum_{y \neq x} S^{\mathfrak{X}}_{\rho_1, \rho_2, \rho_3}(x, y)$.

Now for $\mathfrak{X} \in \{\mathfrak{N}(n), \mathfrak{N}_m(n), \mathfrak{N}'_m(n)\}$, and $0 < \rho < 1$ we define the transition matrix $S_\rho^{\mathfrak{X}^\bullet}$ on $\mathfrak{X}^\bullet$ by $S_\rho^{\mathfrak{X}^\bullet} = S_{0,\rho,1-\rho}^{\mathfrak{X}}$.

Lemma 1. *Let m, n be fixed positive integers, with $m \geq 2$. If $1 > \rho > 0$, $\rho_1 > 0$, $\rho_2 > 0$ and $\rho_3 > 0$, then $S_{\rho_1,\rho_2,\rho_3}^{\mathfrak{N}(n)}$, $S_{\rho_1,\rho_2,\rho_3}^{\mathfrak{N}_m(n)}$ and $S_{\rho_1,\rho_2,\rho_3}^{\mathfrak{N}'_m(n)}$ are irreducible, as well as $S_\rho^{\mathfrak{N}(n)^\bullet}$, $S_\rho^{\mathfrak{N}_m(n)^\bullet}$ and $S_\rho^{\mathfrak{N}'_m(n)^\bullet}$.*

Proof. Without loss of generality, we assume that $Q = \{1, \ldots, n\}$. Let $\mathfrak{X} \in \{\mathfrak{N}(n), \mathfrak{N}_m(n), \mathfrak{N}'_m(n)\}$ and $x \in \mathfrak{X}$. We denote by $\mathcal{A}_0$ the automaton $(Q, \Sigma, \emptyset, Q, Q)$. The automaton $\mathcal{A}_0$ is trim and is in $\mathfrak{X}$. We prove there is a path in $\mathfrak{X}$ from x to $\mathcal{A}_0$. Set $x = (Q, \Sigma, \Delta, I, F)$. Since adding initial or final states to x provides automata that are still in $\mathfrak{X}$, there is a path from x to $y = (Q, \Sigma, \Delta, Q, Q)$ (using $\mathsf{Ch}_{\mathsf{init}}$ and $\mathsf{Ch}_{\mathsf{final}}$). Now, since all states are both initial and final, there is a path from y to $\mathcal{A}_0$ (by deleting all transitions). It follows there is a path in $\mathfrak{X}$ from x to $\mathcal{A}_0$. Since the graph of the Markov chain is symmetric, there is also a path from $\mathcal{A}_0$ to x. Consequently, the Markov chains are irreducible. The proof for $S_\rho^{\mathfrak{N}(n)^\bullet}$, $S_\rho^{\mathfrak{N}_m(n)^\bullet}$ and $S_\rho^{\mathfrak{N}'_m(n)^\bullet}$ are similar.

Lemma 2. *Let m, n be two fixed positive integers. If $1 > \rho > 0$, $\rho_1 > 0$, $\rho_2 > 0$ and $\rho_3 > 0$, then $S_{\rho_1,\rho_2,\rho_3}^{\mathfrak{N}(n)}$, $S_{\rho_1,\rho_2,\rho_3}^{\mathfrak{N}_m(n)}$ and $S_{\rho_1,\rho_2,\rho_3}^{\mathfrak{N}'_m(n)}$ are aperiodic, as well as $S_\rho^{\mathfrak{N}(n)^\bullet}$, $S_\rho^{\mathfrak{N}_m(n)^\bullet}$ and $S_\rho^{\mathfrak{N}'_m(n)^\bullet}$.*

Proof. With the notations of the proof of Lemma 1, there is a path of length n_x from any $x \in \mathfrak{X}$ to $\mathcal{A}_0$. Therefore there is a cycle of length $2n_x$ visiting x.

Now, $\mathsf{Ch}_{\mathsf{init}}(\mathcal{A}_0, 1) \notin \mathfrak{X}$ since 1 is not accessible in $\mathcal{A}_0$. It follows that $S^{\mathfrak{X}}(\mathcal{A}_0, \mathcal{A}_0) \neq 0$. Therefore, there is also a cycle of length $2n_x + 1$ visiting x. Since the gcd of $2n_x$ and $2n_x + 1$ is 1, the chain is aperiodic. The proof for $S_\rho^{\mathfrak{N}(n)^\bullet}$, $S_\rho^{\mathfrak{N}_m(n)^\bullet}$ and $S_\rho^{\mathfrak{N}'_m(n)^\bullet}$ are similar.

Proposition 1. *Let m, n be two fixed positive integers with $m \geq 2$. The Markov chains with matrix $S_{\rho_1,\rho_2,\rho_3}^{\mathfrak{N}(n)}$, $S_{\rho_1,\rho_2,\rho_3}^{\mathfrak{N}_m(n)}$ and $S_{\rho_1,\rho_2,\rho_3}^{\mathfrak{N}'_m(n)}$ are ergodic and their stationary distributions are the uniform distributions.*

Proof. By Lemmas 1 and 2, the chain is ergodic. Since the matrix $S_{\rho_1,\rho_2,\rho_3}^{\mathfrak{N}(n)}$, $S_{\rho_1,\rho_2,\rho_3}^{\mathfrak{N}_m(n)}$ and $S_{\rho_1,\rho_2,\rho_3}^{\mathfrak{N}'_m(n)}$ are symmetric, their stationary distributions are the uniform distributions (over the respective family of automata).

In practice, computing X_{t+1} from X_t is done in the following way: the first step consists in choosing with probabilities ρ_1, ρ_2 and ρ_3 whether we will change either an initial state, a final state or a transition. In a second step and in each case, all the possible changing operations are performed with the same probability. If the obtained automaton is in the corresponding class, X_{t+1} is set to this value. Otherwise, $X_{t+1} = X_t$. Since verifying that an automaton is in the desired class ($\mathfrak{N}(n)$, $\mathfrak{N}_m(n)$ or $\mathfrak{N}'_m(n)$), can be performed in time polynomial in n, computing X_{t+1} from X_t can be done in time polynomial in n.

We define the lazy Markov chain on $\mathfrak{A}(n)$ by $L_{\rho_1,\rho_2,\rho_3}^{\mathfrak{A}(n)}(x,y) = \frac{1}{2}S_{\rho_1,\rho_2,\rho_3}^{\mathfrak{A}(n)}(x,y)$ if $x \neq y$ and $L_{\rho_1,\rho_2,\rho_3}^{\mathfrak{A}(n)}(x,x) = \frac{1}{2} + \frac{1}{2}S_{\rho_1,\rho_2,\rho_3}^{\mathfrak{A}(n)}(x,x)$. It is known that a symmetric Markov chain and its associated lazy Markov chain have similar mixing times.

Proposition 2. *The ε-mixing time $\tau(\varepsilon)$ of $L_{\rho_1,\rho_2,\rho_3}^{\mathfrak{A}(n)}$ satisfies $\tau(\varepsilon) \leq (\frac{1}{\rho_1} + \frac{1}{\rho_2})(n \ln n + \lceil n \ln(\varepsilon^{-1}) \rceil) + \frac{2|\Sigma|^2 n^2}{\rho_3}\left(\ln(|\Sigma|n) + \lceil \ln(\varepsilon^{-1}) \rceil\right)$.*

It follows that $\tau(\varepsilon) = O(n^3)$ when $|\Sigma|$ is considered as a constant. At this stage, we are not able to compute bounds on the mixing times of the other Markov chains. Practical experiments, with various sizes of alphabets, seems to show that about 90 % of the automata generated by the above lazy Markov Chain (using n^3 as mixing bound) are trim. This observation leads us to consider, for other experiments, to move n^3 steps to sample automata. Of course, this is not a proof, just an empirical estimation.

3   Random Generation of Non Deterministic Automata upto Isomorphism

In this section we show how to use the Metropolis-Hastings algorithm to uniformly generate NFAs up to isomorphism and that, for this purpose, it *suffices* to compute the sizes of the automorphism groups of involved NFAs. We prove in Sect. 3.2 that this computation is polynomially equivalent to testing the isomorphism problem for the involving automata. For the classes $\mathfrak{N}_m(n)$, $\mathfrak{N}_m(n)^\bullet$, $\mathfrak{N}'_m(n)$ and $\mathfrak{N}'_m(n)^\bullet$, we show that it can be done in time polynomial in n (if m is fixed). In Sect. 3.4 we show how to practically compute the sizes of automorphism group using labellings techniques. Finally, experimental results are given in Sect. 3.5.

3.1   Metropolis-Hastings Algorithm

For a class $\mathfrak{C}$ of NFAs (closed by isomorphism) and n a positive integer, let $\mathfrak{C}(n)$ be the elements of $\mathfrak{C}$ whose set of states is $\{1, \ldots, n\}$ and let γ_n be the number of isomorphism classes on $\mathfrak{C}(n)$. There are $n!$ possible bijections on $\{1, \ldots, n\}$. If $\mathcal{A} \in \mathfrak{C}(n)$, Let φ_1 and φ_2 be two bijections on $\{1, \ldots, n\}$. One has $\varphi_1(\mathcal{A}) = \varphi_2(\mathcal{A})$ iff $\varphi_2^{-1}\varphi_1(\mathcal{A}) = \mathcal{A}$, iff $\varphi_2^{-1}\varphi_1(\mathcal{A}) \in \text{Aut}(\mathcal{A})$. It follows that the isomorphism classes of $\mathcal{A}$ (in $\mathfrak{C}(n)$) has $\frac{n!}{|\text{Aut}(\mathcal{A})|}$ elements. This leads to the following result.

Proposition 3. *Randomly generating an element x of $\mathfrak{C}(n)$ with probability $\frac{n!}{\gamma_n|\text{Aut}(x)|}$ provides a uniform random generator of the isomorphism classes of $\mathfrak{C}(n)$.*

Proof. Let H be an isomorphism class of $\mathfrak{C}(n)$; H is generated with probability

$$\sum_{x \in H} \frac{n!}{\gamma_n|\text{Aut}(x)|} = \sum_{x \in H} \frac{n!}{\gamma_n|H|} = \frac{1}{\gamma_n|H|}\sum_{x \in H} 1 = \frac{|H|}{\gamma_n|H|} = \frac{1}{\gamma_n}.$$

In order to compute P_ν it is not necessary to compute γ_n, since $\frac{\nu(x)}{\nu(y)} = \frac{|\mathrm{Aut}(y)|}{|\mathrm{Aut}(x)|}$. A direct use of the Metropolis-Hastings algorithm requires to compute all the neighbors of x and the sizes of theirs automorphism groups to move from x. Since a n-state automaton has about $|\Sigma|n^2$ neighbors, it can be a quite huge computation for each move. However, practical evaluations show that in most cases the automorphism group of an automaton is quite small and, therefore, the rejection approach exposed in [CG95] is more tractable. It consists in moving from x to y using $S(x, y)$ (the non-modified chain) and to accept y with probability $\min\left\{1, \frac{\nu(y)}{\nu(x)}\right\}$. If it is not accepted, repeat the process (moving from x to y using S with probability $\min\left\{1, \frac{\nu(y)}{\nu(x)}\right\}$) until acceptance. In practice, we observe a very small number of rejects.

The problem of computing the size of the automorphism group of a NFA is investigated in the next session. Assuming it can be done in a reasonable time, an alternative solution to randomly generate NFAs up to isomorphism may be to use a rejection algorithm: randomly and uniformly generate a NFA $\mathcal{A}$ and keep it with probability $\frac{|\mathrm{Aut}(\mathcal{A})|}{n!}$. This way, each class of isomorphism is picked up with the same probability. However, as we will observe in the experiments, most of automata have a very small group of automorphisms, and the number of rejects will be intractable, even for quite small n's.

3.2   Counting Automorphisms

This section is dedicated to show how to compute $|\mathrm{Aut}(\mathcal{A})|$ by using a polynomial number of calls to the isomorphism problem. It is an adaptation of a corresponding result for directed graphs [Mat79].

Let $\mathcal{A} = (Q, \Sigma, \Delta, I, F)$ be a NFA and $Q' \subseteq Q$. Let σ be an arbitrary bijective function from Q' into $\{1, \ldots, |Q'|\}$, a_0 an arbitrary letter in Σ and $\ell = |Q| + |Q'| + 2$. For each state $r \in Q \backslash Q'$ we denote by $\mathcal{A}_r^{Q'}$ the automaton $(Q_r, \Sigma, \Delta_r, I, F)$ where $Q_r = Q \cup \{(p, i) \mid p \in Q \text{ and } 1 \leq i \leq \ell\}$, and $\Delta_r = \Delta \cup \{(p, a, (p, 1)) \mid p \in Q\} \cup \{((p, i), a_0, (p, i + 1)) \mid p \in Q' \text{ and } 1 \leq i < |Q| + 1 + \sigma(p)\} \cup \{((r, i), a_0, (r, i + 1)) \mid 1 \leq i \leq \ell\} \cup \{((p, i), a_0, (p, i + 1)) \mid p \notin Q' \cup \{r\} \text{ and } 1 < i \leq |Q| + 1\}$. Note that the size of $\mathcal{A}_r^{Q'}$ is polynomial in the size of $\mathcal{A}$.

The two next lemma show how to polynomially reduce the problem of counting automorphisms to the isomorphism problem.

Lemma 3. *Let $\mathcal{A} = (Q, \Sigma, \Delta, I, F)$ be a NFA and Q' a non-empty subset of Q. For every $q, q' \in Q \backslash Q'$, there exists $\phi \in \mathrm{Aut}_{Q'}(\mathcal{A})$ such that $\phi(q) = q'$ iff $\mathcal{A}_q^{Q'}$ and $\mathcal{A}_{q'}^{Q'}$ are isomorphic.*

Lemma 4. *Let $\mathcal{A} = (Q, \Sigma, \Delta, I, F)$ be a NFA and Q' a non-empty subset of Q. For every $q \in Q'$, there exists an integer d such that $|\mathrm{Aut}_{Q' \backslash \{q\}}(\mathcal{A})| = d|\mathrm{Aut}_{Q'}(\mathcal{A})|$. Moreover d can be computed with a polynomial number of isomorphism tests between automata of the form $\mathcal{A}_r^{Q' \backslash \{q\}}$.*

Lemma 4 provides a way to compute sizes of automorphism groups by testing whether two NFAs are isomorphic. Indeed, since $\mathrm{Aut}_Q(\mathcal{A})$ is reduced to the identity, and since $\mathrm{Aut}(\mathcal{A}) = \mathrm{Aut}_\emptyset(\mathcal{A})$, one has, by a direct induction using Lemma 4, $\mathrm{Aut}(\mathcal{A}) = d_1 \ldots d_{|Q|}$, where each d_i can be computed by a polynomial number of isomorphism tests. Therefore, the problem of counting automorphism reduces to test whether two automata are isomorphic.

3.3   Isomorphism Problem for Automata with a Bounded Degree

It is proved (not explicitly) in [Boo78] that the isomorphism problem for deterministic automata is polynomially equivalent to the isomorphism problem for directed finite graphs. We prove (Theorem 2) a similar result for NFAs, by using an encoding preserving some bounds on the output degree. Therefore, combining Theorem 2 and Lemma 4, it is possible to compute the size of the automorphism group of an automaton in $\mathfrak{N}_m(n)$, $\mathfrak{N}'_m(n)$, $\mathfrak{N}_m(n)^\bullet$ and $\mathfrak{N}'_m(n)^\bullet$ in time polynomial in n (assuming that m is a constant).

Theorem 2. *Let m be a fixed integer. The isomorphism problem for automata in $\mathfrak{N}_m$ $\mathfrak{N}'_m$, $\mathfrak{N}_m(n)^\bullet$ and $\mathfrak{N}'_m(n)^\bullet$ can be solved in polynomial time.*

Note that the proof is constructive but the exponents are too huge to provide an efficient algorithm. It will be possible to work on a finer encoding but we prefer, in practice, to use labeling techniques described in the next section and that are practically very efficient on graphs (see [Gal14] for a recent survey).

3.4   Practical Computation Using Labelings

For testing graph isomorphism, the most efficient currently used approach is based on labeling [Gal14] and it works practically for large graphs. Intuitively, if two n-state automata are isomorphic, then they have the same number of initial states and of final states. Rather than testing potential $n!$ possible bijections from the automata to point out an isomorphism, it suffices to test $n_1! + n_2! + n_3! + n_4!$ where n_1 is the number of states that are both initial and final, n_2 the number of final states (that are not initial), n_3 the number of initial states (that are not final), and n_4 is the number of states that are neither initial, nor final. With an optimal distribution, the number of tests falls from $n!$ to $4(n/4)!$. This idea can be generalized by the notion of labeling; the goal is to point out easily computable criteria that are stable by isomorphism to get a partition of the set of states and to reduce the search. The approach can be directly adapted for finite automata. A *labeling* is a computable function τ from $\mathfrak{N}(n) \times \{1, \ldots, n\}$ into a finite set D, such that for $\mathcal{A}_1 = (Q, \Sigma, E_1, I_1, F_1)$ and $\mathcal{A}_2 = (Q, \Sigma, E_2, I_2, F_2)$, if φ is an isomorphism from $\mathcal{A}_1$ to $\mathcal{A}_2$, then, for every $i \in \{1, \ldots, n\}$, $\tau(\mathcal{A}_1, i) = \tau(\mathcal{A}_2, \varphi(i))$. The algorithm consists in looking for functions φ preserving τ. If there exists $\alpha \in D$ such that $|\{i \mid \tau(\mathcal{A}_1, i) = \alpha\}| \neq |\{i \mid \tau(\mathcal{A}_2, i) = \alpha\}|$, then the two automata are not isomorphic. Otherwise, all possible bijections preserving the labeling are tested. In the worst case, there are $n!$ possibilities (the labeling doesn't provide any refinement), but in practice, it works very well. The algorithm is depicted in Fig. 2. Note that if τ_1 and τ_2 are two labellings, then $\tau = (\tau_1, \tau_2)$ is a labeling to, allowing the combination of labeling. In our work, we use the following labellings: the labeling testing whether a state is initial, the one testing whether a state is final, the one testing whether a state is both initial and final, the one returning, for each letter a, the number of outgoing transitions labeled by a, the similar one with ongoing transitions, the one returning the minimal word (in the lexical order) from the state to a final state and the one returning the minimal word (in the lexical order) from an initial state to the given state. Using these labellings the practical computation of the sizes of automorphism groups can be done quite efficiently.

> **Input:** $\mathcal{A}_1, \mathcal{A}_2$ in $\mathfrak{N}(n)$, τ a labelling.
> **Output:** 1 if $\mathcal{A}_1$ and $\mathcal{A}_2$ are isomorphic, 0 otherwise.
> $\quad D := \{d_1, \ldots, d_k\}$ is the image of τ.
> $\quad$ **For** α in D
> $\qquad D_\alpha^1 := \{i \mid \tau(\mathcal{A}_1, i) = \alpha\}$
> $\qquad D_\alpha^2 := \{i \mid \tau(\mathcal{A}_2, i) = \alpha\}$
> $\qquad$ **If** $|D_\alpha^1| \neq |D_\alpha^2|$
> $\qquad\qquad$ **Then Return** 0
> $\qquad$ **EndIf**
> $\quad$ **EndFor**
> $\quad$ **For** φ_{d_1} in the set of bijections from $D_{d_1}^1$ into $D_{d_1}^2$
> $\qquad \ldots$
> $\qquad$ **For** φ_{d_k} in the set of bijections from $D_{d_k}^1$ into $D_{d_k}^2$
> $\qquad\qquad \varphi$ is the bijection such that $\varphi_{|D_j^1} = \varphi_{d_j}$
> $\qquad\qquad$ **If** $\varphi(\mathcal{A}_1) = \varphi(\mathcal{A}_2)$, **Then Return** 1
> $\qquad$ **EndFor**
> $\qquad \ldots$
> $\quad$ **EndFor**
> $\quad$ **Return** 0

Fig. 2. Testing isomorphism using labellings

3.5   Experiments

The experiments were made on a personal computer with processor `IntelCore i3-4150 CPU 3.50GHz x 4`, 7.7 GB of memory and running on a 64 bits Ubuntu 14.04 OS. The implementation is a non optimized prototype written in Python.

The first experimentation consists in measuring the time required to move into the Metropolis chains for $\mathfrak{N}(n)$ and $\mathfrak{N}_m(m)$. Results are reported in Table 1. The labellings used are those described in Sect. 3.4. These preliminary results show that using a 2 or 3-letter alphabet does not seem to have a significant influence. For each generation, the n^3-th elements of the walk is returned, with an arbitrary start. Moreover, bounding or not the degree does not seem to be relevant for the computation time. Note that we do not use any optimization: several computations on labellings may be reused when moving into the chain. Moreover, Python is not an efficient programming language (compared to C or Java). In practice, for directed graphs, the isomorphism problem is tractable for large graphs (see for instance [FPSV09]). Note that the number of moves (n^3) is the major factor for the increasing computation time (relatively to n): the average time for moving a single step is multiplied by about (only) 10 from $n = 20$ to $n = 90$.

Table 1. Average Time (s) to Sample a NFA in $\mathfrak{N}(n)$ (left) and in $\mathfrak{N}'_m(n)$ (right).

n	10	20	50	70	90
$\lvert A\rvert = 2$	0.02	0.43	32.5	166.1	569.9
$\lvert A\rvert = 3$	0.02	0.56	47.1	248.4	848.1

n	10	20	50	70	90
$m = 2, \lvert A\rvert = 2$	0.2	0.43	32.5	166.1	566.8
$m = 2, \lvert A\rvert = 3$	0.2	0.57	47.0	246.7	847.2
$m = 3, \lvert A\rvert = 2$	0.2	0.43	33.0	167.8	561.9
$m = 3, \lvert A\rvert = 3$	0.2	0.57	47.2	248.6	851.3

In the second experiment, we also generate automata with n states by returning the n^3-th element of the walk in the Metropolis Chain. We use the algorithm to estimate the sizes of the automorphism group. By generating 1000 automata on a 2-letter alphabet, with 5, 7, 10, 15, 20 and 50 states, for each case, all the automata have a trivial automorphism group but one or two automata that have an automorphism group of size 2.

For the last experience, we propose to compare our generation for $\mathfrak{N}'_2(n)^\bullet$ with the generator proposed in [TV05] with a density of a-transitions of 2 and 3. The parameter of the algorithm is a probability p_f for final states and a density σ on a-transitions: the set of states of the automaton is $\{1, \ldots, n\}$, only 1 is the initial state, each state is final with a probability p_f and for each p and each a, (p, a, q) is a transition with a probability $\frac{\sigma}{n}$. Therefore for each state and each letter, the expected number of outgoing transitions labeled by this letter is σ. We run this algorithm with $p_f = 0.2$ and $\sigma \in \{1.5, 2, 3\}$. For each size, we compute the average size s of the corresponding minimal automata. We use a two letter alphabet and the average sizes (number of states) are obtained by sampling 1000 automata for each case. Results are reported in Table 2.

Table 2. Average sizes of deterministic and minimal automata corresponding to automata sampling using [TV05] and in $\mathfrak{N}'_2(n)^\bullet$.

$\sigma = 1.5, n =$	5	8	11	14	17	20
s	1.5	4.3	4.7	3.8	3.1	2.7
$\sigma = 2, n =$	5	8	11	14	17	20
s	1.3	3.0	4.8	5.1	4.5	4.0
$\sigma = 3, n =$	5	8	11	14	17	20
s	2.8	4.8	4.7	3.8	3.4	3.0

$\mathfrak{N}'_2(n)^\bullet, n =$	5	8	11	14	17	20
s	3.7	6.1	7.9	10.0	11.5	13.9

One can observe that the generator provides quite different automata. With the Markov chain approach the sizes of the related minimal automata are greater, even if there is no blow-up in both cases.

4 Conclusion

In this paper we proposed a Markov Chain approach to randomly generate non deterministic automata (up to isomorphism) for several classes of NFAs. We showed that moving into these Markov chains can be done quite quickly in practice and, in some interesting cases, in polynomial time. Experiments have been performed whithin a non optimized prototype and, following known experimental results on group isomorphism, they allow us to think that the approach can be used on much larger automata. Implementing such techniques using an efficient programming language is a challenging perspective. Moreover, the proposed approach is very flexible and can be applied to various classes of NFAs. An interesting research direction is to design particular subclasses of NFAs that look like NFAs occurring in practical applications, even if this last notion is hard to define. We think that the classes $\mathfrak{N}'_m(n)^\bullet$ and $\mathfrak{N}_m(n)^\bullet$ constitute first attempts in this direction. Theoretically -as often for Monte-Carlo approach-, computing mixing and strong stationary times are crucial and difficult questions we plan to investigate more deeply.

References

[AMR07]   Almeida, M., Moreira, N., Reis, R.: Enumeration and generation with a string automata representation. Theor. Comput. Sci. **387**(2), 93–102 (2007)

[BN07]   Bassino, F., Nicaud, C.: Enumeration and random generation of accessible automata. Theor. Comput. Sci. **381**(1–3), 86–104 (2007)

[Boo78]   Booth, K.S.: Isomorphism testing for graphs, semigroups, and finite automata are polynomially equivalent problems. SIAM J. Comput. **7**(3), 273–279 (1978)

[CF11]   Carnino, V., De Felice, S.: Random generation of deterministic acyclic automata using markov chains. In: Bouchou-Markhoff, B., Caron, P., Champarnaud, J.-M., Maurel, D. (eds.) CIAA 2011. LNCS, vol. 6807, pp. 65–75. Springer, Heidelberg (2011)

[CF12]   Carnino, V., De Felice, S.: Sampling different kinds of acyclic automata using markov chains. Theor. Comput. Sci. **450**, 31–42 (2012)

[CG95]   Chib, S., Greenberg, E.: Understanding the metropolis-hastings al- gorithm. Am. Stat. **49**, 327–335 (1995)

[CHPZ02] Champarnaud, J-M., Hansel, G., Paranthoën, T., Ziadi, D.: NFAS bitstream-based random generation. In: Fourth International Workshop on Descriptional Complexity of Formal Systems - DCFS (2002), pp. 81–94 (2002)

[CN12]   Carayol, A., Nicaud, C.: Distribution of the number of accessible states in a random deterministic automaton. In: STACS 2012, vol. 14 of LIPIcs, pp. 194–205 (2012)

[CP05] Champarnaud, J.-M., Paranthoën, Th: Random generation of dfas. Theor. Comput. Sci. **330**(2), 221–235 (2005)

[DLW08] Yuval Peres, D.A.L., Wilmer, E.L.: Markov Chain and Mixing Times. American Mathematical Society (2008). http://pages.uoregon.edu/dlevin/ MARKOV/markovmixing.pdf

[FPSV09] Foggia, P., Percannella, G., Sansone, C., Vento, M.: Benchmarking graph-based clustering algorithms. Image Vis. Comput. **27**(7), 979–988 (2009)

[Gal14] Gallian, J.A.: A dynamic survey of graph labeling. The Electronic Journal of Combinatorics. 17 (2014)

[HU79] Hopcroft, J., Ullman, J.: Introduction to Automata Theory: Languages and Computation. Addison-Wesley, Boston (1979)

[Luk82] Luks, E.M.: Isomorphism of graphs of bounded valence can be tested in polynomial time. J. Comput. Syst. Sci. **25**(1), 42–65 (1982)

[Mat79] Mathon, Rudolf: A note on the graph isomorphism counting problem. Inf. Process. Lett. **8**(3), 131–132 (1979)

[MU05] Mitzenmacher, M., Upfal, E.: Probability and Computing. Cambridge University Press, Cambridge (2005)

[Nic09] Nicaud, C.: On the average size of glushkov's automata. In: Dediu, A.H., Ionescu, A.M., Martín-Vide, C. (eds.) LATA 2009. LNCS, vol. 5457, pp. 626–637. Springer, Heidelberg (2009)

[Nic14] Nicaud, C.: Random deterministic automata. In: Csuhaj-Varjú, E., Dietzfelbinger, M., Ésik, Z. (eds.) MFCS 2014, Part I. LNCS, vol. 8634, pp. 5–23. Springer, Heidelberg (2014)

[NPR10] Nicaud, C., Pivoteau, C., Razet, B.: Average analysis of glushkov automata under a bst-like model. In: IARCS Annual Conference on Foundations of Software Technology and Theoretical Computer Science (FSTTCS 2010) LIPIcs, pp. 388–399 (2010)

[TV05] Tabakov, D., Vardi, M.Y.: Experimental evaluation of classical automata constructions. In: Sutcliffe, G., Voronkov, A. (eds.) LPAR 2005. LNCS (LNAI), vol. 3835, pp. 396–411. Springer, Heidelberg (2005)

Random Generation and Enumeration of Accessible Deterministic Real-Time Pushdown Automata

Pierre-Cyrille Héam[(⊠)] and Jean-Luc Joly

FEMTO-ST, CNRS UMR 6174, Université de Franche-Comté, INRIA,
16 Route de Gray, 25030 Besançon Cedex, France
{pheam,jean-luc.joly}@femto-st.fr

Abstract. This paper presents a general framework for the uniform random generation of deterministic real-time accessible pushdown automata. A polynomial time algorithm to randomly generate a pushdown automaton having a fixed stack operations total size is proposed. The influence of the accepting condition (empty stack, final state) on the reachability of the generated automata is investigated.

1 Introduction

Finite automata, of any kind, are widely used for their algorithmic properties in many fields of computer science like model-checking, pattern matching and machine learning. Developing new efficient algorithms for finite automata is therefore a challenging problem still addressed by many recent papers. New algorithms are frequently motivated by improvement of worst cases bound. However, several examples, such as sorting algorithms, primality testing or solving linear problems, show that worst case complexity is not always the right way to evaluate the practical performance of an algorithm. When benchmarks are not available, random testing, with a controlled distribution, represents an efficient mean of performance testing. In this context, the problem of uniformly generating finite automata is a challenging problem.

This paper tackles with the problem of the uniform random generation of real-time deterministic pushdown automata. Using classical combinatorial techniques, we expose how to extend existing works on the generation of finite deterministic automata to pushdown automata. More precisely, we show in Sect. 3 how to uniformly generate and enumerate (in the complete case) accessible real-time deterministic pushdown automata. In Sect. 4, it is shown that using a rejection algorithm it is possible to efficiently generate pushdown automata that don't accept an empty language. The influence of the accepting condition (final state or empty-stack) on the reachability of the generated pushdown automata is also experimentally studied in Sect. 4.

Related Work. The enumeration of deterministic finite automata has been first investigated in [Vys59] and was applied to several subclasses of deterministic

© Springer International Publishing Switzerland 2015
F. Drewes (Ed.): CIAA 2015, LNCS 9223, pp. 153–164, 2015.
DOI: 10.1007/978-3-319-22360-5_13

finite automata [Kor78,Kor86,Rob85,Lis06]. The uniform random generation of accessible deterministic complete automata was initially proposed in [Nic00] for two-letter alphabets and the approach was extended to larger alphabets in [CP05]. Better algorithms can be found in [BN07,CN12]. The random generation of possibly incomplete automata is analyzed in [BDN09]. The recent paper [CF11] presents how to use Monte-Carlo approaches to generate deterministic acyclic automata. As far as we know, the only work focusing on the random generation of deterministic transducers is [HNS10]. This work can be applied to the random generation of deterministic real-time pushdown automata that can be possibly incomplete. However the requirement to fix the size of the stack operation on each transition, represents a major restriction. The reader interested in the random generation of deterministic automata is referred to the survey [Nic14].

2   Formal Background

We assume that the reader is familiar with classical notions on formal languages. For more information on automata theory or on pushdown automata the reader is referred to [HU79] or to [Sak09]. For a general reference on random generation and enumeration of combinatorial structures see [FZC94]. For any word w on an alphabet Σ, $|w|$ denotes its length. The empty word is denoted ε. The cardinal of a finite set X is denoted $|X|$.

Deterministic Finite Automata. A *deterministic finite automaton* on Σ is a tuple $(Q, \Sigma, \delta, q_{init}, F)$ where Q is a finite set of states, Σ is a finite alphabet, $q_{init} \in Q$ is the initial state, $F \subseteq Q$ is the set of final states and δ is a partial function from $Q \times \Sigma$ into Q. If δ is not partial, i.e., defined for each $(q, a) \in Q \times \Sigma$, the automaton is said *complete*. A triplet of the form $(q, a, \delta(p, a))$ is called a *transition*. A finite automaton is graphically represented by a labeled finite graph whose vertices are the states of the automaton and edges are the transitions. A deterministic finite automaton is *accessible* if for each state q there exists a path from the initial state to q. Two finite automata $(Q_1, \Sigma, \delta_1, q_{init1}, F_1)$ and $(Q_2, \Sigma, \delta_2, q_{init2}, F_2)$ are isomorphic if they are identical up to the state's names, formally if there exists a one-to-one function φ from Q_1 into Q_2 such that (1) $\varphi(q_{init1}) = q_{init2}$, (2) $\varphi(F_1) = F_2$, and (3) $\delta_1(q, a) = p$ iff $\delta_2(\varphi(q), a) = \varphi(p)$.

Pushdown Automata. A *real-time deterministic pushdown automaton*, RDPDA for short, is a tuple $(Q, \Sigma, \Gamma, Z_{init}, \delta, q_{init}, F)$ where Q is a finite set of states, Σ and Γ are finite disjoint alphabets, $q_{init} \in Q$ is the initial state, $F \subseteq Q$ is the set of final states, Z_{init} is the initial stack symbol and δ is a partial function from $Q \times (\Sigma \times \Gamma)$ into $Q \times \Gamma^*$. If δ is not partial, i.e., defined for each $(q, (a, X)) \in Q \times (\Sigma \times \Gamma)$, the RDPDA is said to be *complete*. A triplet of the form $(q, (a, X), w, p)$ with $\delta(q, (a, X)) = (p, w)$ is called a *transition* and w is the *output* of the transition. The *output size* of a transition $(q, (a, X), w, p)$ is the length of w. The *output size* of a RDPDA is the sum of the sizes of its

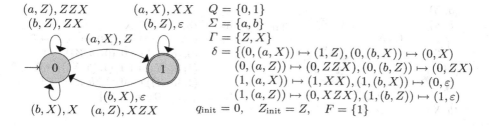

$$\begin{aligned}
&Q = \{0,1\}\\
&\Sigma = \{a,b\}\\
&\Gamma = \{Z,X\}\\
&\delta = \{(0,(a,X)) \mapsto (1,Z),(0,(b,X)) \mapsto (0,X)\\
&\qquad (0,(a,Z)) \mapsto (0,ZZX),(0,(b,Z)) \mapsto (0,ZX)\\
&\qquad (1,(a,X)) \mapsto (1,XX),(1,(b,X)) \mapsto (0,\varepsilon)\\
&\qquad (1,(a,Z)) \mapsto (0,XZX),(1,(b,Z)) \mapsto (1,\varepsilon)\\
&q_{\text{init}} = 0, \quad Z_{\text{init}} = Z, \quad F = \{1\}
\end{aligned}$$

Fig. 1. P_{toy}, a complete RDPDA.

transitions. The *underlying automaton* of an RDPDA, is the finite automaton $(Q, \Sigma \times \Gamma, \delta', q_{\text{init}}, F)$, with $\delta'(q, (a, X)) = p$ iff $\delta((q,(a,X))) = (p,w)$ for some $w \in \Gamma^*$. An RDPDA is *accessible* if its underlying automaton is accessible. A transition whose output is ε is called a *pop transition*. An example of a complete accessible RDPDA is depicted in Fig. 1. The related underlying finite automaton is depicted in Fig. 2.

A *configuration* of a RDPDA is an element of $Q \times \Gamma^*$. The *initial configuration* is $(q_{\text{init}}, Z_{\text{init}})$. Two configurations (q_1, w_1) and (q_2, w_2) are a-consecutive, denoted $(q_1, w_1) \models_a (q_2, w_2)$ if the following conditions are satisfied:

- $w_1 \neq \varepsilon$, and let $w_1 = w_3 X$ with $X \in \Gamma$,
- $\delta(q_1, (X, a)) = (q_2, w_4)$ and $w_2 = w_3 w_4$.

Two configurations are *consecutive* if there is a letter a such that there are a-consecutive. A state p of a RDPDA is *reachable* if there exists a sequence of consecutive configurations $(p_1, w_1), \dots, (p_n, w_n)$ such that (p_1, w_1) is the initial configuration and $p_n = p$. Moreover, if $w_n = \varepsilon$, p is said to be *reachable with an empty stack*. A RDPDA is *reachable* if all its states are reachable. Consider for instance the RDPDA of Fig. 3, where the initial stack symbol is X. State 3 is not reachable since the transition from 0 to 3 cannot be fired. State 1 is reachable with an empty stack. State 2 is reachable, but not reachable with an empty stack. Note that a reachable state is accessible, but the converse is not true in general: accessibility is a notion defined on the underlying finite automaton.

The configurations $(1, XZ)$ and $(2, XXZX)$ are a-consecutive on the RDPDA depicted in Fig. 1.

There are three main kinds of accepting conditions for a word $u = a_1 \dots a_k \in \Sigma^*$ by an RDPDA:

- Under the *empty-stack condition*, u is accepted if there exists configurations $c_1, \dots, c_{k+1}$ such that c_1 is the initial configuration, c_i and c_{i+1} are a_i-consecutive, and c_{k+1} is of the form (q, ε).
- Under the *final-state condition*, u is accepted if there exists configurations $c_1, \dots, c_{k+1}$ such that c_1 is the initial configuration, c_i and c_{i+1} are a_i-consecutive, and c_{k+1} is of the form (q, w), with $q \in F$.

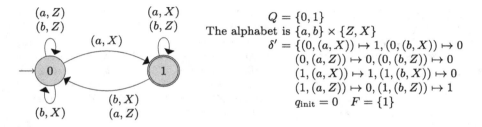

$$Q = \{0, 1\}$$
The alphabet is $\{a, b\} \times \{Z, X\}$
$$\delta' = \{(0, (a, X)) \mapsto 1, (0, (b, X)) \mapsto 0$$
$$(0, (a, Z)) \mapsto 0, (0, (b, Z)) \mapsto 0$$
$$(1, (a, X)) \mapsto 1, (1, (b, X)) \mapsto 0$$
$$(1, (a, Z)) \mapsto 0, (1, (b, Z)) \mapsto 1$$
$$q_{\text{init}} = 0 \quad F = \{1\}$$

Fig. 2. Underlying automaton of P_{toy}.

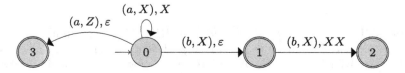

Fig. 3. Acceptance conditions.

– Under the *final-state and empty-stack condition*, u is accepted if there exists configurations $c_1, \ldots, c_{k+1}$ such that c_1 is the initial configuration, c_i and c_{i+1} are a_i-consecutive, and c_{k+1} is of the form (q, ε), with $q \in F$.

Consider for instance the RDPDA of Fig. 3, where the initial stack symbol is X. With the empty-stack condition only the word b is accepted, as well as for the empty-stack and final state condition. With the final state condition, the accepted language is $a^*(b + bb)$.

Two RDPDA $(Q_1, \Sigma, \Gamma, \delta_1, q_{\text{init}1}, F_1)$ and $(Q_2, \Sigma, \Gamma, \delta_2, q_{\text{init}2}, F_2)$ are isomorphic if there exists a one-to-one function φ from Q_1 into Q_2 such that (i) $\varphi(q_{\text{init}1}) = q_{\text{init}2}$, and (ii) $\varphi(F_1) = F_2$, and

$$(iii) \quad \delta_1(q, (a, X)) = (p, w) \quad \text{iff} \quad \delta_2(\varphi(q), (a, X)) = (\varphi(p), w).$$

Note that if two RDPDA are isomorphic, then their underlying automata are isomorphic too.

Generating Functions. A *combinatorial class* is a class $\mathfrak{C}$ of objects associated with a size function $|.|$ from $\mathfrak{C}$ into $\mathbb{N}$ such that for any integer n there are finitely many elements of $\mathfrak{C}$ of size n. The ordinary generating function for $\mathfrak{C}$ is $C(z) = \sum_{c \in \mathfrak{C}} z^{|c|}$. The n-th coefficient of $C(z)$ is exactly the number of objects of size n and is denoted $[z^n]C(z)$. The reader is referred to [FS08] for the general methodology of analytic combinatorics, ans especially the use of generating functions to count objects. The following result [FS08, Theorem VIII.8] will be useful in this paper.

Theorem 1. *Let $C(z)$ be an ordinary generating function satisfying: (1) $C(z)$ is analytic at 0 and have only positive coefficients, (2) $C(0) \neq 0$ and (3) $C(z)$ is*

*aperiodic. Let R be the radius of convergence of $C(z)$ and $T = \lim_{x \to R^-} x \frac{C'(x)}{C(x)}$.
Let $\lambda \in]0, T[$ and ζ be the unique solution of $x \frac{C'(x)}{C(x)} = \lambda$. Then, for $N = \lambda n$ an
integer, one has*

$$[z^N]C(z)^n = \frac{C(\zeta)^n}{\zeta^{N+1}\sqrt{2\pi n \xi}}(1 + o(1)),$$

where $\xi = \frac{d^2}{dz^2}(\log C(z) - \lambda \log(z))|_{z=\zeta}$.

In the above theorem, T is called the *spread* of the $C(z)$.

Rejection Algorithms. A rejection algorithm is a probabilistic algorithm to randomly generate an element in a set X, using an algorithm A for generating an
element of Y in the simple way: repeat A until it returns an element of X. Such
an algorithm is tractable if the expected number of iterations can be kept under
control (for instance is fixed): the probability that an element of Y is in X has
to be large enough.

Random Generation. The theory of Generating Functions provides an efficient
way to randomly and uniformly generate an element of size n of a combinatorial
class $\mathfrak{C}$ using a recursive approach [FS08]. It requires a $O(n^2)$ precomputation
time and each random sample is obtained in time $O(n \log n)$. Another efficient
way to uniformly generate element of $\mathfrak{C}$ is to use Boltzmann samplers [DFLS04]:
the random generation of an object with a size about n (approximate sampling)
is performed in $O(n)$, while the random generation of an object with a size
exactly n is performed in expected time $O(n^2)$ using a rejection algorithm (without precomputation). Boltzmann samplers are quite easy to implement but are
restricted to a limited number of combinatorial constructions. They also require
the evaluation of some generating functions at some values of the variable.

3   Random Generation and Enumeration of RDPDA

In this section, Σ and Γ are fixed disjoint alphabets of respective cardinals α
and β. We denote by $\mathfrak{T}_{s,n,m}$ combinatorial class of the RDPDA (on Σ, Γ) with n
states, s transitions and with an output size of m, up to isomorphism. Note that
this class is well defined since two isomorphic RDPDA have the same output
size. Let $\rho = \alpha\beta$.

3.1   Enumeration of RDPDA

We are interested in the random generation of accessible RDPDA up to isomorphism. The class of all isomorphic classes of accessible automata on $\Sigma \times \Gamma$ with
n states and s transitions is denoted $\mathfrak{A}_{s,n}$. Let ψ be the function from $\mathfrak{T}_{s,n,m}$ into
$\mathfrak{A}_{s,n}$ mapping RDPDA to their underlying automata. The number of elements of
$\psi^{-1}(\mathcal{A})$, where $\mathcal{A}$ is an element of $\mathfrak{A}_{s,n}$, is the number of possible output labelling
of the s transitions of $\mathcal{A}$, which only depends on s and m and is independent of
$\mathcal{A}$. We denote by $c_{s,m}$ this number of labelings. The following proposition is a
direct consequence of the above remark.

Proposition 1. *One has* $|\mathfrak{T}_{s,n,m}| = |\mathfrak{A}_{s,n}| \cdot c_{s,m}$.

Proposition 1 is the base of the enumeration for RDPDA. Let $M(z)$ denote the ordinary generating function of elements of Γ^*. Then, using Proposition 1 and classical constructions on generating functions, one has

$$|\mathfrak{T}_{s,n,m}(\mathcal{M})| = |\mathfrak{A}_{s,n}|.[z^m]M(z)^s. \tag{1}$$

Equation (1) will be exploited for complete finite automata using the following result of [Kor78] – see also [BDN07].

Theorem 2 ([Kor78]).   *There exists a constant* γ_ρ, *such that* $|\mathfrak{A}_{\rho n,n}| \sim n\gamma_\rho \{{\rho n \atop n}\}$, *where* $\{{x \atop y}\}$ *denotes the Stirling numbers of the second kind.*

The case $s = \rho n$ corresponds to complete accessible automata. We will particularly focus throughout this paper on the complete case for a fixed average size λ for the transitions: we assume that there is a fixed $\lambda > 0$ such that $m = \lambda s = \lambda n\rho$. In this context, Eq. (1) becomes

$$|\mathfrak{T}_{s,n,m}(\mathcal{M})| \sim n\gamma_\rho \left\{{\rho n \atop n}\right\}[z^{\lambda n\rho}]M(z)^{n\rho}. \tag{2}$$

Since $M(z)$ is the generating function of words on Γ, one has $M(z) = \frac{1}{1-\beta z}$. Therefore (see [FS08]), one has:

$$[z^m]F_s(z) = \beta^m \frac{s(s+1)(s+2)\ldots(s+m-1)}{m!}. \tag{3}$$

The generating function F_s satisfies the hypotheses of Theorem 1, with an infinite spread. Therefore for any strictly positive λ and any integer $m = \lambda s$, one has

$$[z^{\lambda s}]\,F_s(z) = \frac{C(\zeta)^s}{\zeta^{\lambda s+1}\sqrt{2\pi s\xi}}(1+o(1)), \tag{4}$$

where $C(z) = \frac{1}{1-\beta z}$ and $\zeta = \frac{\lambda}{\beta(\lambda+1)}$ is the unique solution of $x\frac{C'(x)}{C(x)} = \lambda$. It follows that

$$[z^{\lambda s}]F_s(z) = \frac{(\lambda+1)^{(\lambda+1)s+1}\beta^{\lambda s+1}}{\lambda^{\lambda s+1}\sqrt{2\pi s\xi}}(1+o(1)). \tag{5}$$

In addition

$$\xi = \frac{d^2}{dz^2}\left(\log C(z) - \lambda \log(z)\right)|_{z=\zeta} = \frac{(\lambda+1)^3\beta^2}{\lambda}. \tag{6}$$

Consequently Eq. (5) can be rewritten as

$$[z^{\lambda s}]F_s(z) = \frac{(\lambda+1)^{(\lambda+1)\,s-1/2}\,\beta^{\lambda s}}{\lambda^{\lambda s+1/2}\sqrt{2\pi s}}(1+o(1)). \tag{7}$$

The following proposition is a direct consequence of the combination of Eq. (7), Proposition 1 and Theorem 2.

Proposition 2. *The number $f_{\lambda,n}$ of complete accessible RDPDA with n states and with an output size of λn, with $\lambda \geq 1$ a fixed rational number, satisfies*

$$f_{\lambda,n} = \gamma_\rho n \left\{ {\rho n \atop n} \right\} \frac{(\lambda+1)^{(\lambda+1)n\,\rho-1/2}\,\beta^{\lambda\,n\,\rho}}{\lambda^{\lambda\,n\rho+1/2}\sqrt{2\,\pi\,n\,\rho}}(1+o(1)),$$

Note that $f_{\lambda,n} = |\mathfrak{T}_{n\rho,n,\lambda\rho n}|$. The following result can be easily obtained.

Proposition 3. *For RDPDA in $\mathfrak{T}_{s,n,m}$, the average number of pop transitions is $\frac{s(s-1)}{s+m-1}$.*

Proof. We introduce the generating function $G_s(z,u) = \left(\frac{1}{1-\beta z} - 1 + u \right)^s$ counting RDPDA where z counts the output size and u the number of pop transitions. Using [FS08, Proposition III.2], the average number of pop transitions is $\frac{[z^m]\frac{\partial}{\partial u}G_s(z,u)_{|u=1}}{[z^m]G_s(z,1)}$. Since $\frac{\partial}{\partial u}G_s(z,u) = sG_{s-1}(z,u)$, the proposition is a direct consequence of Eq. (3).

3.2   Random Generation

The random generation is also based on Proposition 1; the general schema for the uniform random generation of an element of $\mathfrak{T}_{s,n,m}$ consists of two steps:

1. Generate uniformly an element $\mathcal{A}$ of $\mathfrak{A}_{s,n}$.
2. Generate the output of the transitions of $\mathcal{A}$ such that the sum of their sizes is m.

The first step can be performed in the general case using [HNS10]. For generating complete deterministic RDPDA – when $s = n\alpha$ – faster algorithms are described in [BDN07, CN12]. In the general case, the complexity is $O(n^3)$ and for the complete case, the complexity falls to $O(n^{\frac{3}{2}})$. The second step can be easily done using the classical recursive approach as described in [FZC94] or using Boltzmann samplers.

With a non-optimized Python implementation running on a 2.5 GHz personal computer it is possible to generate 100 complete RDPDA with hundreds of states in few minutes.

4   Influence of the Accepting Condition

Accessibility defined for a RDPDA does not mean that the accessible states can be reached by a calculus. Therefore the random generation may produce semantically RDPDA simpler than wanted. One of the requirements may be to generate RDPDA accepting non empty languages. Another requirement is to produce only reachable states. Finally, if the final state-empty stack accepting condition is chosen, it is frequently required that final states are empty stack reachable.

4.1 Emptiness of Accepted Languages

Proposition 4. *Whatever the selected accepting condition, the probability that an accessible RDPDA with n states, s transitions and an output size of m, accepts a non empty language is greater or equal to $\frac{s-1}{2\beta(s+m-1)}$.*

Proof. Since the considered RDPDA are accessible, there is at least one outgoing transition from the initial state. We will evaluate the probability that this transition is of the form $(q_{\text{init}}, a, Z_{\text{init}}, \varepsilon, p)$ with p final. There is no condition on a. The probability that the stack symbol is Z_{init} is $\frac{1}{\beta}$ since all letters have the same role. The probability that p is final is $\frac{1}{2}$ (see [CP05, BN07]). By Proposition 3 the probability that this transition is a pop transition is $\frac{s-1}{s+m-1}$. It follows that the probability that the transition has the claimed form is $\frac{s-1}{2\beta(s+m-1)}$. If this transition exists, the RDPDA accepts the word a, proving the proposition. □

By Proposition 4, if $m = \lambda s$, for a fixed λ, then RDPDA accepting non-empty languages can be randomly generated by a rejection algorithm, with an expected constant number of rejects. Experiments show that most of the states are reachable (see Tables 3 and 4).

4.2 Empty-Stack Reachability

Proposition 4 shows it is possible to generate complete RDPDA accepting a non-empty language (if m and s are of the same order). However, it doesn't suffice since many states of generated automata can be unreachable. Under the final-state and empty-stack condition of acceptance, a final state that is not reachable with an empty stack is a useless final state, i.e. it cannot be used as a final state to recognize a word – but it can be involved as any state for accepting a word. Using for instance [FWW97], one can decide in polynomial time whether a state is reachable with an empty stack.

Table 1 reports experiments on the average number of empty-stack reachable states. For this experiment, we consider complete and accessible RDPDA with $\alpha = 2$, and $\beta = 2$. Since a state is final with a probability $1/2$, dividing the number by 2 in the table provides the average number of final states reachable with

Table 1. Average number of reachable states with an empty stack, $\alpha = 2$, $\beta = 2$

Number of states →	5	10	15	20	30	40	60	100
$\lambda = 0.5$	3.56	6.14	8.02	11.1	16.26	15	24.7	49.5
$\lambda = 1$	2.6	4.62	4.7	6.16	7.06	7.82	13.85	17.3
$\lambda = 1.5$	2.36	3.05	3.61	3.62	5.2	5.5	5.68	5.8
$\lambda = 2$	2.0	2.6	2.81	3.4	3.02	3.1	3.21	3.89
$\lambda = 3$	1.65	1.8	1.83	1.8	2.26	2.44	2.34	2.6
$\lambda = 5$	1.3	1.41	1.43	1.4	1.42	2.1	1.5	1.5

an empty stack. Experiments show that if λ is greater than 1, then the average number of states reachable with an empty stack is quite small. Remind that λ is the average size of the outputs. For each case, 100 complete RDPDA have been generated. Clearly the random generation of complete accessible RDPDA based on the sizes of the output will not produce enough pop-transitions to empty the stack. Adding a criterion on a minimal number k of pop-transitions may be a solution that can be achieved in the following way:

1. Choose uniformly k transitions of the underlying finite automata that will be pop-transitions.
2. Decorate the $s - k$ other transitions with strings for a total size of m.

The experiments reported in Table 1 has also been done for this procedure to compare the number of rejects. We choose the case $\alpha = \beta = 2$ again with at least 40 % of pop-transitions. The results are reported in Table 2.

Table 2. Average number of reachable states (at least 40 % of pop transitions); $\alpha = \beta = 2$.

Number of states →	5	10	15	20	30	40	60	100
$\lambda = 0.5$	4.05	7.75	11.54	16.09	23.63	30.87	46.83	80.62
$\lambda = 1$	3.09	5.25	8.19	10.78	13.97	22.55	31.23	44.52
$\lambda = 1.5$	2.94	4.44	5.91	8.24	9.06	11.93	18.77	29.58
$\lambda = 2$	2.68	4.44	5.19	6.5	8.53	9.39	12.84	19.91
$\lambda = 3$	2.53	3.29	4.38	4.81	6.37	7.49	8.72	12.18
$\lambda = 5$	2.29	3.31	3.81	4.53	5.09	4.71	5.28	6.42

Imposing a minimal number of pop transitions improves the efficiency (relative to the number of reachable state) for small values of λ. However it is not sufficient when $\lambda \geq 1$.

4.3   Reachability (with No Stack Condition)

If we are now interested in the random generation with the *final state* condition, regardless of the stack, it is interesting to know the number of reachable states (which is on average twice the number of final reachable states). Using [FWW97], the average number of reachable states have been assessed experimentally. Results are reported in Table 3 for $\alpha = 2$ and $\beta = 2$ and in Table 4 for $\alpha = 3$ and $\beta = 5$.

For the random generation of complete accessible RDPDA with a final state accepting condition, our framework seems to be suitable: most of the states are reachable. For the two other accepting conditions, the value of λ has to be small. Otherwise, there will be too few pop transitions to clean out the stack. Proposition 3 confirms this outcome since averagely the number of pop transitions is close to $\frac{\alpha \beta n}{\lambda + 1}$.

Table 3. Average number of reachable states, $\alpha = 2$ and $\beta = 2$.

Number of states →	10	20	30	40	50	60	80	100
$\lambda = 1$	8.29	14.89	21.5	26.93	32.48	35.55	44.4	52.86
$\lambda = 2$	8.73	16.35	25.3	33.45	39.56	47.99	62.32	81.02
$\lambda = 3$	8.84	17.67	27.14	36.19	45.7	54.69	73.35	89.26
$\lambda = 5$	9.23	18.3	28.06	37.47	47.61	56.7	76.11	95.15

Table 4. Average number of reachable states, $\alpha = 3$ and $\beta = 5$.

Number of states →	10	20	30	40	50	60	80	100
$\lambda = 1$	9.72	19.48	29.34	39.71	49.16	59.49	79.6	99.4
$\lambda = 2$	9.9	19.7	29.9	39.8	49.4	59.6	79.7	99.5
$\lambda = 3$	9.93	19.9	29.92	39.9	49.81	59.9	79.6	99.1
$\lambda = 5$	9.95	19.96	29.93	39.9	49.86	59.89	79.79	99.75

Table 5. Average number of random generations to obtain a reachable RDPDA.

	Number of states →	10	20	30	40	50	60	80	100
$\alpha = 2$	$\lambda = 1$	293.1	-	-	-	-	-	-	-
$\beta = 2$	$\lambda = 1.5$	24.6	88.8	278.0	-	-	-	-	-
	$\lambda = 2$	6.9	20.3	14.9	13.2	17.9	25.2	65.9	95.8
	$\lambda = 3$	1.6	1.5	1.8	2.0	1.9	2.2	2.1	2.1
	$\lambda = 5$	1.1	1.2	1.1	1.2	1.1	1.2	1.2	1.2
$\alpha = 4$	$\lambda = 1$	3.8	5.5	9.4	15.7	38.4	39.3	76.9	76.9
$\beta = 2$	$\lambda = 1.5$	1.7	2.0	1.7	1.8	1.9	2.0	2.0	1.9
	$\lambda = 2$	1.3	1.3	1.31	1.3	1.2	1.2	1.1	1.2
	$\lambda = 3$	1.1	1.1	1.1	1.1	1.0	1.1	1.1	1.1
	$\lambda = 5$	1.0	1.0	1.0	1.0	1.0	1.0	1.0	1.0
$\alpha = 2$	$\lambda = 1$	2.3	5.5	16.9	23.8	20.1	44.1	80.4	214.2
$\beta = 4$	$\lambda = 1.5$	1.7	1.8	1.2	2.6	3.0	2.0	1.6	1.9
	$\lambda = 2$	1.4	1.2	1.2	1.1	1.3	1.4	1.3	1.1
	$\lambda = 3$	1.0	1.2	1.2	1.0	1.1	1.1	1.0	1.2
	$\lambda = 5$	1.1	1.0	1.0	1.0	1.0	1.0	1.0	1.0

4.4   A Rejection Algorithm for Reachable Complete RDPDA

A natural question is to consider how to generate a complete accessible RDPDA with exactly n reachable states. An easy way would be to use a rejection approach by generating a complete accessible RDPDA with n states until obtaining a reachable RDPDA. Results presented in Tables 3 and 4 seems to prove that this approach might be fruitful for the parameters of Table 4 but more difficult for

Table 6. Random generation of RDPDA.

Accepting condition	
Empty stack	• General framework does not work: the number of states reachable with an empty-stack is too small
	• Fixing a minimal number of pop transitions (see Sect. 4.2) works for $\lambda < 1$
Final states	• General framework works: a significant number of states are reachable
	• A rejection approach is tractable for generating reachable RDPDA, when both $\lambda \geq 1.5$ and the alphabets are large enough

the parameters of Table 3. Several experiments have been performed to evaluate the average number of rejects and results are reported in Table 5: the average number random generations of RDPDA used to produce 10 reachable RDPDA was reported. When a "−" is reported in the table, it means that after 300 rejects, no such automata was obtained. These results seem to show that the rejection approach is tracktable if $\lambda \geq 2$ and if the alphabets are not too small. With $\alpha = \beta = 2$, it works for $\lambda \geq 3$ and for smaller λ's when the number of states is small.

5   Conclusion

In this paper a general framework for generating accessible deterministic pushdown automata is proposed. We also experimentally showed that with some accepting conditions, it is possible to generate pushdown automata where most states are reachable. The results on the random generation are synthesized in Table 6. In a future work we plan to investigate how to randomly generate real-time deterministic automata, with an empty-stack accepting condition and again, which most states are reachable. We also plan to remove the *real-time* assumption, but it requires a deeper work on the underlying automata.

References

[BDN07] Bassino, F., David, J., Nicaud, C.: REGAL: a library to randomly and exhaustively generate automata. In: Holub, J., Žd'árek, J. (eds.) CIAA 2007. LNCS, vol. 4783, pp. 303–305. Springer, Heidelberg (2007)

[BDN09] Bassino, F., David, J., Nicaud, C.: Enumeration and random generation of possibly incomplete deterministic automata. Pure Math. Appl. **19**, 1–16 (2009)

[BN07] Bassino, F., Nicaud, C.: Enumeration and random generation of accessible automata. Theor. Comput. Sci. **381**(1–3), 86–104 (2007)

[CF11] Carnino, V., De Felice, S.: Random generation of deterministic acyclic automata using Markov chains. In: Bouchou-Markhoff, B., Caron, P., Champarnaud, J.-M., Maurel, D. (eds.) CIAA 2011. LNCS, vol. 6807, pp. 65–75. Springer, Heidelberg (2011)

[CN12] Carayol, A., Nicaud, C.: Distribution of the number of accessible states in a random deterministic automaton. In: STACS 2012, LIPIcs, vol. 14, pp. 194–205 (2012)

[CP05] Champarnaud, J.-M., Paranthoën, T.: Random generation of dfas. Theor. Comput. Sci. **330**(2), 221–235 (2005)

[DFLS04] Duchon, P., Flajolet, P., Louchard, G., Schaeffer, G.: Boltzmann samplers for the random generation of combinatorial structures. Comb. Probab. Comput. **13**(4–5), 577–625 (2004)

[FS08] Flajolet, P., Sedgewick, R.: Analytic Combinatorics. Cambridge University Press, Cambridge (2008)

[FWW97] Finkel, A., Willems, B., Wolper, P.: A direct symbolic approach to model checking pushdown systems (extended abstract). In: INFINITY 1997, Electronic Notes in Theoretical Computer Science, vol. 9 pp. 27–39 (1997)

[FZC94] Flajolet, P., Zimmermann, P., Van Cutsem, B.: A calculus for the random generation of labelled combinatorial structures. Theor. Comput. Sci. **132**(2), 1–35 (1994)

[HNS10] Héam, P.-C., Nicaud, C., Schmitz, S.: Parametric random generation of deterministic tree automata. Theor. Comput. Sci. **411**(38–39), 3469–3480 (2010)

[HU79] Hopcroft, J., Ullman, J.: Introduction to Automata Theory, Languages and Computation. Addison-Wesley, Reading (1979)

[Kor78] Korshunov, D.: Enumeration of finite automata. Problemy Kibernetiki **34**, 5–82 (1978)

[Kor86] Korshunov, A.D.: On the number of non-isomorphic strongly connected finite automata. Elektronische Informationsverarbeitung und Kybernetik **22**(9), 459–462 (1986)

[Lis06] Liskovets, V.A.: Exact enumeration of acyclic deterministic automata. Discret. Appl. Math. **154**(3), 537–551 (2006)

[Nic00] Nicaud, C.: Etude du comportement en moyenne des automate finis et des langages rationnels. Ph.D. thesis, Université Paris VII (2000)

[Nic14] Nicaud, C.: Random deterministic automata. In: Csuhaj-Varjú, E., Dietzfelbinger, M., Ésik, Z. (eds.) MFCS 2014, Part I. LNCS, vol. 8634, pp. 5–23. Springer, Heidelberg (2014)

[Rob85] Robinson, R.: Counting strongly connected finite automata. In: Alavi, Y. (ed.) Graph Theory with Applications to Algorithms and Computer Science, pp. 671–685. Wiley, New York (1985)

[Sak09] Sakarovitch, J.: Elements of Automata Theory. Cambridge University Press, Cambridge (2009)

[Vys59] Vyssotsky, V.: A counting problem for finite automata. Technical report, Bell Telephon Laboratories (1959)

Subword Metrics for Infinite Words

Stefan Hoffmann[1,2] and Ludwig Staiger[3]([✉])

[1] Institut für Mathematik, Martin-Luther-Universität Halle-Wittenberg,
06099 Halle (Saale), Germany
st.hoffmann@student.uni-halle.de
[2] Universität Trier, Fachbereich 4, Informatikwissenschaften, 54286 Trier, Germany
[3] Institut für Informatik, Martin-Luther-Universität Halle-Wittenberg,
06099 Halle (Saale), Germany
staiger@informatik.uni-halle.de

Abstract. The space of one-sided infinite words plays a crucial rôle in several parts of Theoretical Computer Science. Usually, it is convenient to regard this space as a metric space, the Cantor-space. It turned out that for several purposes topologies other than the one of the Cantor-space are useful, e.g. for studying fragments of first-order logic over infinite words or for a topological characterisation of random infinite words.

Continuing the work of [14], here we consider two different refinements of the Cantor-space, given by measuring common factors, and common factors occurring infinitely often. In particular we investigate the relation of these topologies to the sets of infinite words definable by finite automata, that is, to regular ω-languages.

Keywords: Metric spaces · ω-words · Subwords · Shift-invariance · Subword complexity

1 Introduction

The space of one-sided infinite words plays a crucial rôle in several parts of Theoretical Computer Science (see the surveys [18,23]). Usually, it is convenient to regard this space as a topological space provided with the CANTOR topology. This topology can be also considered as the natural continuation of the left topology of the prefix relation on the space of finite words (cf. [3]).

It turned out that for several purposes other topologies on the space of infinite words are also useful [12,16], e.g. for investigations in first-order logic [4], to characterise the set of random infinite words [2] or the set of disjunctive infinite words [20] and to describe the converging behaviour of not necessarily hyperbolic iterative function systems [6,19].

Most of these approaches use topologies on the space of infinite words which are refinements of the CANTOR topology showing a certain kind of shift invariance. In [14] a unified treatment of those shift invariant topologies is given, and here we built on this work, introducing two new topologies arising naturally from the consideration of finite subwords occurring in infinite words.

F. Drewes (Ed.): CIAA 2015, LNCS 9223, pp. 165–175, 2015.
DOI: 10.1007/978-3-319-22360-5_14

2   Notation and Preliminaries

We introduce the notation used throughout the paper. By $\mathbb{N} = \{0, 1, 2, \ldots\}$ we denote the set of natural numbers. Let X be a finite alphabet of cardinality $|X| \geq 2$, and X^* be the set (monoid) of words on X, including the *empty word* e, and X^ω be the set of infinite sequences (ω-words) over X. For $w \in X^*$ and $\eta \in X^* \cup X^\omega$ let $w \cdot \eta$ be their *concatenation*. This concatenation product extends in an obvious way to subsets $W \subseteq X^*$ and $P \subseteq X^* \cup X^\omega$. For a language W let $W^* := \bigcup_{i \in \mathbb{N}} W^i$ be the *submonoid* of X^* generated by W, and by $W^\omega := \{w_1 \cdots w_i \cdots : w_i \in W \setminus \{e\}\}$ we denote the set of infinite strings formed by concatenating words in W. Furthermore $|w|$ is the *length* of the word $w \in X^*$ and $\mathbf{pref}(P)$ ($\mathbf{infix}(P)$) is the set of all finite prefixes (infixes) of strings in $P \subseteq X^* \cup X^\omega$, in particular, $\mathbf{pref}(P) \subseteq \mathbf{infix}(P)$. We shall abbreviate $w \in \mathbf{pref}(\eta)$ ($\eta \in X^* \cup X^\omega$) by $w \sqsubseteq \eta$. If $\xi \in X^\omega$ by $\mathbf{infix}^\infty(\xi) \subseteq \mathbf{infix}(\xi)$ we denote the set of infixes occurring infinitely often in ξ.

Further we denote by $P/w := \{\eta : w \cdot \eta \in P\}$ the *left derivative* or *state* of the set $P \subseteq X^* \cup X^\omega$ generated by the word w. We refer to P as *finite-state* provided the set of states $\{P/w : w \in X^*\}$ is finite. It is well-known that a language $W \subseteq X^*$ is finite state if and only if it is accepted by a finite automaton, that is, it is a regular language.[1]

In the case of ω-languages *regular ω-languages*, that is, ω-languages accepted by finite automata, are the finite unions of sets of the form $W \cdot V^\omega$, where W and V are regular languages (cf. e.g. [18]). Every regular ω-language is finite-state, but, as it was observed in [25], not every finite-state ω-language is regular (cf. also [15]).

It is well-known that the families of regular or finite-state ω-languages are closed under Boolean operations (see [11, 18, 23, 24] or [15]).

3   The CANTOR Topology and Regular ω-Languages

In this section we list some properties of the CANTOR topology on X^ω and regular ω-languages (see [18, 23]).

3.1   Basic Properties of the CANTOR Topology

We consider the space of infinite words (ω-words) X^ω as a metric space with metric ρ defined as follows

$$\rho(\xi, \eta) := \sup\{r^{1-|w|} : w \in \mathbf{pref}(\xi) \,\Delta\, \mathbf{pref}(\eta)\} \tag{1}$$

Here $r > 1$ is a real number[2], Δ denotes the symmetric difference of sets and we set $\sup \emptyset := 0$, that is, $\rho(\xi, \eta) = 0$ if and only of $\xi = \eta$.

---

[1] Observe that the relation $\sim_P$ defined by $w \sim_P v$ iff $P/w = P/v$ is the NERODE right congruence of P.

[2] It is convenient to choose $r = |X|$. Then every ball of radius r^{-n} is partitioned into exactly r balls of radius $r^{-(n+1)}$.

Since $\mathbf{pref}(\xi) \, \Delta \, \mathbf{pref}(\eta) \subseteq \big(\mathbf{pref}(\xi) \, \Delta \, \mathbf{pref}(\zeta)\big) \, \cup \, \big(\mathbf{pref}(\zeta) \, \Delta \, \mathbf{pref}(\eta)\big)$, the metric ρ satisfies the ultra-metric inequality

$$\rho(\xi, \eta) \leq \max\{\rho(\xi, \zeta), \rho(\zeta, \eta)\}\,.$$

A subset $E \subseteq X^\omega$ is *open* if for every $\xi \in E$ there is an $\epsilon > 0$ such that $\eta \in E$ for all η with $\rho(\xi, \eta) < \epsilon$. Complements of open sets are called *closed*. The smallest closed set containing a given set $F \subseteq X^\omega$, $\mathcal{C}(F)$, is referred to as the *closure of F*.

$\mathbf{G}_\delta$-sets are countable intersections of open sets and $\mathbf{F}_\sigma$-sets are countable unions of closed sets. In a metric space every open set is an $\mathbf{F}_\sigma$-set, and every closed set is a $\mathbf{G}_\delta$-set.

We list some further well-known properties of the metric space (X^ω, ρ).

Property 1. The following is true.

1. The non-empty sets $w \cdot X^\omega$ are open balls with radius $r^{-|w|}$ in the metric space (X^ω, ρ).[3] These balls are simultaneously closed.
2. Open sets in (X^ω, ρ) are of the form $W \cdot X^\omega$ where $W \subseteq X^*$.
3. A subset $E \subseteq X^\omega$ is open and closed (clopen) in (X^ω, ρ) if and only if $E = W \cdot X^\omega$ where $W \subseteq X^*$ is finite.
4. A subset $F \subseteq X^\omega$ is closed in (X^ω, ρ) if and only if $F = \{\xi : \mathbf{pref}(\xi) \subseteq \mathbf{pref}(F)\}$.
5. The closure of F satisfies $\mathcal{C}(F) := \{\xi : \xi \in X^\omega \land \mathbf{pref}(\xi) \subseteq \mathbf{pref}(F)\} = \bigcap_{n \in \mathbb{N}} (\mathbf{pref}(F) \cap X^n) \cdot X^\omega$.

The space (X^ω, ρ) is a *complete* space, that is, every sequence[4] $(\xi_i)_{i \in \mathbb{N}}$ where $\rho(\xi_j, \xi_k) < r^{-i}$ whenever $i \leq j, k$ converges to some $\xi \in X^\omega$. Moreover, (X^ω, ρ) is a *compact* space, that is, for every family of open sets $(E_i)_{i \in J}$ such that $\bigcup_{i \in J} E_i = X^\omega$ there is a finite sub-family $(E_i)_{i \in J'}$ satisfying $\bigcup_{i \in J'} E_i = X^\omega$.

3.2 Regular ω-Languages

As a last part of this section we mention some facts on regular ω-languages known from the literature, e.g. [11,18,23]. Regular ω-languages are well-known for being the ω-languages definable by finite automata. We will not refer to this feature, instead we list some basic properties of this family of ω-languages.

The first one shows among other properties the importance of ultimately periodic ω-words. Denote by $\mathsf{Ult} := \{w \cdot v^\omega : w, v \in X^* \setminus \{e\}\}$ the set of ultimately periodic ω-words.

Theorem 1 (Büchi [1]). *The family of regular ω-languages is a Boolean algebra, and if $F \subseteq X^\omega$ is regular, then $u \cdot F$ and F/w are also regular.*

Every non-empty regular ω-language contains an ultimately periodic ω-word, and regular ω-languages $E, F \subseteq X^\omega$ coincide if and only if $E \cap \mathsf{Ult} = F \cap \mathsf{Ult}$.

---

[3] Observe that $e \notin \mathbf{pref}(\xi) \, \Delta \, \mathbf{pref}(\eta)$ and Eq. (1) imply $\rho(\xi, \eta) = \inf\{r^{-|w|} : w \sqsubset \xi \land w \sqsubset \eta\}$.

[4] Those sequences are usually referred to as CAUCHY sequences.

For regular ω-languages we have the following topological characterisations analogous to Property 1.

Property 2. Let $F \subseteq X^\omega$ be regular and $E \subseteq X^\omega$ be finite-state. Then in CANTOR topology the following hold true.

1. F is open if and only if $F = W \cdot X^\omega$ where $W \subseteq X^*$ is a regular language.
2. $F \subseteq X^\omega$ is closed if and only if $F = \{\xi : \mathbf{pref}(\xi) \subseteq \mathbf{pref}(F)\}$ and $\mathbf{pref}(F)$ is regular.
3. $\mathbf{pref}(E)$ is a regular language.
4. $\mathcal{C}(E)$ is a regular ω-language.

Finally, we provide an example of a regular ω-language which is not a $\mathbf{G}_\delta$-set and a necessary and sufficient topological condition when finite-state ω-languages are regular.

Example 1 (Landweber [8]). For $u \in X^* \setminus \{e\}$ the ω-language $X^* \cdot u^\omega$ is regular, an $\mathbf{F}_\sigma$-set but not a $\mathbf{G}_\delta$-set. $\square$

Theorem 2 ([15]). *Every finite-state ω-language in the class $\mathbf{F}_\sigma \cap \mathbf{G}_\delta$ is a Boolean combination of regular ω-languages open in (X^ω, ρ), thus, in particular, a regular ω-language.*

4   Topologies Defined by Subword Metrics

It was shown that regular ω-languages are closely related to the (asymptotic) subword complexity of infinite words (cf. [17, Sect. 5] and [21]). Therefore, as other refinements of the CANTOR topology we introduce two topologies defined via metrics on X^ω which are based on the sets of subwords occurring or occurring infinitely often in the ω-words, respectively.

Definition 1 (Subword metrics)

$$\rho_I(\xi, \eta) := \sup\{r^{1-|w|} : w \in (\mathbf{pref}(\xi) \,\Delta\, \mathbf{pref}(\eta)) \cup (\mathbf{infix}(\xi) \,\Delta\, \mathbf{infix}(\eta))\}$$
$$\rho_\infty(\xi, \eta) := \sup\{r^{1-|w|} : w \in (\mathbf{pref}(\xi) \,\Delta\, \mathbf{pref}(\eta)) \cup (\mathrm{infix}^\infty(\xi) \,\Delta\, \mathrm{infix}^\infty(\eta))\}$$

These metrics respect except for the length of a shortest non-common prefix of ξ and η also the length of a shortest non-common subword (non-common subword occurring infinitely often). Thus

$$\rho_I(\xi, \eta) \geq \rho(\xi, \eta) \text{ and } \rho_\infty(\xi, \eta) \geq \rho(\xi, \eta), \tag{2}$$
$$\rho_I(\xi, \eta) = \max\{\rho(\xi, \eta), \sup\{r^{1-|u|} : u \in \mathbf{infix}(\xi) \,\Delta\, \mathbf{infix}(\eta)\}\}, \text{ and} \tag{3}$$
$$\rho_\infty(\xi, \eta) = \max\{\rho(\xi, \eta), \sup\{r^{1-|u|} : u \in \mathrm{infix}^\infty(\xi) \,\Delta\, \mathrm{infix}^\infty(\eta)\}\}. \tag{4}$$

Similar to the case of ρ one can verify that ρ_I and ρ_∞ satisfy the ultra-metric inequality. Therefore, balls in the metric spaces (X^ω, ρ_I) and (X^ω, ρ_∞) are simultaneously open and closed. Moreover, Eq. (2) shows that both topologies refine the CANTOR topology of X^ω, that is, ω-languages open (closed) in CANTOR topology are likewise open (closed, respectively) in both spaces (X^ω, ρ_I) and (X^ω, ρ_∞).

4.1   Shift-Invariance

We call a metric space (X^ω, ρ') *shift invariant* if for every open set $E \subseteq X^\omega$ and every word $w \in X^*$ the sets $w \cdot E$ and E/w are also open. In this part we show that the metric spaces (X^ω, ρ_∞) and (X^ω, ρ_I) are shift-invariant. According to Corollary 2 of [14] this property guarantees that the closure of a finite-state ω-language is again finite-state (cf. the stronger Property 2.4 for the CANTOR topology).

To this end we derive some simple properties of the metrics.

Lemma 1. *Let $u \in X^*$ and $v, w \in X^m$. Then*

$$\rho_\infty(u \cdot \xi, u \cdot \eta) \leq \rho_\infty(\xi, \eta), \tag{5}$$

$$\rho_\infty(\xi, \eta) \leq r^m \cdot \rho_\infty(w \cdot \xi, v \cdot \eta), \tag{6}$$

$$\rho_I(u \cdot \xi, u \cdot \eta) \leq \rho_I(\xi, \eta), \text{ and} \tag{7}$$

$$\rho_I(\xi, \eta) \leq r^m \cdot \rho_I(w \cdot \xi, v \cdot \eta). \tag{8}$$

Proof. All inequalities are trivially satisfied if $\xi = \eta$. So, in the following, we may assume $\xi \neq \eta$.

As $\mathrm{infix}^\infty(\xi) = \mathrm{infix}^\infty(u \cdot \xi)$, Eqs. (5) and (6) follow from Eq. (4) and the respective properties of the metric ρ of the CANTOR topology $\rho(u \cdot \xi, u \cdot \eta) \leq \rho(\xi, \eta)$ and $\rho(w \cdot \xi, v \cdot \eta) \geq \rho(w \cdot \xi, w \cdot \eta) = r^{-|w|} \cdot \rho(\xi, \eta)$.

Let $\rho_I(\xi, \eta) = r^{-n}$, that is, $\mathrm{infix}(\xi) \cap X^n = \mathrm{infix}(\eta) \cap X^n$ and $w \sqsubset \xi$ and $w \sqsubset \eta$ for some $w \in X^n$. Then, obviously, $v \sqsubset u \cdot \xi$ and $v \sqsubset u \cdot \eta$ for some $v \in X^n$. Moreover, $\mathrm{infix}(u \cdot \xi) \cap X^n = (\mathrm{infix}(u \cdot w) \cap X^n) \cup (\mathrm{infix}(\xi) \cap X^n) = \mathrm{infix}(u \cdot \eta) \cap X^n$. This proves Eq. (7).

If $w \neq v$ then in view of $\rho(w \cdot \xi, v \cdot \eta) \geq r^{-(m-1)}$, Eq. (8) is obvious. Let $w = v$ and $\rho_I(\xi, \eta) = r^{-n}$ for some $n \in \mathbb{N}$. We have to show that $\rho_I(w \cdot \xi, w \cdot \eta) \geq r^{-(n+m)}$.

If $\rho(\xi, \eta) = r^{-n}$ then $\rho(w \cdot \xi, w \cdot \eta) = r^{-(n+m)}$ and Eq. (3) proves $\rho_I(w \cdot \xi, w \cdot \eta) \geq r^{-(n+m)}$.

If $\rho(\xi, \eta) < r^{-n}$ in view of $\rho_I(\xi, \eta) = r^{-n}$ we have $(\mathrm{infix}(\xi) \triangle \mathrm{infix}(\eta)) \cap X^{n+1} \neq \emptyset$, that is, $u \in (\mathrm{infix}(\xi) \triangle \mathrm{infix}(\eta)) \cap X^{n+1}$ for some $u \in \mathrm{infix}(\xi)$, say. Now, it suffices to show $(\mathrm{infix}(w\xi) \triangle \mathrm{infix}(w\eta)) \cap X^{n+m+1} \neq \emptyset$. Assume $v'u \notin \mathrm{infix}(w\xi) \triangle \mathrm{infix}(w\eta)$ for all $v' \in X^m$. Then $u \in \mathrm{infix}(\xi)$ implies $v'u \in \mathrm{infix}(w\xi) \cap \mathrm{infix}(w\eta)$ for some $v' \in X^m$. Since $|w| = |v'| = m$, we have $u \in \mathrm{infix}(\eta)$, a contradiction. □

As a consequence we obtain our result.

Corollary 1. *The topologies (X^ω, ρ_I) and (X^ω, ρ_∞) are shift invariant.*

Proof. We use the fact that, in view of Lemma 1, the mappings Φ_u and Φ_m defined by $\Phi_u(\xi) := u \cdot \xi$ and $\Phi_m(w \cdot \xi) := \xi$ for $w \in X^m$ are continuous w.r.t. the metrics ρ_I and ρ_∞, respectively.

Thus, if $F \subseteq X^\omega$ is open in (X^ω, ρ_I) or (X^ω, ρ_∞) the preimage $\Phi_u^{-1}(F) = F/u$ and, for $m = |w|$, also $w \cdot F = \Phi_m^{-1}(F) \cap w \cdot X^\omega$ are open sets. □

4.2   Balls in (X^ω, ρ_I) and (X^ω, ρ_∞)

Denote by $K_I(\xi, r^{-n})$ and $K_\infty(\xi, r^{-n})$ the open balls[5] of radius r^{-n} around ξ in the spaces (X^ω, ρ_I) and (X^ω, ρ_∞), respectively. For $w \sqsubseteq \xi$ with $|w| = n + 1$ and $W := X^{n+1} \cap \mathbf{infix}(\xi)$, $V := X^{n+1} \cap \mathrm{infix}^\infty(\xi)$, $\overline{W} := X^{n+1} \setminus \mathbf{infix}(\xi)$ and $\overline{V} := X^{n+1} \setminus \mathrm{infix}^\infty(\xi)$ we obtain the following description of balls via regular ω-languages.

$$K_I(\xi, r^{-n}) = w \cdot X^\omega \cap \bigcap_{u \in W} X^* \cdot u \cdot X^\omega \setminus \bigcup_{u \in \overline{W}} X^* \cdot u \cdot X^\omega, \text{ and} \qquad (9)$$

$$K_\infty(\xi, r^{-n}) = w \cdot X^\omega \cap X^* \cdot \left(\left(\prod_{u \in V} X^* \cdot u \right)^\omega \setminus \bigcup_{u \in \overline{V}} X^* \cdot u \cdot X^\omega \right). \qquad (10)$$

In Eq. (10) the order of the words $u \in V$ can be arbitrarily chosen. In particular, Eqs. (9) and (10) show that balls in (X^ω, ρ_I) and (X^ω, ρ_∞) are regular ω-languages. Thus every non-empty open subset in each of the spaces contains an ultimately periodic ω-word.

An immediate consequence of the representations in Eqs. (9) and (10) is the following relation between the space (X^ω, ρ_I) and the CANTOR space (X^ω, ρ).

Lemma 2

1. *Every ball $K_I(\xi, r^{-n})$ is a Boolean combination of regular ω-languages open in (X^ω, ρ), therefore, simultaneously an $\mathbf{F}_\sigma$- and a $\mathbf{G}_\delta$-set in CANTOR topology.*
2. *Every open set in (X^ω, ρ_I) is an $\mathbf{F}_\sigma$-set in CANTOR topology.*

Proof

1. It is well-known know that open sets in a metric space are simultaneously $\mathbf{F}_\sigma$- and $\mathbf{G}_\delta$-sets. Then, according to Property 1, the set $K_I(\xi, r^{-n})$ is simultaneously an $\mathbf{F}_\sigma$- and $\mathbf{G}_\delta$-set in the CANTOR topology.
2. is a consequence of 1 and the fact that there are only countably many open balls in (X^ω, ρ_I).   □

Equations (9) and (10) and Lemma 2 show a connection between certain regular ω-languages and the open sets in (X^ω, ρ_I). It would be interesting if we could characterise some regular ω-languages open in (X^ω, ρ_I) using CANTOR topology. The next example considering the simple case of closed sets, however, shows that not every regular ω-language closed in CANTOR topology is open in (X^ω, ρ_I).

Example 2 ([7]). Consider the regular ω-language $F = \{1, 00\}^\omega \subseteq \{0, 1\}^\omega$ which is closed in the CANTOR topology. Assume F to be open in (X^ω, ρ_I). Then $\eta = \prod_{i \in \mathbb{N}} 10^{2i} \in F$ and, therefore, $K_I(\eta, r^{-n}) \subseteq F$ for some $n \in \mathbb{N}$, $n \geq 1$.

Consider $\xi = \prod_{i=0}^{n} 10^{2i} \cdot \prod_{i=2n+1}^{\infty} 10^i \notin F$. Then we have $\prod_{i=0}^{n} 10^{2i} \sqsubseteq \eta$, $\prod_{i=0}^{n} 10^{2i} \sqsubseteq \xi$ and, moreover,

$$\mathbf{infix}(\xi) \cap \{0, 1\}^{2n} = \left(\mathbf{infix}(\prod_{i=0}^{n} 10^{2i}) \cup 0^* \cdot 1 \cdot 0^* \cup 0^* \right) \cap \{0, 1\}^{2n}$$
$$= \mathbf{infix}(\eta) \cap \{0, 1\}^{2n}.$$

It follows $\rho_I(\xi, \eta) \leq r^{-2n}$, that is, $\xi \in K_I(\eta, r^{-n}) \subseteq \{1, 00\}^\omega$, a contradiction.   □

---

[5] They are also closed balls of radius $r^{-(n+1)}$.

Using the Morse-Hedlund Theorem (cf. also the proof of Theorem 1.3.13 of [9]) one obtains special representations of small balls containing ultimately periodic ω-words. To this end we derive the following lemma.

Lemma 3. *Let* $w, u \in X^*, u \neq e$ *and* $\xi \in X^\omega$. *Then* $w \cdot u \sqsubset \xi$ *and* $\mathbf{infix}(\xi) \cap X^{|w \cdot u|} = \mathbf{infix}(w \cdot u^\omega) \cap X^{|w \cdot u|}$ *imply* $\xi = w \cdot u^\omega$.

Proof. First observe that $|\mathbf{infix}(w \cdot u^\omega) \cap X^{|w \cdot u|}| = |\mathbf{infix}(w \cdot u^\omega) \cap X^{|w \cdot u|+1}|$. Thus, for every $v \in \mathbf{infix}(w \cdot u^\omega) \cap X^{|w \cdot u|}$, there is a unique $v' \in \mathbf{infix}(w \cdot u^\omega) \cap X^{|w \cdot u|}$ such that $v \sqsubset a \cdot v'$ for some $a \in X$. Consequently, the ω-word $\xi \in X^\omega$ with $w \cdot u \sqsubset \xi$ and $\mathbf{infix}(\xi) \cap X^{|w \cdot u|} = \mathbf{infix}(w \cdot u^\omega) \cap X^{|w \cdot u|}$ is uniquely specified.

Lemma 4. *Let* $w \cdot u^\omega \in X^\omega$ *where* $|w| \leq |u|, |u| > 0$, *and let* $m > |w| + |u|$ *and* $n > |u|$. *Then*

$$K_I(w \cdot u^\omega, r^{-m}) = \{w \cdot u^\omega\}, \ and \tag{11}$$

$$K_\infty(w \cdot u^\omega, r^{-n}) = w' \cdot X^* \cdot u^\omega \ where \ w' \sqsubset w \cdot u^n \ and \ |w'| = n. \tag{12}$$

Proof. Every $\xi \in K_I(w \cdot u^\omega, r^{-m})$ satisfies $w \cdot u \sqsubset \xi$ and $\mathbf{infix}(\xi) \cap X^m = \mathbf{infix}(w \cdot u^\omega) \cap X^m$, and the assertion of Eq. (11) follows from Lemma 3.

If $\xi \in K_\infty(w \cdot u^\omega, r^{-n})$ then there is a tail ξ' of ξ such that $u \sqsubset \xi'$ and $\mathrm{infix}^\infty(\xi) \cap X^n = \mathbf{infix}(\xi') \cap X^n = \mathbf{infix}(u^\omega) \cap X^n$ whence, again by Lemma 3, $\xi' = u^\omega$. $\qquad\square$

This allows us to state the following property concerning isolated points[6] in the spaces (X^ω, ρ_I) and (X^ω, ρ_∞). The additional Item 3 in connection with Lemma 2.2 shows a further difference between both spaces.

Corollary 2

1. *The set of isolated points of the space* (X^ω, ρ_I) *is* Ult.
2. *The space* (X^ω, ρ_∞) *has no isolated points and all sets of the form* $X^* \cdot u^\omega$ *are simultaneously closed and open.*
3. *In the space* (X^ω, ρ_∞) *there are open sets which are not* $\mathbf{F}_\sigma$-*sets in* CANTOR *topology.*

Proof. Since every non-empty open subset of (X^ω, ρ_I) and also (X^ω, ρ_∞) contains an ultimately periodic ω-word, every isolated point has to be ultimately periodic. Now Eq. (11) shows that every $w \cdot u^\omega$ is an isolated point in (X^ω, ρ_I), and Eq. (12) proves that (X^ω, ρ_∞) has no isolated points. The remaining part of Item 2 follows from Eq. (12) and $X^* \cdot u^\omega = \bigcup_{w \in X^n} w \cdot X^* \cdot u^\omega$.

Finally, it is known that $X^\omega \setminus X^* \cdot u^\omega$ is not an $\mathbf{F}_\sigma$-set in CANTOR topology (cf. Example 1). $\qquad\square$

---

[6] A point ξ is referred to as *isolated* if $\rho'(\xi, \eta) \geq \epsilon_\xi$ for all $\eta \neq \xi$. Here the distance $\epsilon_\xi > 0$ may depend on ξ.

4.3   Non-Preservation of Regular ω-Languages

In this section we investigate whether similar to the CANTOR topology the closure of a finite-state ω-language is always regular in the spaces (X^ω, ρ_I) and (X^ω, ρ_∞).

In contrast to the CANTOR topology it is, however, not true that the closure of finite-state ω-languages are regular. We can even show that in both spaces (X^ω, ρ_I) and (X^ω, ρ_∞) there are regular ω-languages with non-regular closures.

Since we do not have a characterisation like the Property 1.5 for the closures $\mathcal{C}_I$ and $\mathcal{C}_\infty$ in the spaces (X^ω, ρ_I) and (X^ω, ρ_∞), respectivly, we circumvent this obstacle by presenting examples where the closure $\mathcal{C}_I(F)$ or $\mathcal{C}_\infty(F)$ of a regular ω-language F is shown to be larger than F but does not contain more ultimately periodic ω-words than F. In view of Theorem 1 this implies that the closures cannot be regular ω-languages.

For the closure $\mathcal{C}_I$ we use that, according to Example 1 the ω-language $\{0,1\}^* \cdot 0^\omega$ is no $\mathbf{G}_\delta$-set in the CANTOR topology, thus in view of Lemma 2.2 not closed in (X^ω, ρ_I).

Example 3. We show that $\mathcal{C}_I(\{0,1\}^* \cdot 0^\omega) \cap \mathsf{Ult} = \{0,1\}^* \cdot 0^\omega$. Let $w \cdot u^\omega \notin \{0,1\}^* \cdot 0^\omega$. Then $u \notin \{0\}^*$ and $0^{|w \cdot u|} \notin \mathbf{infix}(w \cdot u^\omega)$. Now Eq. (9) yields $K_I(w \cdot u^\omega, r^{-|w \cdot u|}) \cap X^* \cdot 0^{|w \cdot u|} \cdot X^\omega = \emptyset$. Thus $\rho_I(w \cdot u^\omega, v \cdot 0^\omega) \geq r^{-|w \cdot u|}$ for all $v \in X^*$ whence $w \cdot u^\omega \notin \mathcal{C}_I(\{0,1\}^* \cdot 0^\omega)$. The other inclusion being trivial.

Assume $\mathcal{C}_I(\{0,1\}^* \cdot 0^\omega)$ were a regular ω-language. Then Theorem 1 implies $\mathcal{C}_I(\{0,1\}^* \cdot 0^\omega) = \{0,1\}^* \cdot 0^\omega$, that is, $\{0,1\}^* \cdot 0^\omega$ is closed in (X^ω, ρ_I), a contradiction to Lemma 2.2 ☐

Since $\{0,1\}^* \cdot 0^\omega$ is closed in (X^ω, ρ_∞), we cannot use this ω-language in that case.

Example 4. Let $F := \{0,1\}^* \cdot ((00)^*1)^\omega$. As explained above, it suffices to show that $\mathcal{C}_\infty(F) \cap \mathsf{Ult} = F \cap \mathsf{Ult}$ and $\mathcal{C}_\infty(F) \supset F$.

Let $w \cdot u^\omega \in \mathcal{C}_\infty(F)$. Then there is a $\xi \in F$ such that $\rho_\infty(w \cdot u^\omega, \xi) < r^{-|wu|}$. According to Lemma 4 we have $\xi \in w \cdot X^* \cdot u^\omega$. Thus $u^\omega = u' \cdot \eta$ where $\eta \in ((00)^*1)^\omega$ whence $w \cdot u^\omega = w \cdot u' \cdot \eta \in F$.

Finally, consider $\zeta = \prod_{j=0}^\infty 10^j = 110100 \cdots$. Since ζ has infinitely many infixes 10^j1 where j is odd, $\zeta \notin F$. Moreover, $\mathbf{infix}^\infty(\zeta) \cap X^n = \{0^n\} \cap \{0^j \cdot 1 \cdot 0^{n-j-1} : 0 \leq j < n\}$. Consider the ω-words $\xi_i := \prod_{j=0}^{2i} 10^j \cdot (1 \cdot 0^{2i})^\omega \in F$. It holds $\mathbf{pref}(\xi_i) \cap X^n = \mathbf{pref}(\zeta) \cap X^n$ and $\mathbf{infix}^\infty(\xi_i) \cap X^n = \mathbf{infix}^\infty(\zeta) \cap X^n$ for $n \leq 2i + 1$. This implies $\rho_\infty(\xi_i, \zeta) \leq r^{-2i}$, that is, $\lim_{i \to \infty} \xi_i = \zeta \in \mathcal{C}_\infty(F)$ in (X^ω, ρ_∞). ☐

5   Completeness and Compactness

Here we show that the spaces (X^ω, ρ_I) and (X^ω, ρ_∞) are neither complete nor compact.

To show that they are not complete we consider the sequence $(\xi_i)_{i \in \mathbb{N}}$ where $\xi_i := \prod_{j=i}^\infty 0^j1$. This sequence converges in CANTOR topology to the limit point

0^ω. Since (X^ω, ρ_I) and (X^ω, ρ_∞) refine (X^ω, ρ), the limit points, if they exist, should be the same. But $\mathbf{infix}(\xi_i)$ and $\mathrm{infix}^\infty(\xi_i)$ both contain the word 1 which is not in $\mathbf{infix}(0^\omega) = \mathrm{infix}^\infty(0^\omega)$. Thus $\rho_I(\xi_i, 0^\omega) = \rho_\infty(\xi_i, 0^\omega) = 1$.

It remains to show that the sequence $(\xi_i)_{i \in \mathbb{N}}$ fulfils the CAUCHY property. To this end we observe that for $j \geq i$ we have $0^i \sqsubset \xi_j$ and $\mathbf{infix}(\xi_j) \cap X^i = \mathrm{infix}^\infty(\xi_j) \cap X^i = \{0^i\} \cup \{0^m 10^{i-m-1} : 0 \leq m < i\}$. Thus $\rho_I(\xi_j, \xi_k) \leq r^{-i}$ and $\rho_\infty(\xi_j, \xi_k) \leq r^{-i}$ for $j, k \geq i$.

In general it holds that no topology refining the CANTOR topology is compact. A proof uses Corollary 3.1.14 in [5]. Here we provide the more illustrative and seemingly stronger examples of partitions of the whole space X^ω into infinitely many open subsets.

Example 5. Let $X = \{0, 1\}$. Then the sets $0^i 1 \cdot X^\omega$ for $i \in \mathbb{N}$ are open in the CANTOR topology, hence open in (X^ω, ρ_I) and according to Corollary 2.1 the set $\{0^\omega\}$ is also open (X^ω, ρ_I).

Then $\{\{0^\omega\}\} \cup \{0^i 1 \cdot X^\omega : i \in \mathbb{N}\}$ is a partition of X^ω into sets open in (X^ω, ρ_I). □

Example 6. Let $X = \{0, 1\}$. Then the sets $0^i 1 \cdot X^\omega$ for $i \in \mathbb{N}$ are open in the CANTOR topology, hence open in (X^ω, ρ_∞) and according to Corollary 2.2 the set $X^* \cdot 0^\omega$ is open and closed in (X^ω, ρ_∞).

Then $\{X^* \cdot 0^\omega\} \cup \{0^i 1 \cdot X^\omega \setminus X^* \cdot 0^\omega : i \in \mathbb{N}\}$ is a partition of X^ω into sets open in (X^ω, ρ_∞). □

6 Subword Complexity

In Sect. 4 we mentioned that regular ω-languages are closely related to the (asymptotic) subword complexity of infinite words. Adapting the metrics ρ_I and ρ_∞ to subwords we may draw some connections to the level sets $F_\gamma^{(\tau)}$ of the asymptotic subword complexity (see [17, 21]).

First we introduce the concept of asymptotic subword complexity.

Definition 2 (Asymptotic subword complexity). $\tau(\xi) := \lim\limits_{n \to \infty} \dfrac{\log_{|X|} |\mathbf{infix}(\xi) \cap X^n|}{n}$

Using the inequality $|\mathbf{infix}(\xi) \cap X^{n+m}| \leq |\mathbf{infix}(\xi) \cap X^n| \cdot |\mathbf{infix}(\xi) \cap X^m|$ it is easy to see that the limit in Definition 2 exists and

$$\tau(\xi) = \inf\left\{ \frac{\log_{|X|} |\mathbf{infix}(\xi) \cap X^n|}{n} : n \in \mathbb{N} \wedge n \geq 1 \right\}. \tag{13}$$

Equation (5.2) of [17] shows that in Definition 2 and Eq. (13) one can replace the term $\mathbf{infix}(\xi)$ by $\mathrm{infix}^\infty(\xi)$.

Let, for $0 < \gamma \leq 1$, $F_\gamma^{(\tau)} := \{\xi : \xi \in X^\omega \wedge \tau(\xi) < \gamma\}$ be the *lower level sets* of the asymptotic subword complexity. For $\gamma = 0$ we set $F_0^{(\tau)} := \mathsf{Ult}$ (instead of $F_0^{(\tau)} = \emptyset$). We want to show that these sets are open in (X^ω, ρ_I) and (X^ω, ρ_∞). As a preparatory result we derive the subsequent Lemma 5.

Let $E_n(\xi) := \{\eta : \mathbf{infix}(\eta) \cap X^n \subseteq \mathbf{infix}(\xi)\}$ and $E'_n(\xi) := \{\eta : \mathrm{infix}^\infty(\eta) \cap X^n \subseteq \mathrm{infix}^\infty(\xi)\}$ be the sets of ω-words having only infixes or infixes occurring infinitely often of length n of ξ, respectively. These sets can be equivalently described as

$$E_n(\xi) = X^\omega \setminus X^* \cdot (X^n \setminus \mathbf{infix}(\xi)) \cdot X^\omega \quad \text{and}$$
$$E'_n(\xi) = X^* \cdot \left(X^\omega \setminus X^* \cdot (X^n \setminus \mathrm{infix}^\infty(\xi)) \cdot X^\omega\right), \quad \text{respectively}$$

which resembles in some sense the characterisation of open balls in Eqs. (9) and (10). In fact, it appears that the sets $E_n(\xi)$ and $E'_n(\xi)$ are open in the respective spaces (X^ω, ρ_I) and (X^ω, ρ_∞).

Lemma 5. *Let $\xi \in X^\omega$. Then $\xi \in E_n(\xi) \cap E'_n(\xi)$, the set $E_n(\xi)$ is open in (X^ω, ρ_I) and the set $E'_n(\xi)$ is open in (X^ω, ρ_∞).*

Proof. The first assertion is obvious. For a proof of the second one we show that $\eta \in E_n(\xi)$ implies that the ball $K_I(\eta, r^{-n})$ is contained in $E_n(\xi)$.

Let $\eta \in E_n(\xi)$ and $\zeta \in K_I(\eta, r^{-n})$. Then, $\rho_I(\eta, \zeta) < r^{-n}$, that is, in particular, $\mathbf{infix}(\eta) \cap X^n = \mathbf{infix}(\zeta) \cap X^n$, whence $\zeta \in E_n(\xi)$.

The proof for $E'_n(\xi)$ is similar. $\qquad\square$

This much preparation enables us to show that the level sets are open sets.

Theorem 3. *Let $0 \le \gamma \le 1$. Then the sets $F_\gamma^{(\tau)}$ are open in (X^ω, ρ_I) and (X^ω, ρ_∞).*

Proof. For $\gamma = 0$ we have $F_\gamma^{(\tau)} = \mathsf{Ult}$ which, according to Corollary 2, is open in (X^ω, ρ_I) as well as in (X^ω, ρ_∞).

Let $\gamma > 0$ and $\tau(\xi) < \gamma$. We show that then $E_n(\xi) \subseteq F_\gamma^{(\tau)}$ and $E'_n(\xi) \subseteq F_\gamma^{(\tau)}$ for some $n \in \mathbb{N}$. Together with Lemma 5 this shows that $F_\gamma^{(\tau)}$ contains, with every ξ, open sets containing this ξ.

If $\tau(\xi) < \gamma$ then in view of Eq. (13) we have $\frac{\log_{|X|} |\mathbf{infix}(\xi) \cap X^n|}{n} < \gamma$ for some $n \in \mathbb{N}$. Then for every $\eta \in E_n(\xi)$ it holds $\tau(\eta) \le \frac{\log_{|X|} |\mathbf{infix}(\xi) \cap X^n|}{n} < \gamma$ and, consequently, $E_n(\xi) \subseteq F_\gamma^{(\tau)}$.

The proof for (X^ω, ρ_∞) is similar using $\mathrm{infix}^{(\infty)}$ instead of $\mathbf{infix}$ and the respective modification of Eq. (13) whose validity was mentioned above. $\qquad\square$

The proof shows also that $\xi \in F_\gamma^{(\tau)}$ implies that $X^\omega \setminus X^* \cdot (X^n \setminus \mathbf{infix}(\xi)) \cdot X^\omega \subseteq F_\gamma^{(\tau)}$ for some $n > 0$. Thus $F_\gamma^{(\tau)}$ is a countable union of regular ω-languages closed in CANTOR topology, hence an $\mathbf{F}_\sigma$-set in CANTOR topology. The sets $F_\gamma^{(\tau)}$ are finite-state[7] non-regular ω-languages because their complement $X^\omega \setminus F_\gamma^{(\tau)}$ is non-empty and does not contain any ultimately periodic ω-word. Thus, in view of Theorem 2, they are not $\mathbf{G}_\delta$-sets in CANTOR-space and they are examples of sets open in (X^ω, ρ_I) and (X^ω, ρ_∞) which are non-regular $\mathbf{F}_\sigma$-sets in CANTOR-space.

---

[7] In particular, they satisfy $F_\gamma^{(\tau)}/w = F_\gamma^{(\tau)}$ for all $w \in X^*$.

References

1. Büchi, J.R.: On a decision method in restricted second order arithmetic. In: Proceedings of the 1960 International Congress for Logic, pp. 1–11. Stanford Univ. Press, Stanford (1962)
2. Calude, C.S., Marcus, S., Staiger, L.: A topological characterization of random sequences. Inform. Process. Lett. **88**, 245–250 (2003)
3. Calude, C.S., Jürgensen, H., Staiger, L.: Topology on words. Theoret. Comput. Sci. **410**, 2323–2335 (2009)
4. Diekert, V., Kufleitner, M.: Fragments of first-order logic over infinite words. In: Albers, S., Marion, J.-Y. (eds.) Proceedings of the STACS 2009, pp. 325–336. Schloss Dagstuhl - Leibniz-Zentrum für Informatik (2009)
5. Engelking, R.: General Topology. Państwowe wydawnictwo naukowe, Warszawa (1977)
6. Fernau, H., Staiger, L.: Iterated function systems and control languages. Inform. Comput. **168**, 125–143 (2001)
7. Hoffmann, S.: Metriken zur Verfeinerung des Cantor-Raumes auf X^ω. Diploma thesis, Martin-Luther-Universität Halle-Wittenberg (2014)
8. Landweber, L.H.: Decision problems for ω-automata. Math. Syst. Theory **3**, 376–384 (1969)
9. Lothaire, M.: Algebraic Combinatorics on Words. Cambridge University Press, Cambridge (2002)
10. McNaughton, R.: Testing and generating infinite sequences by a finite automaton. Inform. Control **9**, 521–530 (1966)
11. Perrin, D., Pin, J.-E.: Infinite Words. Elsevier, Amsterdam (2004)
12. Redziejowski, R.R.: Infinite word languages and continuous mappings. Theoret. Comput. Sci. **43**, 59–79 (1986)
13. Rozenberg, G., Salomaa, A. (eds.): Handbook of Formal Languages. Springer, Berlin (1997)
14. Schwarz, S., Staiger, L.: Topologies refining the Cantor topology on X^ω. In: Calude, C.S., Sassone, V. (eds.) Theoretical Computer Science. IFIP, vol. 323, pp. 271–285. Springer, Berlin (2010)
15. Staiger, L.: Finite-state ω-languages. J. Comput. Syst. Sci. **27**, 434–448 (1983)
16. Staiger, L.: Sequential mappings of ω-languages. ITA **21**, 147–173 (1987)
17. Staiger, L.: Kolmogorov complexity and Hausdorff dimension. Inf. Comput. **103**, 159–194 (1993)
18. Staiger, L.: ω-languages. In: [13], vol. 3, pp. 339–387
19. Staiger, L.: Weighted finite automata and metrics in Cantor Space. J. Automata Lang. Comb. **8**, 353–360 (2003)
20. Staiger, L.: Topologies for the set of disjunctive ω-words. Acta Cybern. **17**, 43–51 (2005)
21. Staiger, L.: Asymptotic subword complexity. In: Bordihn, H., Kutrib, M., Truthe, B. (eds.) Languages Alive. LNCS, vol. 7300, pp. 236–245. Springer, Heidelberg (2012)
22. Staiger, L., Wagner, K.: Automatentheoretische und automatenfreie Charakterisierungen topologischer Klassen regulärer Folgenmengen. Elektronische Informationsverarbeitung und Kybernetik **10**, 379–392 (1974)
23. Thomas, W.: Automata on infinite objects. In: Van Leeuwen, J. (ed.) Handbook of Theoretical Computer Science, vol. B, pp. 133–191. Elsevier, Amsterdam (1990)
24. Thomas, W.: Languages, automata, and logic. In: [13], vol. 3, pp. 389–455
25. Trakhtenbrot, B.A.: Finite automata and monadic second order logic. Sibirsk. Mat. Ž. **3**, 103–131 (1962). (Russian; English translation: AMS Transl. 59, 23–55, (1966))

From Two-Way to One-Way Finite Automata—Three Regular Expression-Based Methods

Mans Hulden[(✉)]

University of Colorado Boulder, Boulder, USA
mans.hulden@colorado.edu

Abstract. We describe three regular expression-based methods to characterize as a regular language the language defined by a two-way automaton. The construction methods yield relatively simple techniques to directly construct one-way automata that simulate the behavior of two-way automata. The approaches also offer conceptually uncomplicated alternative equivalence proofs of two-way automata and one-way automata, particularly in the deterministic case.

1 Introduction

An early result in automata theory is that of the equivalence of two-way and one-way finite automata. Rabin and Scott [13] outlined a proof of this equivalence by analyzing the so-called *crossing sequences* that occur during the acceptance of a string by a two-way automaton. This proof was slightly simplified by Shepherdson [15]. Later, Vardi [16] has shown equivalence through a subset construction that is used to characterize the complement of the language accepted by a two-way nondeterministic automaton. The *crossing sequences* proof and construction is more involved if one wants to include non-deterministic two-way automata (2NFA) in addition to deterministic ones (2DFA). While these methods in principle allow for the construction of the equivalent one-way automaton, the calculations involved are rather complex. In the crossing sequences approach, this calls for the analysis of the possible crossing sequences for possible prefixes of strings, and the complement construction requires laborious bookkeeping. While two-way automata have been analyzed intensely, especially as regards theoretical size bounds in conversions from two-way to one-way automata and state complexity of operations [3,4,7,8,10–12], practical conversion algorithms have received less attention. The intricacies involved in previous conversion methods may also be reflected in the paucity of actual implementations for converting arbitrary two-way to one-way automata.

In this paper, we describe a new method to characterize the language accepted by some two-way automata (2DFA/2NFA). Our approach is very direct: we model the set of accepting computation sequences of a two-way automaton as strings in a regular language that includes annotations about the behavior of a two-way automaton. Following this, a homomorphism is applied to delete the

© Springer International Publishing Switzerland 2015
F. Drewes (Ed.): CIAA 2015, LNCS 9223, pp. 176–187, 2015.
DOI: 10.1007/978-3-319-22360-5_15

annotations, yielding the set of actual strings accepted by a two-way automaton. Central to the modeling is a compact simulation of the accepting sequences of a given 2DFA/2NFA. Apart from providing a construction method, the approach also gives an alternative to the equivalence proofs of one-way and two-way automata customarily provided in most textbooks on automata (e.g. [5,9,14]).

2 Notation and Definitions

We define a 2NFA M the standard way as a 5-tuple $(\Sigma, Q, Q_0, \delta, F)$, where Σ denotes the alphabet, Q the finite set of states, a set of initial states $Q_0 \subseteq Q$. The transition function is denoted by $\delta : Q \times \Sigma \to 2^{Q \times \{L,S,R\}}$, and the set of final states by $F \subseteq Q$. If $|\delta(q,a)| \leq 1$ for all $q \in Q$ and all $a \in \Sigma$ and $|Q_0| = 1$, the automaton is deterministic. We say M accepts a string w whenever there is a transition path in M from an initial state to a final state such that M at each state moves its read head in the direction specified by the transition function (to the left (L), right (R), or staying (S)) and ends up at the right edge of w. Formally, we can define acceptance as a specific series of configurations using strings of the format $\Sigma^* Q \Sigma^*$. A string wqx describes the circumstance where the input string is wx and q is the current state when M is scanning the first symbol of x. We say $wpax \vdash waqx$ is a permitted change of configuration if $(q, R) \in \delta(p, a)$, as is $wbpax \vdash wpbax$ if $(q, L) \in \delta(p, a)$ and $|b| = 1$, and $wpax \vdash wqax$ if $(q, S) \in \delta(p, a)$. We say M accepts w if there exists some choice of a sequence of configuration changes such that $pw \vdash \ldots \vdash wq$, where $p \in Q_0$ and $q \in F$.

In the following, we also make use of extended regular expressions to denote operations on regular languages. We will make use of the following standard notational devices in regular expressions: a is a single symbol drawn from an alphabet, ϵ is the empty string, and $\emptyset$ the empty language. Additionally we use the following operators: $L_1 L_2$ (denoting concatenation), $L_1 \cup L_2$ (union), $L_1 \cap L_2$ (intersection), $\neg L_1$ (complement), $L_1 - L_2$ (subtraction), L_1^* (Kleene closure) and L_1^+ (Kleene plus). We also make use of the fact that regular languages are closed under homomorphisms $h : \Gamma^* \to \Sigma^*$.

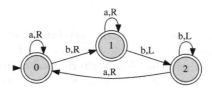

Fig. 1. Example (deterministic) 2DFA M with initial state 0. The language described is $(a|ba)^*(b|\epsilon)$.

3   Overview

The general idea behind the three methods given below is to simulate accepting or nonaccepting move sequences of a two-way automaton by a specific string representation, the correctness of which is locally checkable, that is, verifiable by a one-way automaton. The verification can be modeled through regular expressions.

We model the set of accepting computations of a two-way automaton M operating over the alphabet Σ as a regular set of strings over an auxiliary alphabet Γ and the original alphabet Σ. In particular, strings in this language consist of symbols from Σ (the alphabet of M), interspersed with auxiliary substrings that characterize relevant movements of the 2DFA/2NFA. The auxiliary alphabet Γ consists of symbols representing states in Q as well as symbols corresponding to possible moves $\{L, S, R, C\}$, i.e. $\Gamma = \{q_0, \ldots, q_n, L, S, R, C\}$. The symbol C is a move that models a crash (only used in the second construction method). We enforce that these auxiliary substrings always come in triplet-size chunks where the three-symbol sequence encodes a move by M in the order (1) the source state, (2) the target state, and (3) the direction of movement (left, right, stay).

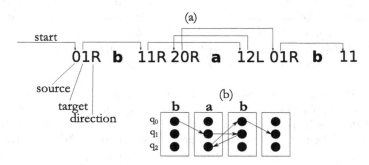

Fig. 2. Example of an accepting sequence for the 2DFA in Fig. 1 in our string encoding (a). The arrows show which source-target-direction triplets license the presence of others in the first construction method. The figure (b) illustrates the computation of the 2DFA the the string effectively encodes.

Figure 2 shows such a sequence in our encoding. The intuition behind the constructions is that, out of all possible sequences over $(\Gamma \cup \Sigma)^*$, we want to characterize all and only those that fulfill criteria that correspond to acceptance or non-acceptance of a string over Σ by some two-way automaton M. We do so by providing additional local constraints on these strings to simulate the legal movements in a 2DFA/2NFA.

In such an string, we say that any subsequence of symbols from Γ is in the same *position* as another subsequence from Γ so long as there are no intervening symbols from Σ between the two. For example, in Fig. 2, 11R and 20R are in the same *position*, while 20R is in the position preceding 12L.

4   Method 1: 2DFA to 1NFA/1DFA

In this construction method, which applies to deterministic two-way automata only, the idea is to declare a language over $(\Gamma \cup \Sigma)^*$ in such a way that the first Γ-triplet at the left edge corresponds to movement from an initial state, and that any other two-symbol sequences representing state pairs present at any position need to all be 'licensed' by some previous move from a previous position. We also require that all strings end in a qq sequence, reflecting a halt in a final state at the right edge of a string.

More formally, the conditions for well-formedness of a string w in our encoding for an accepting sequence by a 2DFA M can be specified as follows:

(1) The string w is of the form $(T^+ a)^*\, T_{end}$, where T is any three-symbol sequence pqD representing transitions of M where p and q correspond to a valid transition $p \to q$ in M between states in Q using symbol a, and $D \in \{L, S, R\}$ denotes the corresponding direction of movement in M. T_{end} is a two-symbol string qq, where q is any final state in M.

(2) Additionally, when w contains a two-symbol sequence pq, then at least one the following holds:

    (i) p corresponds to the initial state Q_0 in the definition of M and is at the left edge of w.

    (ii) there is a substring pS in the same *position* in w.

    (iii) there is a substring pL in the following *position* in w.

    (iv) there is a substring pR in the preceding *position* in w.

For example, in Fig. 2, the first occurrence of 01 is permitted since it occurs at the left edge of the word and 0 is an initial state in M (by condition (2i)), while the second occurrence is permitted because the preceding position contains 0R (by condition (2iv)). Likewise, 20 is permitted because it is followed by 2L in the following position (condition (2iii)), etc.

Note that the constraints above say nothing about the order in which the triplets themselves are permitted. There may also be arbitrary repetitions of the same triplets within a position, so long as their presence is allowed by conditions (1) and (2i-iv); these two questions are irrelevant for purposes of the encoding. By the same token, the encoding says nothing about the specific order in which the moves actually occur when M accepts a word w—only that each substring representing a move or halting be *licensed* by some other substring representing another move, save for the base case of the initial state symbol, which is always allowed as the first symbol in the string.

The sets of strings that satisfy (1) and (2i-iv) is regular, and an automaton that accepts the sets is easily constructed (see below for precise regular expressions).

If we call the language where all strings fulfill condition (1) L_{base} and the language where all strings fulfill conditions (2i-iv) $L_{license}$, we have, for a homomorphism that deletes symbols in Γ, $h(a) = \epsilon$ for all a in Γ:

$$L_1 = h(L_{base} \cap L_{license}) \tag{1}$$

Theorem 1. *M accepts a word w iff $w \in L_1$.*

Proof. First, consider the case where M accepts w. By induction on the number of steps in the computation of M, we see that $(L_{base} \cap L_{license})$ then contains a string ending in qq, for some final state q, and hence that $w \in L_1$. In the other direction: all strings in $(L_{base} \cap L_{license})$ end in the sequence qq (by definition). Now, such a string qq is only permitted by, the presence of some other move encoding at some position which in turn is permitted by some previous move, etc. (by 2ii-iv), forming a sequence of position-state pairs $(p_1, q_1) \leftarrow (p_2, q_2) \ldots \leftarrow (p_k, q_k)$, tracing the computation backward. In such a sequence no position-state pair may repeat, since—by assumption—the two-way automaton is deterministic. In other words, repetition of a state-position pair (p_i, q_j) with symbol a at position p_i would imply that $|\delta(q_i, a)| > 1$. Since there are only $|Q|n$ possible unique position-state pairs for a string of length n, this sequence must terminate, which is only possible by (2i). Hence, the final substring qq must ultimately be licensed by the initial state and the sequence describes a legitimate accepting path in M.   □

5   Construction Details

The construction described above can be immediately implemented for an arbitrary 2DFA M.

We use the alphabets:

- Σ (of M)
- $\Gamma = \{q_0, \ldots, q_n, L, S, R\}$
- $\Delta = \Sigma \cup \Gamma$ (as shorthand)

The language L_{base}, which enforces the general structure of the string, can be defined as:

$$L_{base} = \left(T_{a_1}^+ a_1 \cup \ldots \cup T_{a_n}^+ a_n\right)^* L_{end} \tag{2}$$

for symbols $a_1, \ldots, a_n \in Q$. Here, T_{a_i} contains all three-symbol strings pqD corresponding to M's transitions $p \to q$ reading symbol a_i and moving in the direction D. Formally, $pqD \in T_{a_i}$ iff $(q, D) \in \delta(p, a_i)$.

L_{end} is the set of two-symbol strings qq, where q is a halting state in M. That is, $qq \in L_{end}$ iff $q \in F$.

To describe $L_{license}$, we make use of the regular expression idiom

$$\neg(\neg S \; T \; \neg U)$$

to convey the idea that strings drawn from the set T must either be preceded by some string from the set S or followed by some string from U. This allows us to express the relevant parts of (2i-iv) above concisely. We assume that we have a set of single symbols $\mathcal{Q} \subset \Gamma$, representing the states of M. Now conditions (2i-iv) for some state q are expressed as:

$$L_{licence_q} = \neg(\neg(\mathcal{Z} \cup \underbrace{\Delta^* qS\Gamma^*}_{} \cup \underbrace{\Delta^* qR\Gamma^* \Sigma\Gamma^*}_{}) \, q\mathcal{Q} \neg(\underbrace{\Gamma^* \Sigma\Gamma^* qL\Delta^*}_{} \cup \underbrace{\Gamma^* qS\Delta^*}_{}) \,)$$

'stay' move in same position to the left	'right' move in previous position	'left' move in following position	'stay' move in same position to the right

$$\tag{3}$$

Here,

$$\mathcal{Z} = \begin{cases} \epsilon & \text{if } q \in Q_0 \\ \emptyset & \text{otherwise} \end{cases} \tag{4}$$

In other words, a sequence $q\mathcal{Q}$ (the symbol for state q followed by any other state symbol), must be preceded by right move in the previous position or a stay move in the same position with target state q, or followed by a stay move in the same position or a left move in the following position with target state q. Note that the 'stay' move is brought up twice in the expression because within a position, the moves listed are in arbitrary order, and we must therefore account for the possibility that an S-move can occur either to the left or the right of the relevant state symbol q. Additionally, symbols representing an initial state may always occur initially in the string (modeled by $\mathcal{Z}$).

For a 2DFA M with states $Q = q_0, \ldots, q_n$, the language L_1 is then

$$h(L_{base} \cap L_{license_0} \cap \ldots \cap L_{license_n}) \tag{5}$$

6 Method 2: 2NFA to 1DFA by Complement Construction

The previous method cannot be used to convert a nondeterministic two-way automaton to a 1DFA as is seen from the correctness argument which hinges on the two-way automaton being deterministic. However, we can use the same string encoding to create a similar setup where we model as a regular set all and only the words the are rejected by some 2NFA, i.e. the complement of acceptance.

The only minor change to the encoding used previously is in the auxiliary alphabet, which now becomes $\Gamma = \{q_0, \ldots, q_n, L, S, R, C\}$. the symbols L, S, R are as before, and the symbol C is an extra arbitrary symbol we use to denote a 'crash' configuration—either a state that has no outgoing transitions with some symbol, or a nonfinal state at the right edge.

In this construction, the idea is to capture all possible failing paths as a regular language by (1) insisting that all initial states be present as source states in the first position of the string encoding, and (2) requiring that each move encoding be *followed* by another move encoding or a crash—note that this enforcement is different from method 1 where we *permitted* a state-pair in the string if it resulted from a legitimate previous move; here we *require* a subsequent move for any state-pair. In other words, the presence of each transition triplet pqD

requires the presence of all legal transition triplets qrD in the following, preceding, or the current positions (for moves right, left, stay). For any state q without an outgoing transition with the symbol in that position, this requirement will also be satisfied by a triplet qqC. Such crash triplets do not themselves require a follow-up move. This encoding ensures that if a valid path through a 2NFA M exists, that path cannot be encoded in our string representation, since we lack a halting configuration. Conversely, the encoding contains all invalid paths.

Here, we have the following requirements on the well-formedness of a string w in the encoding:

(1) The string w is of the form $(T_1 \ldots T_k a_i)^* \, T_{end}$, where the Ts are three-symbol transition sequences of the form $p_1 q_1 D_1, \ldots, p_k q_k D_k$ corresponding to all transitions from p in M using symbol a, and moving to q, and $D \in \{L, S, R, C\}$ denotes the corresponding direction of movement in M. In case a state has no transition with a, ppC may be present. T_{end} is a three-symbol string qqC, where q is any nonfinal state in M. Also, the first position in string w contains all sequences pq where $p \in Q_0$ and some $q \in Q$.

(2) Additionally, when w contains a two-symbol sequence pD where $p \in Q$ (representing a state in Q) and $D \in \{L, S, R\}$, then the following holds:
   (i) if D is L there is a substring pq in the preceding *position* in w, where $q \in Q$, or pD is in the leftmost position.
   (ii) if D is R there is a substring pq in the following *position* in w, where $q \in Q$.
   (iii) if D is S there is a substring pq in the current *position* in w, where $q \in Q$.

Condition (1) enforces the general well-formedness of the strings, assuring that each symbol in Σ is surrounded by sequences of triplets corresponding to valid transitions in M. Also, if one triplet pqD is present, all other possible outgoing transitions from p also need to be listed. It also sets up the base case that all initial states are represented at the first position. Conditions (1) and (2) also ensure that any transition modeled is followed by all possible outgoing transitions from the target state, or, in the case that the target state has no valid outgoing transitions and is nonfinal, that fact is marked by a ppC, where p is the state with no outgoing transitions with the symbol at hand.

Again, the conditions (1)–(3) are all local and easily testable by DFA(s) and hence the set of strings that fulfill all conditions is a regular set.

We now claim, using the same pattern as before, that the language where all strings fulfill condition (1) L_{base} and the language where all strings fulfill conditions (2i-iii) $L_{license}$, we have for a homomorphism $h(a) = \epsilon$ for all a in Γ:

$$L_2 = \Sigma^* - h(L_{base} \cap L_{license}) \tag{6}$$

Theorem 2. *A 2NFA M accepts a word w iff $w \in L_2$.*

Proof. Suppose M accepts w. Then the accepting path through M will be modeled by (1) and (2i-iii) in $L_{base} \cap L_{license}$ with the exception of the accepting move which is never permitted, and so w is not in $h(L_{base} \cap L_{license})$. M can reject a word w if all paths in the computation eventually lack a transition for the symbol being read, end up at the right edge of a word in a nonfinal state, or try to transition left at the left edge. All such configurations are accepted by $L_{base} \cap L_{license}$, and hence w is in the language $h(L_{base} \cap L_{license})$. $\qquad\square$

Details of the actual construction are very similar to that of the first method and are omitted here.

6.1   A Note on the Construction

This approach bears similarities to the method suggested by Vardi [16]. In that work, a type of subset construction is used that directly constructs the states in the complement language accepted by a 2NFA. That construction relies on the following lemma:

Lemma 1 (Vardi, 1989). *Let $M = (\Sigma, Q, Q_0, \delta, F)$ be a two-way automaton, and $w = a_0, \dots, a_n$ be a word in Σ^*. M does not accept w if and only if there exists a sequence $T_0, \dots, T_{n+1}$ of subsets of Q such that the following conditions hold:*

1. $Q_0 \subseteq T_0$
2. $T_{n+1} \cap F = \emptyset$
3. *for $0 \leq i \leq n$, if $q \in T_i$, $(q', k) \in \delta(q, a)$, and $i + k > 0$, then $q' \in T_{i+k}$*

It is assumed here that k is an integer $\{-1, 0, 1\}$ corresponding to the directions of movement in the transition function ($\{L, S, R\}$ in our notation).

One of the consequences of this more abstract construction is that it cannot directly be used to model the set of strings *accepted* by a 2NFA, and requires the complement construction.[1] Our regular language 2NFA-1DFA construction, however, can be modified to do precisely that which is alluded to in [16]; we present the details of this additional construction method below.

7   Method 3: 2NFA to 1DFA Directly

With the 2NFA-1DFA construction above, is it not possible to directly model the set of accepting sequences by a 2NFA M, instead of modeling the complement? That is, can one not combine the techniques in method 1 and method 2 and construct a language that contains the same triplets that mark transitions in such a way as to only contain valid computation sequences of M that end in a final state. This would mean, in addition to enforcing the overall format of the strings,

---

[1] "It may be tempting to think that it is easy to get a similar condition to acceptance of w by A. It seems that all we have to do is to change the second clause in [the lemma] to $T_{n+1} \cap F \neq \emptyset$. Unfortunately, this is not the case; to characterize acceptance we also have to demand that the T_i's be minimal. While the conditions in the lemma are local, and therefore checkable by a finite-state automaton, minimality is a global condition." [16], p. 3.

requiring that all well-formed strings have initial states represented at the left edge, and that each transition pqD 'require' that a subsequent transition from q be present in the appropriate position, except for a final qq, at the right edge. Additionally, any 'crash' configuration would of course not be in the language. The problem with such an idea is exactly what is touched upon in [16]—that one must also require that any accepting path be minimal. This is illustrated in Fig. 3. Here (a) exemplifies a nonminimal path that leads to acceptance since the path of computation from the left edge fulfills the criteria by ending in a loop. Additionally, there is a spurious transition triplet before y leading to acceptance. In (b) we see the corresponding minimal path induced by the same string, showing a case of non-acceptance.

Fig. 3. Illustration of a nonminimal path starting from the initial state with a spurious path which causes nonminimality, together with the corresponding minimal path.

The idea behind this third construction is to modify the construction so that only minimal paths are in the simulation. To do this, consider the language $L = L_{base} \cap L_{license}$ that contains strings over Σ (with interspersed path descriptions) if M accepts, but that also includes spurious nonminimal paths. Now, consider the homomorphism $h(a) = \epsilon$ for all a in Γ. Define an operation $insert(L)$: $\{y \mid x \in L \wedge h^{-1}(x) = y \wedge |x| < |y|\}$, i.e. the inverse homomorphism with the additional requirement that at least one symbol from Γ is inserted. If L is regular, so is obviously $insert(L)$. In practice, we model this by composition of the identity transducer for L with a transducer Ins (see Fig. 4) that inserts at least one symbol from Γ, and reconvert to an automaton by taking the output projection: $proj_2(Id(L) \circ Ins(\Gamma))$.

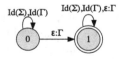

Fig. 4. Illustration of insertion transducer Ins.

The insert-operation can be used to remove the nonminimal paths in some language L that represents computations in the string encoding, and we can define the set of accepting strings by a 2NFA directly as:

$$L_3 = h(L - Insert(L)) \tag{7}$$

Taking advantage of this, we can define the conditions for any w as follows, and then use our ability to enforce minimality.

Here, we have the following requirements on the well-formedness of a string w in the encoding:

(1) The string w is of the form $(T^*a)^*$ $(T_{end}\cup\epsilon)$, where T is a set of three-symbol transition sequences of the form pqD corresponding to *some* transitions p in M using symbol a, and moving to q, and $D \in \{L, S, R\}$ denotes the corresponding direction of movement in M. T_{end} is a set of two-symbol strings qq, where q is any final state in M. Also, the first position in string w contains a sequence pq for all $p \in Q_0$ and some $q \in Q$

(2) Additionally, when w contains a two-symbol sequence pD where $p \in Q$ (representing a state in Q) and $D \in \{L, S, R\}$, then the following holds:

(i) if D is L there is a substring pq in the preceding *position* in w, where $q \in Q$.

(ii) if D is R there is a substring pq in the following *position* in w, where $q \in Q$.

(iii) if D is S there is a substring pq in the current *position* in w, where $q \in Q$.

In essence, we have modified method 2 to remove the possibility of including any 'crash' moves, and added the possibility of having qq substrings at the right edge to signal what would be an accepting path in M. We have also removed the requirement of follow-up states to moves carrying *all* possible transitions, i.e. we're not exploring paths in parallel with the model.

Again, call the language that conforms to (1) and (2) L, which is obviously regular, and we may construct the following language:

$$L_3 = h((L - Insert(L)) \cap (\Delta^*QQ)) \qquad (8)$$

Theorem 3. *M accepts a word w iff $w \in L_3$.*

Proof. Suppose M accepts w. Then, by induction we see that L contains a string ending in qq and so $w \in L_3$. Conversely, if the language L contains a string u that ends in qq, then either (1) M accepts $h(u)$ or (2) running M on $h(u)$ would end in a nonterminating loop, and additional symbols are present in u that model another path ending in qq that does not start from an initial state. But then, in the latter case, L also accepts a shorter string u' that does not contain the subpath ending in qq. But this implies that u is not in $h((L - Insert(L)) \cap (\Delta^*QQ))$, and that M accepts w.   □

8   Implementations

The methods above are practical and relatively straightforward to implement in very little space, assuming one has access to a compiler for regular expressions. We have developed a simple conversion tool that reads descriptions of 2DFAs/2NFAs and converts them into regular expressions as defined above, which can then be compiled into one-way automata.[2] For the implementation we rely on the regular expression formalism supported by the *PARC Finite-State Tool* [1] and the finite-state toolkit *foma* [6].

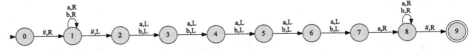

Fig. 5. Two-way automaton.

Table 1. Illustration of the growth in states when intersecting the sublanguages in suboptimal order (right) and the more efficient order that includes the L_{base} (left) with method 1, compiling the 2DFA in Fig. 5. The final 1DFA has 66 states.

k	size($L_{base} \cap L_{license_0} \cap \ldots \cap L_{license_k}$)	size($L_{license_0} \cap \ldots \cap L_{license_k}$)
0	33	58
1	77	1,394
2	112	29,634
3	166	589,570
4	204	11,271,170
5	226	NF
6	210	NF
7	131	NF
8	138	NF
9	181	NF

9   Practical Concerns

In an actual implementation it is important to calculate the intersections $L_{base} \cap L_{license_0} \cap \ldots \cap L_{license_k}$ in left-to-right order to avoid undesired exponential growth in the number of states. The $L_{license}$-languages (except for the 0-case) are symmetrical and therefore of the same size (n states) and so, in the worst case, the size of the minimal DFA result of intersection is n^k [2]. Separately constructing L_{base} and $L_{license_0} \cap \ldots \cap L_{license_n}$ is suboptimal in practice and quickly leads to unnecessary growth in the result, which would often be curbed had the general structure of L_{base} been imposed first. This is illustrated in Table 1. There we also see that the maximal partial result in the example is not substantially larger than the resulting final minimized DFA, if intersection is done in the proposed order. It is, of course, also advisable to minimize partial results through standard DFA-minimization. Additional optimizations not presented above for the sake of clarity include constraining the positions between the symbols from Σ to not contain repetitions of triplets representing transitions.

---

[2] Available at https://github.com/mhulden/2nfa.

10   Conclusion

We have presented three variants of a basic approach to converting 2DFA/2NFA to one-way automata. The construction methods offer a way to leverage the existence of efficient tools for compiling extended regular expressions into one-way automata, and thus makes it practicable to integrate two-way specifications into practical applications. We expect that the simulation method can be extended to cover more specific and constrained variants of two-way automata and two-way transducers.

References

1. Beesley, K.R., Karttunen, L.: Finite State Morphology. CSLI Publications, Stanford (2003)
2. Birget, J.C.: Intersection and union of regular languages and state complexity. Inf. Process. Lett. **43**(4), 185–190 (1992)
3. Birget, J.C.: State-complexity of finite-state devices, state compressibility and incompressibility. Math. Syst. Theor. **26**(3), 237–269 (1993)
4. Chrobak, M.: Finite automata and unary languages. Theoret. Comput. Sci. **47**, 149–158 (1986)
5. Hopcroft, J.E., Ullman, J.D.: Introduction to Automata Theory, Languages, and Computation. Addison-Wesley, Reading (1979)
6. Hulden, M.: Foma: a finite-state compiler and library. In: Proceedings of the 12th Conference of the European Chapter of the Association for Computational Linguistics, pp. 29–32. Association for Computational Linguistics (2009)
7. Kapoutsis, C.A.: Removing bidirectionality from nondeterministic finite automata. In: Jedrzejowicz, J., Szepietowski, A. (eds.) MFCS 2005. LNCS, vol. 3618, pp. 544–555. Springer, Heidelberg (2005)
8. Kapoutsis, C.A.: Size complexity of two-way finite automata. In: Diekert, V., Nowotka, D. (eds.) DLT 2009. LNCS, vol. 5583, pp. 47–66. Springer, Heidelberg (2009)
9. Kozen, D.C.: Automata and Computability. Springer, New York (1997)
10. Kunc, M., Okhotin, A.: Describing periodicity in two-way deterministic finite automata using transformation semigroups. In: Mauri, G., Leporati, A. (eds.) DLT 2011. LNCS, vol. 6795, pp. 324–336. Springer, Heidelberg (2011)
11. Kunc, M., Okhotin, A.: State complexity of union and intersection for two-way nondeterministic finite automata. Fundamenta Informaticae **110**(1), 231–239 (2011)
12. Mereghetti, C., Pighizzini, G.: Optimal simulations between unary automata. SIAM J. Comput. **30**(6), 1976–1992 (2001)
13. Rabin, M., Scott, D.: Finite automata and their decision problems. IBM J. **3**(2), 114–125 (1959)
14. Shallit, J.: A Second Course in Formal Languages and Automata Theory. Cambridge University Press, Cambridge (2008)
15. Shepherdson, J.C.: The reduction of two-way automata to one-way automata. IBM J. Res. Dev. **3**, 198–200 (1959)
16. Vardi, M.Y.: A note on the reduction of two-way automata to one-way automata. Inf. Process. Lett. **30**(5), 261–264 (1989)

Describing Homing and Distinguishing Sequences for Nondeterministic Finite State Machines via Synchronizing Automata

Natalia Kushik[✉] and Nina Yevtushenko

Tomsk State University, Tomsk, Russia
ngkushik@gmail.com, yevtushenko@sibmail.com

Abstract. There is a long standing problem of the study of homing and distinguishing sequences for deterministic and nondeterministic Finite State Machines (FSMs) which are widely used in many applications. A homing sequence allows establishing the state of the given FSM after applying the sequence while a distinguishing sequence allows learning the state of the given FSM before the sequence is applied. On the other hand, other sequences, namely, synchronizing sequences, have been thoroughly studied for finite automata. For a synchronizing automaton, there is a state such that a synchronizing sequence takes the automaton from any state to this state. There are many papers reported on such automata as well as on the complexity of synchronizing sequences. In this paper, given a complete nondeterministic FSM, we propose a method for deriving a corresponding finite automaton such that the set of all homing (or distinguishing) sequences coincides with the set of all synchronizing sequences of the derived automaton.

Keywords: Nondeterministic finite state machines · Homing sequence (homing word) · Distinguishing sequence (distinguishing word) · Synchronizing sequence (synchronizing word)

1 Introduction

There is a long standing problem of the study of homing and distinguishing sequences for deterministic and nondeterministic Finite State Machines (FSMs) which are widely used in many applications (see, for example, [1,4,13,17]). A homing sequence allows establishing the state of the given FSM after applying the sequence while a distinguishing sequence allows learning the state of the given FSM before the sequence is applied. The problem has been well studied for complete deterministic FSMs where for each pair 'state/input' there is exactly one pair 'output/next_state'. A number of papers have been published where homing and distinguishing (also called separating) sequences are derived for nondeterministic FSMs [1,14,16,20]. On the other hand, other sequences, namely,

The work is partially supported by RFBR grant No. 15-58-46013 CT_a.

F. Drewes (Ed.): CIAA 2015, LNCS 9223, pp. 188–198, 2015.
DOI: 10.1007/978-3-319-22360-5_16

synchronizing sequences, have been thoroughly studied for finite automata [15] that can be somehow considered as FSMs with no outputs. For a synchronizing automaton, there is a state such that a synchronizing sequence takes the automaton from any state to this state. There are many papers reported on such automata, the length of synchronizing sequences and the complexity of methods for checking the existence of such sequences (see, for example, [19]). In this paper, we fill a gap between homing and distinguishing sequences (words) for nondeterministic FSMs and synchronizing sequences for finite automata. In fact, given a complete nondeterministic FSM, we propose a method for deriving a corresponding finite automaton such that the set of all homing (or distinguishing) sequences of the FSM coincides with the set of all synchronizing sequences of the derived automaton and thus, the complexity of checking the existence of a homing (or a distinguishing) sequence for a complete nondeterministic observable FSM coincides with that for synchronizing sequences for partial nondeterministic automata. The automaton has the same set of inputs as the initial FSM and the number of states of the automaton is at most $n(n-1)/2+1$ for an FSM with n states. Therefore, despite the fact that the length of a homing (or a distinguishing) sequence can be exponential w.r.t. the number of FSM states [9], the problem of checking the existence of a distinguishing sequence is PSPACE. The completeness of this problem is directly implied by the fact that the problem of checking the existence of a preset distinguishing sequence for deterministic FSMs is PSPACE-complete [11]. For a homing sequence the same result for nondeterministic FSMs was obtained in a different way in [10]. However, the coincidence of the set of all homing (or distinguishing) sequences with the set of all synchronizing sequences of the corresponding automaton allows to use all the knowledge about synchronizing sequences for deriving homing and distinguishing sequences for complete nondeterministic FSMs. Moreover, to the best of our knowledge, there have not been obtained any results regarding such coincidence for homing sequences, neither for deterministic, nor for nondeterministic FSMs.

The paper is organized as follows. Section 2 contains preliminaries. Sections 3 and 4 present an idea behind the approach how to synthesize an automaton for a nondeterministic FSM such that the set of all homing/distinguishing sequences for the FSM coincides with the set of all synchronizing sequences of the automaton. Section 5 concludes the paper.

2   Preliminaries

In this section, we address the classical notions of homing, distinguishing and synchronizing sequences mostly taken from [9] and [8].

A *(non-initialized) Finite State Machine (FSM)* **S** is a 4-tuple (S, I, O, h_S), where S is a finite set of states; I and O are finite non-empty disjoint sets of inputs and outputs; $h_S \subseteq S \times I \times O \times S$ is a *transition relation* where a 4-tuple $(s, i, o, s') \in h_S$ is a *transition*.

An FSM **S** $= (S, I, O, h_S)$ is *complete* if for each pair $(s, i) \in S \times I$ there exists a pair $(o, s') \in O \times S$ such that $(s, i, o, s') \in h_S$; otherwise, the machine is *partial*.

Given a partial FSM $\mathbf{S}$, an input i is a *defined* input at state s if there exists a pair $(o, s') \in O \times S$ such that $(s, i, o, s') \in h_S$. FSM $\mathbf{S}$ is *nondeterministic* if for some pair $(s, i) \in S \times I$, there exist at least two transitions (s, i, o_1, s_1), $(s, i, o_2, s_2) \in h_S$, such that $o_1 \neq o_2$ or $s_1 \neq s_2$. FSM $\mathbf{S}$ is *observable* if for each two transitions (s, i, o, s_1), $(s, i, o, s_2) \in h_S$ it holds that $s_1 = s_2$. All machines considered in this paper are assumed to be complete. They may be nondeterministic, but are assumed to be observable. A *trace* of $\mathbf{S}$ at state s is a sequence of input/output pairs of sequential transitions starting from state s. Given a trace $i_1 o_1 \ldots i_k o_k$ at state s, $i_1 \ldots i_k$ is the *input projection* of the trace while $o_1 \ldots o_k$ is an *output response* to the input sequence $i_1 \ldots i_k$ at state s. The set of all traces of $\mathbf{S}$ at state s including the empty trace is $Tr(\mathbf{S}/s)$. As usual, for state s and a sequence $\gamma \in (IO)^*$ of input/output pairs, the γ-*successor* of state s is the set of all states that are reached from s by trace γ. If γ is not a trace at state s then the γ-successor of state s is empty. For an observable FSM $\mathbf{S}$, the cardinality of the γ-successor of state s is at most one for any string $\gamma \in (IO)^*$. Given a nonempty subset S' of states of the FSM $\mathbf{S}$ and $\gamma \in (IO)^*$, the γ-successor of the set S' is the union of γ-successors over all $s \in S'$. An input sequence α is a *homing* sequence for FSM $\mathbf{S}$ if for each trace $\gamma \in (IO)^*$ with the input projection α the γ-successor of the set S is empty or is a singleton $\{s'\}$. The set of all homing sequences of $\mathbf{S}$ is denoted $L_{home}(\mathbf{S})$. If an FSM has a homing sequence then the FSM is *homing*. An input sequence α is a *distinguishing* sequence (also called a *separating* sequence for nondeterministic FSMs) for FSM $\mathbf{S}$ if the sets of output responses to this input sequence at any two different states of $\mathbf{S}$ do not intersect. The set of all distinguishing sequences of S is denoted $L_{dist}(\mathbf{S})$. If an FSM has a distinguishing sequence then the FSM is *distinguishing*.

The notion of an FSM is very close to the automaton model [18] that does not support output responses, i.e., automaton transitions are labeled by actions that are not divided into inputs and outputs and usually these actions are called inputs or, simply, *letters* [12]. One may eliminate outputs at each transition of a given FSM in order to get the underlying automaton that can be nondeterministic for a nondeterministic FSM. In this paper, similar to non-initialized FSMs, we consider non-initialized finite automata. A sequence α is a *synchronizing* sequence (a *synchronizing word*) for a given automaton A if there exists a state a' such that α takes the automaton from any state to state a'.

If an automaton has a synchronizing sequence then the automaton is *synchronizing* [19]. The set of all synchronizing sequences is denoted as $L_{synch}(\mathrm{A})$. Synchronizing sequences for automata have been well studied. In particular, it has been shown that the length of a synchronizing word for a complete deterministic automaton with n states is bounded by a polynomial of degree three while for a nondeterministic automaton it is of the order 2^n [2,5,6]. For partial automata, as well as for nondeterministic automata the length of a synchronizing sequence (if it exists) is also of the order 2^n [5,12]. The complexity of the problem of checking whether a given automaton is synchronizing is PSPACE-complete independently whether the automaton is complete or not [12,15].

We show how the known results for synchronizing automata can be applied to efficiently check if a nondeterministic complete observable FSM has a homing or a distinguishing sequence as it is done for deterministic FSMs in [3]. For this reason, we briefly sketch the algorithm for deriving a homing / distinguishing sequence (HS / DS). Since both algorithms are almost the same we describe the algorithm for deriving a HS making then a corresponding remark about the DS derivation.

Algorithm 1. for deriving a shortest HS for an FSM

Input : $\mathbf{S} = (S, I, O, h_S)$

Output: A shortest HS for $\mathbf{S}$ or the message " the FSM $\mathbf{S}$ is not homing"

Derive a truncated successor tree for the FSM $\mathbf{S}$. The root of the tree is labeled with the set of the pairs $\overline{s_p, s_q}$, where $s_p, s_q \in S$, $s_p \neq s_q$; the nodes of the tree are labeled by sets of pairs of the set S. Edges of the tree are labeled by inputs and there exists an edge labeled by i from a node P of level j, $j \geq 0$, to a node Q such that a pair $\overline{s_p, s_q} \in Q$ if this pair is an io-successor of some pair from P. The set Q contains a singleton if io-successors of some pair of P coincide for some $o \in O$. If the input i distinguishes each pair of states of P, then the set Q is empty.

Given a node P at the level k, $k > 0$, the node is *terminal* if one of the following conditions holds.

Rule-1: P is the empty set.

Rule-2: P contains a set R without singletons that labels a node at a level j, $j < k$.

Rule-3: P has only singletons.

if *the successor tree has no nodes labeled with a set of singletons or with the empty set, i.e., is not truncated using Rules 1 or 3* **then**

| Return the message "FSM $\mathbf{S}$ is not homing".

else

| Determine a path with minimal length from the root to a node labeled with a set of singletons or with the empty set;

| Return HS as the input sequence α that labels the selected path.

end

By construction of the successor tree in Algorithm 1, the following statements can be established by induction [7].

Proposition 1. *Given a path of the truncated successor tree derived by Algorithm 1 from the root to a node labeled with the set P of state pairs, let the path carry an input sequence α. The set P contains the pair $\overline{s_p, s_q}$ if and only if there exists a trace $\gamma \in (IO)^\star$ with the input projection α such that s_p and s_q are γ-successors of some state pair of the FSM $\mathbf{S}$.*

Proposition 2. *Given a path of the truncated successor tree derived by Algorithm 1 from the root to a node labeled with the set of singletons or with the* *empty set, let the path carry an input sequence* α. *The sequence* α *is a HS for the FSM* **S**.

Proposition 3. *Each HS of FSM* **S** *is the prolongation of an input sequence that labels a path of the successor tree from the root to a node labeled with the set of singletons or with the empty set.*

An example of a truncated successor tree can be found in Fig. 1 for an FSM **S** taken from [9]. Table 1 has transitions of FSM **S** with four states 0,1,2,3, the set $I = \{i_0, i_1, i_2\}$ of inputs, and the set $O = \{(j,k), (j < k)\&(j,k \in \{0,1,2,3\})\}$ of outputs.

Table 1. FSM S

Input/State	0	1	2	3
i_0	$3/(0,3)$	$1/(0,3)$	$2/(0,3)$	$3/(0,3)$
i_1	$1/(0,2)$	$0/(0,2)$	$2/(0,2)$	$3/(0,3)$
	$1/(0,3)$	$0/(0,3)$	$2/(1,2)$	$3/(1,3)$
	$1/(2,3)$		$2/(2,3)$	$3/(2,3)$
i_2	$2/(0,1)$	$2/(0,1)$	$0/(0,1)$	$3/(0,3)$
	$2/(0,3)$	$2/(0,3)$	$0/(0,3)$	$3/(1,3)$
	$2/(1,3)$	$2/(1,3)$	$1/(1,3)$	$3/(2,3)$
				$1/(0,1)$

When Algorithm 1 is used for deriving a DS, Rule 3 and the above propositions are slightly modified in the following way.

Given a node P at the level k, $k > 0$, the node is *terminal* if one of the following conditions holds.

Rule-1: P is the empty set.
Rule-2: P contains the subset R' of all pairs of the set that labels a node at a level j, $j < k$.
Rule-3′: P has a singleton.

If the successor tree has no nodes labeled with the empty set, i.e., is not truncated using Rule 1 then return the message "FSM **S** is not distinguishing". Otherwise, determine a path with minimal length from the root to a node labeled with the empty set; Return DS as the input sequence α that labels the selected path.

Correspondingly, the following propositions hold.

Proposition 4. *Given a path of the truncated successor tree derived by modified Algorithm 1 from the root to a node labeled with the empty set, let the path carry an input sequence* α. *The sequence* α *is a DS for the FSM* **S**.

Proposition 5. *Each DS of FSM* **S** *is the prolongation of an input sequence that labels a path of the truncated successor tree derived by modified Algorithm 1 from the root to a node labeled with the empty set.*

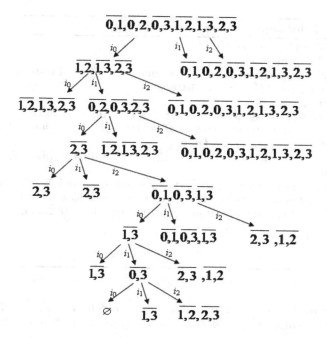

Fig. 1. The truncated tree for an FSM with a flow table in Table 1

The successor tree derived by Procedure 1 can be naturally considered for deriving a transition diagram of a special automaton where states are pairs of states of the FSM **S** together with the designated state *sink*. The transitions of the automaton are defined according to the tree branches where the state *sink* is reached under input i when a corresponding pair is distinguished by the input i or for some $o \in O$, the *io*-successor is a singleton. According to Proposition 2, a homing sequence exists for the FSM **S** if and only if the obtained automaton has a synchronizing sequence to the *sink* state.

When deriving such automaton for the set of all distinguishing sequences, the automaton can become partial when, given a pair $\overline{s_p, s_q}$, for some input i there exists o such that the *io*-successor of $\overline{s_p, s_q}$ is a singleton (Rule 3').

3 Describing the Set of All Homing Sequences of a Nondeterministic FSM via a Synchronizing Automaton

Consider a nondeterministic FSM $\mathbf{S} = (S, I, O, h_S)$, $S = \{s_1, \ldots, s_n\}$, derive an automaton $S^2{}_{home}$ such that the set of all synchronizing sequences of this automaton coincides with the set of all homing sequences of FSM **S**, i.e., $L_{home}(S) = L_{synch}(S^2{}_{home})$.

By construction, the automaton $S^2{}_{home}$ has the following features.

Algorithm 2. for deriving the automaton $S^2{}_{home}$

Input   : Complete observable FSM $\mathbf{S} = (S, I, O, h_S)$
Output: Automaton $S^2{}_{home}$
States of $S^2{}_{home}$ are pairs $\overline{s_j, s_k}$, $j < k$, and the designated state $sink$ while
actions are inputs of the FSM $\mathbf{S}$ (letters);
for *each input $i \in I$* **do**
$\quad$ **for** *each state $\overline{s_j, s_k}$ of the automaton $S^2{}_{home}$* **do**
$\quad\quad$ **if** *$\{s_p, s_t\}$ is the io-successor of the set $\{s_j, s_k\}$ for some output*
$\quad\quad$ *$o \in O$, $p < t$ and $j < k$* **then**
$\quad\quad\quad$ | $\quad$ Add to the automaton $S^2{}_{home}$ the transition $(\overline{s_j, s_k}, i, \overline{s_p, s_t})$
$\quad\quad$ **end**
$\quad\quad$ **if** *for each output $o \in O$ the io-successor of the pair $\overline{s_j, s_k}$ is a*
$\quad\quad$ *singleton or states s_j and s_k are distinguished by the input i* **then**
$\quad\quad\quad$ | $\quad$ Add to the automaton $S^2{}_{home}$ the transition $(\overline{s_j, s_k}, i, sink)$
$\quad\quad$ **end**
$\quad$ **end**
end

Proposition 6. *Given a complete observable nondeterministic FSM $\mathbf{S} = (S, I, O, h_S)$, the automaton $S^2{}_{home}$ derived by Algorithm 2 can be nondeterministic.*

Proof. The automaton $S^2{}_{home}$ can be nondeterministic, since given a pair $\overline{s_j, s_k}$, $j < k$, of states of FSM $\mathbf{S}$, there can be transitions to different pairs under the same input i according to different io_1-successors and io_2-successors, $o_1 \neq o_2$, of the pair $\overline{s_j, s_k}$.

Proposition 7. *Given a sequence $\alpha \in I^\star$, a pair $\overline{s_p, s_q}$, $s_p, s_q \in S$, $s_p \neq s_q$, α takes the automaton $S^2{}_{home}$ from the pair $\overline{s_p, s_q}$ to the sink state if and only for each trace $\gamma \in (IO)^\star$ with the input projection α, the γ-successor of $\overline{s_p, s_q}$ is the empty set or a singleton.*

The statement can be easily proven by induction on the length of the sequence α.

The next theorem that is based on Proposition 7, shows that the set of synchronizing sequences of the automaton $S^2{}_{home}$ coincides with the set of all homing sequences of the FSM $\mathbf{S}$.

Theorem 1. *An input sequence α is a homing sequence for the FSM $\mathbf{S}$ if and only if α is a synchronizing sequence for $S^2{}_{home}$.*

Proof. By construction, state $sink$ has a self-loop, and thus, if the automaton $S^2{}_{home}$ has a synchronizing sequence then this sequence takes the automaton from each state to the state $sink$. Moreover, given a sequence $\alpha \in I^\star$ and a pair $\overline{s_p, s_q}$, $s_p, s_q \in S$, $s_p < s_q$, α takes the automaton $S^2{}_{home}$ from the pair $\overline{s_p, s_q}$ to the $sink$ state if and only for any trace $\gamma \in (IO)^\star$ with the input projection α, the γ-successor of $\overline{s_p, s_q}$ is the empty set or a singleton. Thus, due to Proposition 2, α is a synchronizing sequence for the automaton $S^2{}_{home}$ if and only if α is a homing sequence for FSM $\mathbf{S}$.

Corollary 1. *The set of all homing sequences of the FSM S is equal to the set of all synchronizing sequences of the automaton $S^2{}_{home}$, i.e., $L_{home}(S) = L_{synch}(S^2{}_{home})$.*

Corollary 2. *FSM S is homing if and only if the automaton $S^2{}_{home}$ is synchronizing.*

As an example, consider again an FSM from [9], for which a homing sequence is a prolongation of a shortest homing sequence with the length that is exponential w.r.t. the number of FSM states. The flow table of the automaton $S^2{}_{home}$ returned by Algorithm 2 is shown in Table 2.

Table 2. The flow table of the automaton $S^2{}_{home}$ for the FSM S from Table 1

Input/State	$\overline{0,1}$	$\overline{0,2}$	$\overline{0,3}$	$\overline{1,2}$	$\overline{1,3}$	$\overline{2,3}$	sink
i_0	$\overline{1,3}$	$\overline{2,3}$	sink	$\overline{1,2}$	$\overline{1,3}$	$\overline{2,3}$	sink
i_1	$\overline{0,1}$	$\overline{1,2}$	$\overline{1,3}$	$\overline{0,2}$	$\overline{0,3}$	$\overline{2,3}$	sink
i_2	sink	$\overline{0,2}$	$\overline{2,3}$	$\overline{0,2}$	$\overline{1,2}$	$\overline{0,1}$	sink
		$\overline{1,2}$		$\overline{1,2}$	$\overline{2,3}$	$\overline{0,3}$	
						$\overline{1,3}$	

For example, consider transitions from state $\overline{2,3}$. For inputs i_0 and i_1, there are transitions to state $\overline{2,3}$; there are transitions to states $\overline{0,1}$, $\overline{1,3}$, and $\overline{0,3}$ under input i_2. By direct inspection, one can assure that for the automaton $S^2{}_{home}$, a shortest synchronizing sequence is the sequence $i_0i_1i_0i_2i_0i_1i_0$, and this sequence is a shortest homing sequence for FSM S [9]. Any prolongation of this sequence is a synchronizing sequence for the automaton $S^2{}_{home}$ and a homing sequence for the FSM S.

4   Describing the Set of All Distinguishing Sequences of a Nondeterministic FSM via a Synchronizing Automaton

In order to describe the set of all distinguishing sequences we remind that the procedure for deriving a distinguishing sequence is also based on a truncated successor tree that is very close to that derived when using Algorithm 1 . The only difference is that when for some $o \in O$, the io-successors of states of some pair of P coincide there is no edge from the node labeled by i to the next tree level. For this reason, we can derive the automaton almost in the same manner: the only difference is that the automaton has undefined transitions, i.e., the automaton can become partial.

Similar to the statements of the previous section, the following results can be established.

Theorem 2. *An input sequence α is a distinguishing sequence for the FSM S if and only if α is a synchronizing sequence for $S^2{}_{dist}$.*

Algorithm 3. for deriving the automaton $S^2{}_{dist}$

Input   : Complete observable FSM $\mathbf{S} = (S, I, O, h_S)$
Output: Automaton $S^2{}_{dist}$
States of $S^2{}_{dist}$ are pairs $\overline{s_j, s_k}$, $j < k$, and the designated state $sink$ while
actions are inputs of the FSM $\mathbf{S}$;
for *each input* $i \in I$ **do**
⎸ **for** *each state state* $\overline{s_j, s_k}$ *of the automaton* $S^2{}_{dist}$ **do**
⎸ ⎸ **if** *states* s_j *and* s_k *are separated by input* i **then**
⎸ ⎸ ⎸ Add to the automaton $S^2{}_{dist}$ the transition $(\overline{s_j, s_k}, i, sink)$
⎸ ⎸ **end**
⎸ ⎸ **if** *for each* $o \in O$, *the io-successors of states* s_j *and* s_k *do not*
⎸ ⎸ *coincide and* $\{s_p, s_t\}$ *is the io'-successor of the set* $\{s_j, s_k\}$ *for*
⎸ ⎸ *some* $o' \in O$, $p < t$ *and* $j < k$ **then**
⎸ ⎸ ⎸ Add to the automaton $S^2{}_{dist}$ the transition $(\overline{s_j, s_k}, i, \overline{s_p, s_t})$
⎸ ⎸ **end**
⎸ **end**
⎸ Add to the automaton $S^2{}_{dist}$ the transition $(sink, i, sink)$;
end

Corollary 3. *The set of all distinguishing sequences of the FSM $\boldsymbol{S}$ coincides with the set of all synchronizing sequences of the automaton $S^2{}_{dist}$, i.e., $L_{dist}(\boldsymbol{S}) = L_{synch}(S^2{}_{dist})$.*

Given a complete observable nondeterministic FSM $\mathbf{S}$, the automaton $S^2{}_{dist}$ derived by Algorithm 3 can be not only nondeterministic but also partial, since for some pair $\overline{s_j, s_k}$, $j < k$, of states of FSM $\mathbf{S}$, there can be no transition under input i if states s_j and s_k have the same nonempty io-successor for some output o.

Theorem 3. *The problem of checking the existence of a distinguishing sequence is PSPACE-complete for complete observable nondeterministic FSMs.*

Proof. By definition, the problem of deriving the automaton $S^2{}_{dist}$ is in P, and this automaton has a polynomial number of transitions. Indeed, the maximum number of states of $S^2{}_{dist}$ equals $n(n-1)/2 + 1$, where n is the number of states of the FSM $\mathbf{S}$. Since $\mathbf{S}$ is observable, the maximum number of transitions at each state under each input is at most $n(n-1)/2$. Correspondingly, the number of transitions of the automaton does not exceed $|I| \cdot n^2(n-1)^2$, and the problem of checking the existence of a distinguishing sequence for an observable complete nondeterministic FSM can be reduced in a polynomial time to the one of synchronizing sequence for (partial) nondeterministic automaton. The latter is PSPACE, while the problem of checking the existence of a distinguishing sequence is PSPACE-complete for deterministic FSMs [11]. Therefore, the problem of checking the existence of a distinguishing sequence is PSPACE-complete for observable complete nondeterministic FSMs as well.

As an example, consider an FSM with Table 3 as the flow table. Given FSM **S**, we derive the corresponding automaton S^2_{dist} (Table 4). By direct inspection, one can assure that there exists a shortest synchronizing sequence $i_0 i_1 i_2 i_2 i_1$ for S^2_{dist} and thus, this sequence is a shortest distinguishing sequence for the initial FSM **S**.

Table 3. FSM **S**

Input/State	0	1	2	3
i_0	$3/o_1, 0/o_5$	$1/o_1, 0/o_7$	$2/o_5, 2/o_7$	$3/o_5, 3/o_7$
i_1	$1/o_2, o_6$	$0/o_3$	$2/o_6, o_5, o_3$	$3/o_5$
i_2	$2/o_1, 2/o_6$	$2/o_3, 3/o_6$	$0/o_4, o_6, 1/o_5$	$3/o_3, 2/o_5$

Table 4. The flow table of the automaton S^2_{dist} for the FSM **S** from Table 3

Input/State	$\overline{0,1}$	$\overline{0,2}$	$\overline{0,3}$	$\overline{1,2}$	$\overline{1,3}$	$\overline{2,3}$	$sink$
i_0		$\overline{1,3}$	$\overline{0,2}$	$\overline{0,3}$	$\overline{0,2}$	$\overline{0,3}$	$\overline{2,3}$
i_1	$sink$	$\overline{1,2}$	$sink$	$\overline{0,2}$	$sink$	$\overline{2,3}$	$sink$
i_2	$\overline{2,3}$	$\overline{0,2}$	$sink$	$\overline{0,3}$	$\overline{2,3}$	$\overline{1,2}$	$sink$

5 Conclusion

In this paper, when deriving homing or distinguishing sequences for a nondeterministic FSM, we reduce this problem to a problem of deriving a synchronizing sequence for a nondeterministic, possibly, partial automaton. We have shown that the set of all homing (distinguishing) sequences of a given FSM is equal to the set of (all) synchronizing sequences of an appropriate finite automaton that has at most $n(n-1)/2 + 1$ states for an FSM with n states and the same input alphabet. Using corresponding results for synchronizing automata and deterministic FSMs we can conclude that, similar to homing sequences, the problem of checking the existence of a distinguishing sequence for a nondeterministic complete observable FSM is PSPACE-complete.

References

1. Alur, R., Courcoubetis, C., Yannakakis, M.: Distinguishing tests for nondeterministic and probabilistic machines. In: Proceedings of the 27th ACM Symposium on Theory of Computing, pp. 363–372 (1995)
2. Cern'y, H.: Pozn'amka k homog'ennym eksperimentom s konecn'ymi avtomatami. Mat. Fyz. Cas. Slovensk. Akad. Vied. **14**, 208–216 (1964). (in Slovak)

3. Güniçen, C., İnan, K., Türker, U.C., Yenigün, H.: The relation between preset distinguishing sequences and synchronizing sequences. Formal Aspects Comput. **26**(6), 1153–1167 (2014)
4. Hierons, R.M., Jourdan, G.V., Ural, H., Yenigun, H.: Using adaptive distinguishing sequences in checking sequence constructions. In: Proceedings of the 2008 ACM Symposium on Applied Computing, pp. 682–687 (2008)
5. Ito, M., Shikishima-Tsuji, K.: Some results on directable automata. In: Karhumäki, J., Maurer, H., Păun, G., Rozenberg, G. (eds.) Theory Is Forever. LNCS, vol. 3113, pp. 125–133. Springer, Heidelberg (2004)
6. Klyachko, A.A., Rystsov, I.K., Spivak, M.A.: In extremal combinatorial problem associated with the bound on the length of a synchronizing word in an automaton. Cybernetics **23**, 165–171 (1987)
7. Kushik, N.: Methods for deriving homing and distinguishing experiments for nondeterministic FSMs. Ph.D. thesis, Tomsk State University (2013)
8. Kushik, N., El-Fakih, K., Yevtushenko, N., Cavalli, A.: On adaptive experiments for nondeterministic finite state machines. Int. J. Softw. Tools Technol. Transf. (2014) (in press)
9. Kushik, N., Yevtushenko, N.: On the length of homing sequences for nondeterministic finite state machines. In: Proceedings of the 18th International Conference on Implementation and Application of Automata. pp. 220–231 (2013)
10. Kushik, N.G., Kulyamin, V.V., Evtushenko, N.V.: On the complexity of existence of homing sequences for nondeterministic finite state machines. Program. Comput. Softw. **40**, 333–336 (2014)
11. Lee, D., Yannakakis, M.: Testing finite-state machines: state identification and verification. IEEE Trans. Comput. **43**(3), 306–320 (1994)
12. Martugin, P.V.: Lower bounds for the length of the shortest carefully synchronizing words for two- and three-letter partial automata. J. Appl. Ind. Math **4**(15), 44–56 (2008). (in Russian)
13. Milner, R.: A Calculus of Communicating Systems. Springer-Verlag, Berlin (1980)
14. Petrenko, A., Yevtushenko, N.: Adaptive testing of deterministic implementations specified by nondeterministic FSMs. In: Wolff, B., Zaïdi, F. (eds.) ICTSS 2011. LNCS, vol. 7019, pp. 162–178. Springer, Heidelberg (2011)
15. Sandberg, S.: 1 homing and synchronizing sequences. In: Broy, M., Jonsson, B., Katoen, J.-P., Leucker, M., Pretschner, A. (eds.) Model-Based Testing of Reactive Systems. LNCS, vol. 3472, pp. 5–33. Springer, Heidelberg (2005)
16. Spitsyna, N., El-Fakih, K., Yevtushenko, N.: Studying the separability relation between finite state machines. Softw. Test. Verification Reliab. **17**(4), 227–241 (2007)
17. Starke, P.: Abstract Automata. American Elsevier, North-Holland (1972)
18. Trahtenbrot, B., Barzdin, J.: Finite Automata: Behavior and Synthesis. Nauka, Moscow (1970)
19. Volkov, M.V.: Synchronizing automata and the Černý conjecture. In: Martín-Vide, C., Otto, F., Fernau, H. (eds.) LATA 2008. LNCS, vol. 5196, pp. 11–27. Springer, Heidelberg (2008)
20. Zhang, F., Cheung, T.: Optimal transfer trees and distinguishing trees for testing observable nondeterministic finite-state machines. IEEE Trans. Softw. Eng. **19**(1), 1–14 (2003)

Expressive Capacity of Concatenation Freeness

Martin Kutrib$^{(\boxtimes)}$ and Matthias Wendlandt

Institut für Informatik, Universität Giessen, Arndtstr. 2, 35392 Giessen, Germany
{kutrib,matthias.wendlandt}@informatik.uni-giessen.de

Abstract. The expressive capacity of regular expressions without concatenation, but with complementation and a finite set of words as literals is studied. In particular, a characterization of unary concatenation-free languages by the Boolean closure of certain sets of languages is shown. The characterization is then used to derive regular languages that are not concatenation free. Closure properties of the family of concatenation-free languages are derived. Furthermore, the position of the family in the subregular hierarchy is considered and settled for the unary case. In particular, there are concatenation-free languages that do not belong to all of the families in the hierarchy. Moreover, except for comets, all of the families in the subregular hierarchy considered are strictly included in the family of concatenation-free languages.

1 Introduction

The investigation of regular expressions originates in [8]. They allow a set-theoretic characterization of languages accepted by finite automata. Compared to automata, regular expressions may be better suited for human users and therefore are often used as interfaces to specify certain patterns or languages. For example, regular(-like) expressions can be found in many software tools, where the syntax used to represent them may vary, but the concepts are very much the same everywhere. The leading idea is to describe languages by using constants and operator symbols. Classically, the constants are literals from the underlying alphabet and the symbol for the empty set, together with the operations union, concatenation, and Kleene star. However, the regular languages are closed under many more operations. So, adding these operations to regular expressions cannot increase their expressive power. On the other hand, removing an operation or replacing it by another may decrease the expressive capacity. For example, replacing the star by complementation yields the well-known and important subregular family of star-free (or regular non-counting) languages [4]. This family obeys nice characterizations, for example, in terms of aperiodic syntactic monoids [12], permutation-free DFA [9], and loop-free alternating finite automata [11]. See [6] for a recent survey on the complexity of regular(-like) expressions.

Here we study the expressive power of regular expressions without concatenation, but with complementation. In analogy with the star-free expressions we call them concatenation-free expressions. However, in order to allow non-trivial languages to be expressed, we allow any finite set of words as literals.

© Springer International Publishing Switzerland 2015
F. Drewes (Ed.): CIAA 2015, LNCS 9223, pp. 199–210, 2015.
DOI: 10.1007/978-3-319-22360-5_17

The paper is organized as follows. In the next section, we present the basic notations and definitions, and provide an introductory example. Section 3 is devoted to explore the limits of the expressive capacity of concatenation-free expressions. The basic question is whether or not they capture the regular languages. The question is answered negatively. To this end, the unary concatenation-free languages are characterized by the Boolean closure of certain sets of languages. The characterization is then used to derive regular languages that are not concatenation free. Furthermore, it is shown that just one concatenation operation suffices to describe all unary regular languages. The obvious closure properties of the family of concatenation-free languages are complemented in Sect. 4. The properties are summarized in Table 1. Finally, in Sect. 5 the position of the family of concatenation-free languages in the hierarchy of several subregular language families (see Fig. 2) is considered. It turns out that there are concatenation-free languages that do not belong to any other of the subregular families. For the special case of unary languages the precise position can be settled. In detail, though unary concatenation-free expressions are not as expressive as general regular expressions, they yield a language family that strictly includes all of the families in the subregular hierarchy depicted in Fig. 2, except for the (two-sided) comets to which it is incomparable.

2   Preliminaries and Definitions

We write Σ^* for the set of all words over the finite alphabet Σ. The *empty word* is denoted by λ. For the *length* of w we write $|w|$. We use $\subseteq$ for *inclusions* and $\subset$ for *strict inclusions*. The *complement* of a language L over alphabet Σ is again a language over alphabet Σ which is denoted by $\overline{L}$. The *family of finite languages* is denoted by FIN.

The *regular expressions* over an alphabet Σ and the languages they describe are defined inductively in the usual way: $\emptyset$ and every word (of length one) $v \in \Sigma$ are regular expressions, and when s and t are regular expressions, then $(s \cup t)$, $(s \cdot t)$, and $(s)^*$ are also regular expressions. The language $L(r)$ defined by a regular expression r is defined as follows: $L(\emptyset) = \emptyset$, $L(v) = \{v\}$, $L(s \cup t) = L(s) \cup L(t)$, $L(s \cdot t) = L(s) \cdot L(t)$, and $L(s^*) = L(s)^*$.

Since the regular languages are closed under many more operations, the approach to add operations like intersection ($\cap$), complementation ($^-$), or squaring (^{2}) does not increase the expressive power of regular expressions. However, replacing operations by others may decrease the expressive power. So, in general, $\mathrm{RE}(\Sigma, \Lambda, \Phi)$, where $\Lambda \subset \Sigma^*$ is a finite set of initial words, and Φ is a set of (regularity preserving) operations, denotes all regular(-like) expressions over Λ using only operations from Φ. Hence $\mathrm{RE}(\Sigma, \Sigma, \{\cup, \cdot, *\})$ refers to the set of all ordinary regular expressions, and $\mathrm{RE}(\Sigma, \Sigma, \{\cup, \cdot, ^-\})$ defines the star-free languages.

Here we study the expressive power of regular expressions without concatenation but with complementation, that is, $\mathrm{RE}(\Sigma, \Lambda, \{\cup, *, ^-\})$, and call the family of languages represented by such expressions *concatenation-free languages*. This definition nicely complements the definition of star-free languages except for the set of initial words. Since in the presence of concatenation, every word in Λ can be obtained by concatenating letters from Σ, the set Λ can be created for free. Here, however, we do not have concatenation and, thus, provide initially a *finite set* of words. For convenience, parentheses in regular expressions are sometimes omitted, where it is understood that the unary operations complementation and star have a higher priority than union.

In order to clarify our notion, we continue with an example.

Example 1. Let $L \subseteq \{a, b\}^*$ be the language of words that either begin with an a and have at least two consecutive b, or begin with a b.

Language L is described by the concatenation-free expression $r = \overline{(a \cup ab)^*}$. The subexpression $(a \cup ab)^*$ gives all words over the alphabet $\{a, b\}$ beginning with an a that do not have the factor bb. The complement of r describes L. ∎

3 Limits of Concatenation-Free Expressions

In this section, we investigate the expressive limits of concatenation-free expressions. The basic question is whether or not all regular languages can be described by such an expression. We are going to answer the question negatively. To this end, first the representable unary languages are characterized. We recall a well-known useful fact on unary languages, which is related to number theory:

Lemma 2. *Let $p, q \geq 1$ be two integers which are relatively prime, that is, the greatest common divisor $\gcd(p, q)$ equals 1. Then the biggest integer that cannot be written as a linear combination of these two integers is $pq - p - q$.*

A *cofinite unary* language L is stretched in the following sense. For $m \geq 1$, we set

$$L_{(m)} = \{ w \mid |w| = m \cdot |v|, \text{ for some } v \in L \}.$$

Now the family of all unary languages that are either finite or have a representation as cofinite language stretched by $m \geq 1$ joint with a finite language is denoted by $\mathscr{U}_{(m)}$. The union of all of these families is $\mathscr{U} = \bigcup_{m \geq 1} \mathscr{U}_{(m)}$.

In the following, we are particularly interested in the family UKF of languages that is defined to be the Boolean closure of $\mathscr{U}$. So, UKF is the *least family of languages which contains all members of $\mathscr{U}$ and is closed under complementation and union (and, thus, under intersection).*

Each language $L \in$ UKF has a representation

$$\bigcup_{1 \leq i \leq k} \bigcap_{1 \leq j \leq l_i} L_{i,j}, \text{ where } k, l_1, l_2, \ldots, l_k \geq 0 \text{ and } L_{i,j} \in \mathscr{U} \text{ or } \overline{L}_{i,j} \in \mathscr{U}.$$

First we consider the closure of UKF under star.

Lemma 3. *The family UKF is closed under star.*

Proof. Let $L \subseteq \{a\}^*$ be a language from the family UKF. If $L = \emptyset$ or $L = \{\lambda\}$, then in each case $L^* = \{\lambda\}$ is a finite language that belongs to UKF as well. We distinguish three further cases.

First, assume that $L = \{a^m\}$, for some $m \geq 1$, is a singleton. Then we obtain $L^* = \{a^{m \cdot n} \mid n \geq 0\} \in \mathcal{U}_{(m)}$ and, thus, $L^* \in$ UKF.

Second, if L contains two words whose lengths p and q are relatively prime, then $L^* \supseteq \{a^n \mid n > pq - p - q\}$ follows by Lemma 2. This implies $L^* \in \mathcal{U}_{(1)}$ and, thus, $L^* \in$ UKF.

For the last case we assume that the lengths of each two different words in L are *not* relatively prime. Moreover, we denote the greatest common divisor of the lengths of *all* words in L by m, where it may happen that $m = 1$. Now we consider the language of words of L whose lengths are divided by m: $L_{\langle m \rangle} = \{a^n \mid a^{m \cdot n} \in L\}$. A simple example is $L = \{a^{12}, a^{20}, a^{30}\}$, where $m = 2$ and $L_{\langle m \rangle} = \{a^6, a^{10}, a^{15}\}$. In general, if $L_{\langle m \rangle}$ contains two words w_1 and w_2 so that $\gcd(|w_1|, |w_2|) = 1$ (which is always true when $L_{\langle m \rangle}$ contains only two words), then $L^*_{\langle m \rangle} \supseteq \{a^n \mid n > |w_1||w_2| - |w_1| - |w_2|\}$ follows by Lemma 2. This implies $L^*_{\langle m \rangle} \in \mathcal{U}_{(1)}$ and, therefore, $L^* = \{a^{m \cdot n} \mid a^n \in L_{\langle m \rangle}\}^* \in \mathcal{U}_{(m)}$. This implies $L^* \in$ UKF.

Finally, let also the lengths of each two different words in $L_{\langle m \rangle}$ be *not* relatively prime. Then $L_{\langle m \rangle}$ contains always three different words w_1, w_2, and w_3 so that $\gcd(|w_1|, |w_2|, |w_3|) = 1$. Now we denote the greatest common divisor of $|w_1|$ and $|w_2|$ by d (which is greater than 1 by assumption). So, we have the representation $|w_1| = x \cdot d$ and $|w_2| = y \cdot d$, for some $x, y \geq 1$. Moreover, x and y are relatively prime. By Lemma 2, for any power d^k of d greater than $xy - x - y$ there are i and j so that $ix + jy = d^k$. Since $w_1^i w_2^j \in L^*_{\langle m \rangle}$ and $i|w_1| + j|w_2| = i \cdot x \cdot d + j \cdot y \cdot d = d^{k+1}$, the word $a^{d^{k+1}}$ belongs to $L^*_{\langle m \rangle}$. From $\gcd(|w_1|, |w_2|, |w_3|) = 1$ we derive that d and $|w_3|$ are relatively prime. So, d^{k+1} and $|w_3|$ are relatively prime and both words belong to $L^*_{\langle m \rangle}$. We conclude that $L^*_{\langle m \rangle} \supseteq \{a^n \mid n > d^{k+1}|w_3| - d^{k+1} - |w_3|\}$. As before, this implies $L^*_{\langle m \rangle} \in \mathcal{U}_{(1)}$ and, therefore, $L^* \in \mathcal{U}_{(m)}$. This shows $L^* \in$ UKF. $\square$

Now we are prepared to prove the characterization of unary concatenation-free languages.

Theorem 4. *A unary language is concatenation free if and only if it belongs to the family UKF, that is, to the Boolean closure of $\mathcal{U}$.*

Proof. Let L be a unary concatenation-free language given by a regular expression r from $\mathrm{RE}(\Sigma, \Lambda, \{\cup, *, ^-\})$. Then r is built from elements of Λ by finitely many applications of the operations union, star, and complementation. Since all finite sets belong to $\mathcal{U}$ and, thus, to UKF, and the family UKF is by definition closed under union and complementation and by Lemma 3 closed under star, language L represented by r belongs to UKF as well.

Now consider a unary language from UKF having a representation of the form $\bigcup_{1 \leq i \leq k} \bigcap_{1 \leq j \leq l_i} L_{i,j}$, where $k, l_1, l_2, \ldots, l_k \geq 0$ and $L_{i,j} \in \mathcal{U}$ or $\overline{L}_{i,j} \in \mathcal{U}$. Since the intersection can be simulated by union and complementation, this representation can immediately be converted into a concatenation-free regular expression, provided the languages $L_{i,j}$ are concatenation free. So, it remains to be shown that any language from $\mathscr{U}$ is concatenation free.

If $L \in \mathscr{U}$, then there exists an $m \geq 1$, so that $L \in \mathscr{U}_{(m)}$. Since any finite language belongs to $\mathscr{U}_{(m)}$, we consider L to be a cofinite language stretched by m joint with a finite language L_f, say $L = L_s \cup L_f$. Without loss of generality, we assume $\Sigma = \{a\}$. The stretching of L_s is undone by setting $L'_s = \{\, a^i \mid a^{m \cdot i} \in L_s \,\}$. Let l be the length of the longest word *not* belonging to L'_s. Then all words with lengths up to l are removed from L'_s thus obtaining a cofinite language L''_s containing *all* words whose lengths are at least $l + 1$ and no shorter words. The finitely many words removed from L'_s are stretched again and are included in L_f thus obtaining a finite language L'_f. The stretching of L''_s is now implemented as $r = \overline{\{\, a^j \mid j < m \cdot (l + 1) \,\}} \cap \{a^m\}^*$.

Altogether we have $L = L(r) \cup L'_f$. Since expression r contains only finite languages and the intersection can be implemented by union and complementation, we obtain a concatenation-free expression for L. □

Next we turn to show that there are regular languages which are not concatenation-free. So, the concatenation-free languages form a strict subregular family. We use languages of the form $\{\, a^n \mid n \equiv x \pmod{y} \,\}$ where $1 \leq x < y$ and $x \neq \frac{y}{2}$ as witnesses. Clearly, the later restriction is always met for odd y.

In the following, for $i \geq 1$, the ith prime number is denoted by p_i, where $p_1 = 2$.

Lemma 5. *Let $1 \leq x < y$ be two integers with $x \neq \frac{y}{2}$. Then the language $L = \{\, a^n \mid n \equiv x \pmod{y} \,\}$ is not concatenation free.*

Proof. In contrast to the assertion assume that L is concatenation free. Then it belongs to UKF and has a representation of the form $\bigcup_{1 \leq i \leq k} L_i$ with languages $L_i = \bigcap_{1 \leq j \leq l_i} L_{i,j}$, for $k, l_1, l_2, \ldots, l_k \geq 0$ and $L_{i,j} \in \mathscr{U}$ or $\overline{L}_{i,j} \in \mathscr{U}$.

For any prime number $p \geq y$, the word of length $x + x \cdot p!$ belongs to L. We choose one of the languages $L_i = L_{i,1} \cap L_{i,2} \cap \cdots \cap L_{i,t} \cap L_{i,t+1} \cap L_{i,t+2} \cap \cdots \cap L_{i,l_i}$ with $L_{i,j} \in \mathscr{U}$ for $1 \leq j \leq t$, and $\overline{L}_{i,j} \in \mathscr{U}$ for $t + 1 \leq j \leq l_i$, that contains infinitely many of these words. In particular, this implies that all languages $L_{i,j}$, $1 \leq j \leq l_i$, contain infinitely many of these words.

First we consider the languages $L_{i,j}$ belonging to $\mathscr{U}$. Since they are infinite they must be representable by the union of a finite language and a cofinite language stretched by some $m_j \geq 1$. Moreover, for infinitely many prime numbers p greater than y, and m_j, the word of length $x + x \cdot p!$ belongs to $L_{i,j}$. The word length $x + x \cdot p! = x(p! + 1)$ is divisible only by x and possibly by other numbers having only additional prime factors greater than p which, in turn, is greater than m_j. However, all but finitely many word lengths in $L_{i,j}$ are multiples of m_j. This implies that m_j is a divisor of x. Therefore, all but finitely many multiples of x are word lengths in the intersection $L_{i,1} \cap L_{i,2} \cap \cdots \cap L_{i,t}$.

Since $L_i \subseteq L$, words whose lengths are multiples of x that do not belong to L have to be excluded by intersection with $L_{i,t+1} \cap L_{i,t+2} \cap \cdots \cap L_{i,l_i}$, that is, the languages $L_{i,j}$ whose complements belong to $\mathcal{U}$. In particular, the words of length $x(p! - 1)$ have to excluded, where p is any prime number large enough. Since $p!$ is divisible by y we have $x(p! - 1) \equiv y - x \pmod{y}$ which is not equal to x due to $x \neq \frac{y}{2}$. So, these words do not belong to L. Therefore, there is at least one of the languages $L_{i,t+1}, L_{i,t+2}, \dots, L_{i,l_i}$, say $L_{i,s}$, so that infinitely many of these words are not contained in $L_{i,s}$. We consider the complement $\overline{L}_{i,s}$ which contains infinitely many of these words and belongs to $\mathcal{U}$. So, it is representable by the union of a finite language and a cofinite language stretched by some $m_s \geq 1$. Similar as before, we argue that the word length $x(p! - 1)$ is divisible only by x and possibly by other numbers having only additional prime factors greater than p which, in turn, is greater than m_s. However, all but finitely many word lengths in $\overline{L}_{i,s}$ are multiples of m_s. This implies that m_s is a divisor of x. Therefore, all but finitely many multiples of x are word lengths in $\overline{L}_{i,s}$ and, in turn, only finitely many multiples of x are word lengths in $L_{i,s}$. This implies that only finitely many multiples of x are word lengths in L_i. Since L_i has been chosen to contain infinitely many words with length of the form $x + x \cdot p!$, we obtain a contradiction. □

The next examples show that the conditions $1 \leq x < y$ and $x \neq \frac{y}{2}$ of Lemma 5 are in fact necessary.

Example 6. Let $y \geq 1$ be an integer. Then the concatenation-free regular expression $r = (a^y)^*$ represents the language $L(r) = \{ a^n \mid n \equiv 0 \pmod{y} \}$. ∎

Example 7. Let $1 \leq x < y$ be two integers with $x = \frac{y}{2}$. Then the concatenation-free regular expression $r = \overline{(a^x)^*} \cup (a^y)^*$ represents the language

$$L(r) = \{ a^n \mid n \equiv x \pmod{y} \}.$$

∎

Lemma 5 immediately implies the next theorem.

Theorem 8. *The (unary) concatenation-free languages are strictly included in the (unary) regular languages.*

Since the family of concatenation-free languages is properly included in the regular languages, it can be extended by allowing concatenations. But how many applications of concatenations are necessary to obtain all regular languages. Next we show that one concatenation suffices to obtain all unary regular languages. To this end, we use the following notation. For any fixed integer $k \geq 0$, the set of extended concatenation-free expressions that may contain at most k applications of the operation concatenation is referred to as concatenation-free expressions of degree k. The family of languages represented by such expressions is referred to by concatenation-free languages of degree k.

Theorem 9. *Every unary regular language can be represented by a concatenation-free expression of degree 1.*

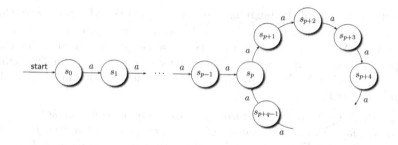

Fig. 1. General structure of a DFA accepting a unary language

Proof. Let L be a unary regular language. If L is finite it is concatenation free. So, we assume that L is infinite. A DFA accepting L has the general structure as depicted in Fig. 1. After an initial tail of $p \geq 0$ states, it runs through a cycle of some length $q \geq 1$. Any state may be accepting or rejecting.

Let M be a DFA accepting L, where s_0 is the initial state, F denotes the set of accepting states, $S_1 = \{s_1, s_2, \ldots, s_{p-1}\}$ are the states of the initial tail, and $S_2 = \{s_p, s_{p+1}, \ldots, s_{p+q-1}\}$ are the states of the cycle (see Fig. 1). The set of accepting states on the initial tail $S_1 \cap F$ are denoted by F_1 and the set of remaining accepting states $S_2 \cap F$ by F_2.

Two concatenation-free expressions r_1 and r_2 are constructed as follows. Expression r_1 describes the finitely many words from L that are accepted on the initial tail. That is, $L(r_1) = \{a^i \mid s_i \in F_1\}$.

Next, for any $s_i \in F_2$, we define $r(s_i) = \{a^i\} \cdot \{a^{\ell \cdot q} \mid \ell \geq 0\}$. The union of all $L(r(s_i))$ can be written as

$$r_2 = (a^q)^* \cdot \bigcup_{s_i \in F_2} \{a^i\}.$$

So, the expression $r_1 \cup r_2$ is concatenation free of degree 1.

We claim that L is the language $L(r_1 \cup r_2)$. It follows immediately from the construction that r_1 represents exactly the finitely many words in L whose length is at most $p - 1$. Therefore, it remains to show the claim for words whose length is at least p.

Let $w \in L(M)$ with $|w| \geq p$. Then w is accepted by M with a final state from F_2. Therefore, w belongs to the set $\{a^{i+\ell \cdot q} \mid s_i \in F_2, \ell \geq 0\}$. In particular, there is a $s_{i_0} \in F_2$ so that $w \in \{a^{i_0 + \ell \cdot q} \mid \ell \geq 0\}$. The latter set can be written as $(a^q)^* \cdot \{a^{i_0}\}$. By construction, we derive that w belongs to $L(r(s_{i_0}))$. Therefore, w belongs to $L(r_2)$ and, hence, to $L(r_1 \cup r_2)$.

Next, let $w \in L(r_1 \cup r_2)$ with $|w| \geq p$. Since all words in $L(r_1)$ are shorter than p, we have $w \in L(r_2)$. So, there is a state $s_{i_1} \in F_2$ such that w is of the form $(a^q)^* \cdot \{a^{i_1}\}$. We conclude that a computation of M on input w ends in state $s_{i_1} \in F_2$ and the word w is accepted by M. □

We conclude the section by discussing another property of concatenation-free languages. The question is whether additional symbols in the alphabet, that never occur in a word of the language, can help to increase the expressive capacity. For example, let $r = \overline{(a \cup b)^* \cup \overline{(a \cup c)^*}}$ be a concatenation-free expression over alphabet $\Sigma = \{a, b, c\}$ so that $L(r)$ is a language over a strict subset $\Sigma' = \{a\} \subset \Sigma$ only. In general, is there always a concatenation-free expression r' over alphabet Σ' describing $L(r)$? Or else, can the additional symbols be reasonably utilized?

Lemma 10. *Let r be a concatenation-free expression over alphabet Σ, so that $L(r)$ is a language over a strict subset Σ' of Σ. Then a concatenation-free expression r' over alphabet Σ' describing $L(r)$ can effectively be constructed.*

Proof. Let $r \in \mathrm{RE}(\Sigma, \Lambda, \{\cup, *, \overline{}\})$ be a concatenation-free regular expression, and $\Sigma' \subset \Sigma$ so that $L(r) \subseteq \Sigma'^*$. The construction of a concatenation-free regular expression r' in $\mathrm{RE}(\Sigma', \Lambda', \{\cup, *, \overline{}\})$ is as follows.

We have $L(r) = L(r) \cap \Sigma'^*$ and consider the intersection with Σ'^* as operation. This operation commutes with union and star:

$$(L_1 \cup L_2) \cap \Sigma'^* = (L_1 \cap \Sigma'^*) \cup (L_2 \cap \Sigma'^*) \quad \text{and} \quad L^* \cap \Sigma'^* = (L \cap \Sigma'^*)^*.$$

For complementation, we provide the rule $\overline{L} \cap \Sigma'^* = \overline{L \cup \overline{\Sigma'^*}}$.

So, we start with $L(r) \cap \Sigma'^*$ and apply these rules repeatedly to move the intersections with Σ'^* towards the words from Λ and possibly the symbol $\emptyset$. Then the intersections are applied yielding either empty sets or the words from Λ unchanged. In this way the language described is not changed, and the regular expression r' thus constructed belongs to $\mathrm{RE}(\Sigma', \Lambda', \{\cup, *, \overline{}\})$.   □

4   Closure Properties

Since some of the closure properties of concatenation-free languages follow already from their definition, this short section is devoted to summarize and complement the properties.

Theorem 11. *The family of concatenation-free languages is closed under inverse λ-free homomorphism.*

Proof. Let $r \in \mathrm{RE}(\Sigma, \Lambda, \{\cup, *, \overline{}\})$ be a concatenation-free regular expression, and $h : \Gamma^* \to \Sigma^*$ be a λ-free homomorphism. We are going to construct a concatenation-free regular expression r' for the language $h^{-1}(L(r))$. The inverse homomorphism h^{-1} commutes with each of the operations union, star, and complementation. This means, we have $h^{-1}(L_1 \cup L_2) = h^{-1}(L_1) \cup h^{-1}(L_2)$, $h^{-1}(L^*) = (h^{-1}(L))^*$, and $h^{-1}(\overline{L}) = \overline{h^{-1}(L)}$.

Since any concatenation-free regular expression is built from words of Λ and the symbol $\emptyset$ by a finite number of applications of union, star, and complementation, we may start with $h^{-1}(L(r))$ and apply these rules repeatedly to move the application of the inverse homomorphism towards the words from Λ and the symbol $\emptyset$. Then h^{-1} is applied to the occurrences of words from Λ and $\emptyset$ thus obtaining sets $h^{-1}(w) = \{u \in \Gamma^* \mid h(u) = w\}$ for all $w \in \Lambda$, and $h^{-1}(\emptyset) = \emptyset$. Since h is λ-free all these sets are finite. Finally, every such set is replaced by the union of its elements. This concludes the construction of r'.   □

Theorem 12. *The family of concatenation-free languages is closed under injective homomorphism.*

Proof. Let $r \in \mathrm{RE}(\Sigma, \Lambda, \{\cup, *, \overline{}\})$ be a concatenation-free regular expression, and $h : \Sigma^* \to \Gamma^*$ be an injective homomorphism. Similar as in the proof of Theorem 11 we construct a concatenation-free regular expression r' for the language $h(L(r))$. The injective homomorphism h commutes with the operations union and star: $h(L_1 \cup L_2) = h(L_1) \cup h(L_2)$ and $h(L^*) = (h(L))^*$.

In general, even injective homomorphisms do not commute with complementation. However, we claim the following rule: $h(\overline{L}) = \overline{h(L)} \cap (h(\Sigma))^*$.

To give evidence of the claim, let $w \notin L$. Then $h(w) \notin h(L)$ which implies $h(w) \in \overline{h(L)}$. Since $h(w) \in (h(\Sigma))^*$ is trivial one direction of the claim follows. Now let $v \in (h(\Sigma))^*$. Since h is injective the preimage $w = h^{-1}(v)$ is unique. If, in addition, $v \in \overline{h(L)}$ then $h(w) \in \overline{h(L)}$ which in turn implies $h(w) \notin h(L)$. We conclude $w \notin L$ and, thus, $w \in \overline{L}$. Therefore, we have $h(w) \in h(\overline{L})$ and, finally, $v \in h(\overline{L})$. So, the other direction and, hence, the claim follows.

The rest of the construction is as in the proof of Theorem 11 using the transformed rule $h(\overline{L}) = \overline{h(L)} \cap (h(\Sigma))^* = \overline{h(L) \cup \overline{(h(\Sigma))^*}}$ for complementations.   □

Theorem 13. *The family of concatenation-free languages is not closed under concatenation.*

Proof. The singleton language $\{a\}$ is clearly concatenation free. Let $y \geq 3$ be some integer. Example 6 shows that $\{a^n \mid n \equiv 0 \pmod{y}\}$ is concatenation free as well. However, the concatenation $\{a\} \cdot \{a^n \mid n \equiv 0 \pmod{y}\}$ is equal to $\{a^n \mid n \equiv 1 \pmod{y}\}$ which is not concatenation free by Lemma 5.   □

Theorem 14. *The family of concatenation-free languages is closed under reversal.*

Proof. Since the reversal operation commutes with union, star, and complementation, the claim follows along the line of the proof of Theorem 11.   □

Theorem 15. *The family of concatenation-free languages is not closed under intersection with regular languages.*

Proof. The language a^* is concatenation free. Its intersection with any unary regular language $L \subseteq \{a\}^*$ gives L. Since the unary concatenation-free languages are properly included in the unary regular languages the non-closure follows.   □

5   Relations with Other Subregular Families

Relations between several subregular language families are studied in [5]. These subfamilies are well motivated by their representations as finite automata or regular expressions. Just to mention a few of them: finite languages (are accepted by acyclic finite automata), definite languages [1,10] (can be realized by a register and a combinational circuit), star-free languages or regular non-counting

Table 1. Closure properties of the family of concatenation-free languages

$-$	$\cup$	$\cap$	$\cap_{REG}$	$\cdot$	$*$	$h_{injective}$	h_λ	h	h_λ^{-1}	R
yes	yes	yes	no		no	yes	yes	?	?	yes	yes

languages [4,9] (which can be described by regular(-like) expressions using only union, concatenation, and complement and which have nice characterizations in terms of aperiodic monoids and permutation-free DFA [9]), combinational languages (are accepted by automata modeling combinational circuits), locally testable languages (where the set of factors of a given length obtained from a word uniquely determines whether or not the word belongs to the language), star languages [2], and (two-sided) comet languages [3]. The hierarchy of these and some other subregular language families is depicted in Fig. 2.

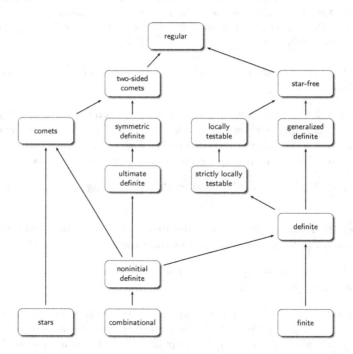

Fig. 2. Hierarchy of subregular language families under investigation. The inclusions are strict, where for stars the inclusion does not apply to the language $\{\lambda\}$.

Next we turn to discuss the position of the family of concatenation-free languages in that hierarchy. For the special case of unary languages the precise position can be settled. On the right branch, we obtain a family properly in between the regular and star-free languages. Since a unary language is star free if and only if it is either finite or cofinite [7], the unary concatenation-free regular expressions are strictly more expressive the unary star-free regular expressions.

Theorem 16. *The family of unary star-free languages is strictly included in the family of unary concatenation-free languages.*

Proof. Every unary language is star-free if and only if it is either finite or cofinite. Since all finite and cofinite languages belong to UKF, they are concatenation free (Theorem 4). On the other hand, Example 6 gives a unary concatenation-free language, that is neither finite nor cofinite. □

A language $L \subseteq \Sigma^*$ is a *star language* if and only if it can be written as $L = G^*$, for some regular language $G \subseteq \Sigma^*$, $L \subseteq \Sigma^*$ is a *comet language* if and only if it can be represented as concatenation $G^* H$ of a regular star language $G^* \subseteq \Sigma^*$ and a regular language $H \subseteq \Sigma^*$, such that $G \neq \{\lambda\}$ and $G \neq \emptyset$, and $L \subseteq \Sigma^*$ is a *two-sided comet language* if and only if $L = EG^* H$, for a regular star language $G^* \subseteq \Sigma^*$ and regular languages $E, H \subseteq \Sigma^*$, such that $G \neq \{\lambda\}$ and $G \neq \emptyset$.

Theorem 17. *The families of unary concatenation-free languages and unary (two-sided) comet languages are incomparable.*

Proof. It follows immediately from the definitions that any (two-sided) comet language that contains at least one nonempty word is infinite. On the other hand, any finite language is concatenation free.

Conversely, by Lemma 5 the language $L = \{ a^n \mid n \equiv 1 \pmod 3 \}$ is not concatenation free. On the other hand, it is described by $\{a^3\}^* \{a\}$ and, thus, a comet language. □

The next theorem completes the comparisons with the families on the left branch of the hierarchy. Its proof follows immediately from the proof of Lemma 3. In fact, the proof of the lemma does not utilize any property of the given language except for being unary. It is well known that the star of any unary language is regular. The proof of Lemma 3 shows that it is even concatenation free.

Theorem 18. *The family of unary star languages is strictly included in the family of unary concatenation-free languages.*

Concerning the middle branch of the hierarchy we know already that the family of concatenation-free languages is incomparable with the family of two-sided comet languages. One aspect of the definition of symmetric definite languages is to relax the condition of finiteness to arbitrary regular languages. A language $L \subseteq \Sigma^*$ is *symmetric definite* if and only if $L = G\Sigma^* H$, for some regular languages $G, H \subseteq \Sigma^*$.

Theorem 19. *The family of unary symmetric definite languages is strictly included in the family of unary concatenation-free languages.*

Proof. Let $L = G\{a\}^* H$ be symmetric definite. If G or H is the empty language, then L is empty and, thus, concatenation free. Otherwise, L contains all words whose length is at least the sum of the lengths of the shortest words in G and H. So, L is cofinite and, therefore, concatenation free. □

The results show that unary concatenation-free expressions are very powerful. Though they are not as expressive as general regular expressions, they yield a language family that strictly includes all of the families in the subregular hierarchy depicted in Fig. 2, except for the (two-sided) comets to which it is incomparable.

References

1. Brzozowski, J.A.: Canonical regular expressions and minimal state graphs for definite events. In: Mathematical Theory of Automata, pp. 529–561. Polytechnic Institute of Brooklyn (1962)
2. Brzozowski, J.A.: Roots of star events. J. ACM **14**, 466–477 (1967)
3. Brzozowski, J.A., Cohen, R.S.: On decompositions of regular events. J. ACM **16**, 132–144 (1969)
4. Cohen, R.S., Brzozowski, J.A.: Dot-depth of star-free events. J. Comput. Syst. Sci. **5**, 1–16 (1971)
5. Havel, I.M.: The theory of regular events II. Kybernetica **6**, 520–544 (1969)
6. Holzer, M., Kutrib, M.: The complexity of regular(-like) expressions. Int. J. Found. Comput. Sci. **22**, 1533–1548 (2011)
7. Holzer, M., Kutrib, M., Meckel, K.: Nondeterministic state complexity of star-free languages. Theoret. Comput. Sci. **450**, 68–80 (2012)
8. Kleene, S.C.: Representation of events in nerve nets and finite automata. In: Shannon, C.E., McCarthy, J. (eds.) Automata Studies, pp. 3–42. Princeton University Press, Princeton (1956)
9. McNaughton, R., Papert, S.: Counter-Free Automata. Research Monographs, vol. 65. MIT Press, Cambridge (1971)
10. Perles, M., Rabin, M.O., Shamir, E.: The theory of definite automata. IEEE Trans. Electr. Comput. **EC-12**, 233–243 (1963)
11. Salomaa, K., Yu, S.: Alternating finite automata and star-free languages. Theoret. Comput. Sci. **234**, 167–176 (2000)
12. Schützenberger, M.P.: On finite monoids having only trivial subgroups. Inform. Control **8**, 190–194 (1965)

The Membership Problem for Linear and Regular Permutation Languages

Grzegorz Madejski[✉]

Institute of Informatics, University of Gdańsk,
Wita Stwosza 57, 80-952 Gdańsk, Poland
gmadejsk@inf.ug.edu.pl

Abstract. We analyze the complexity of membership problem for two subclasses of permutation languages: *PermReg* and *PermLin*. These are languages generated by regular and linear grammars respectively, extended by rules that allow to permute symbols in derivation, such as $abX \rightarrow bXa$. We prove two NP-hardness results and analyze parameterized complexity of the problem.

1 Introduction

Context-free and context-sensitive languages are one of the most thoroughly studied language classes in the formal language theory. There are, however, such phenomena that cannot be modeled with a good accuracy or speed by either of these two. Therefore, many families of languages were introduced that extend context-free languages, but are contained within context-sensitive class.

One of such extensions is the family of permutation languages $PermCF$, introduced in [11] and also studied in [10,12]. This class consists of languages generated by context-free grammars extended with special interchange rules, that allow to permute symbols in the process of deriving a word. This class is particularly interesting because of its relation to concurrency theory. Indeed, permutation rules, as we see in upcoming proofs, can be used to model shuffle expressions, which are known to be useful in the interleaving of processes [4–6]. In addition, a subclass of permutation languages called partially-commutative context-free languages (PCCFL) was studied in terms of finding a good substitute for the Process Algebra [2]. Finally, permutation grammars could be studied in relation with natural languages with a relatively free word order, such as Finnish [9].

To be of practical use, a class of languages (or grammars) must have several good properties. One of the most important property of a grammar is to be parsable in polynomial time. Therefore, it is one of the priorities to investigate the complexity of the uniform membership problem:

Input: a language $L \subseteq A^*$, a word $w \in A^*$,
Question: $w \in L$?

where A is an alphabet.

© Springer International Publishing Switzerland 2015
F. Drewes (Ed.): CIAA 2015, LNCS 9223, pp. 211–223, 2015.
DOI: 10.1007/978-3-319-22360-5_18

The uniformity of this problem comes from the fact that the representation of language L is also given as input. One could also consider a simpler membership problem, called non-uniform, where language $L \subseteq A^*$ is fixed and its representation is not given as input:

Input: a word $w \in A^*$,
Question: $w \in L$?

Notice that investigating complexity of the above problem actually corresponds to the question: what is the complexity class of L? Another important observation is that the uniform version is harder (or to be precise: not easier) to solve than the non-uniform. There are classes, such as shuffle languages, for which the non-uniform membership problem is solvable in polynomial time [8] but the uniform version is NP-complete [13].

Unfortunately these problems seem to be hard for the permutation languages. There exist languages in $PermCF$ that are NP-hard which can be shown using the results in [13], or [1] as was mentioned in [10]. In our effort to make the languages recognizable in polynomial time we restrict the context-free rules only to be linear or regular. Thus, two subclasses of $PermCF$ emerge: $PermLin$ and $PermReg$. We find a language in $PermLin$ for which the non-uniform membership is NP-complete and prove that the uniform membership problem for $PermReg$ is NP-hard. Only the non-uniform version for $PermReg$ is not studied and remains an open problem. The results are summarized in the table together with numbers of the theorems.

Membership Problem	$PermLin$	$PermReg$
Non-Uniform	**NP-complete**[Theorem 1]	?
Uniform	**NP-hard**	**NP-hard**[Theorem 2]

The runtime of the above problems is measured in terms of the input size only. It would be interesting to know if the runtime is polynomial in the input size and exponential or worse in a parameter k, where k is the number of occurrences of permutation rules in derivation. With time bounded by a function $f(k)p(|input|)$, for some polynomial $p : \mathbb{N} \to \mathbb{N}$ and computable function $f : \mathbb{N} \to \mathbb{N}$ this problem would be fixed-parameter tractable. We could then try to minimize $f(k)$ to try to make the problem solvable not in polynomial, but yet in decent time. We show however that this problem is $W[1]$-hard which contradicts the above statements, unless $W[1] = FTP$.

The paper is organized as follows. In Sect. 2 we give some preliminary definitions and examples. In Sect. 3 we prove the NP-completeness of the non-uniform membership problem for $PermLin$. Section 4 contains the NP-hardness proof for the uniform membership problem for $PermReg$. In Sect. 5 the $W[1]$-hardness of the uniform problem for $PermReg$ is proved. We conclude the paper with Sect. 6 containing some final remarks and open problems.

2   Preliminaries

We assume the reader is familiar with the basics of formal language theory. If not, please consider reading [7] beforehand.

We quickly recall the definition of a Parikh mapping $\Psi : A^* \to \mathbb{N}^{|A|}$ over an alphabet $A = \{t_1, ..., t_n\}$. For each word in A^* this function returns a vector of occurrences of letters in the word $\Psi(w) = (|w|_{t_1}, |w|_{t_2}, ..., |w|_{t_n})$, where $|w|_t$ is the number of occurrences of letter t in w. For example for alphabet $\{a, b, c, d\}$: $\Psi(abad) = (2, 1, 0, 1)$.

Let $G = (N, A, P, S)$ be a grammar with N as the set of variables, A as the alphabet, P as the set of rules and $S \in N$ as a start symbol. In this paper we will only consider the following rules.

- A context-free rule is of the form $X \to \alpha$, where $X \in N$, $\alpha \in (A \cup N)^*$.
- A linear rule is of the form $X \to uYv$ or $X \to u$, where $X, Y \in N$, $u, v \in A^*$.
- A regular rule is of the form $X \to uY$ or $X \to u$, where $X, Y \in N$, $u \in A^*$.
- A permutation (or interchange) rule is of the form $\alpha \to \beta$, where $\alpha, \beta \in (N \cup A)^* N (N \cup A)^*$, $\alpha \neq \beta$, $\Psi(\alpha) = \Psi(\beta)$ for a Parikh mapping Ψ over the alphabet $N \cup A$. Also, two permutation rules $\alpha \to \beta$ and $\beta \to \alpha$ will be shortly denoted as $\alpha \leftrightarrow \beta$.

Let $PermCF$ stand for the family of permutation languages, i.e. languages generated by permutation grammars which contain context-free and permutation rules. The linear permutation languages $PermLin$ and regular permutation languages $PermReg$ are defined by analogy using linear permutation grammars (linear and permutation rules) and regular permutation grammars (regular and permutation rules) respectively.

Observe that from the above definitions the following containment is obvious:

$$PermReg \subseteq PermLin \subseteq PermCF.$$

We also recall the notion of a shuffle and shuffle closure operations (see [8]). We define the shuffle operation of two words inductively

$$\lambda \odot \lambda = \{\lambda\}, \quad \lambda \odot u = u \odot \lambda = \{u\}, \quad au \odot bv = a(u \odot bv) \cup b(au \odot v)$$

where $u, v \in A^*$, $a, b \in A$. A shuffle of two languages is a set $L \odot K = \{u \odot v : u \in L, v \in K\}$, whereas by $L^{\odot i}$ we denote a language $L \odot L^{\odot(i-1)}$ and $L^{\odot 0} = \{\lambda\}$. A shuffle closure of a language is a set $L^{\otimes} = \bigcup_{i \in \mathbb{N}} L^{\odot i}$.

For example for the alphabet $\{a, b, c\}$, we denote the language $L = \{w \in \{a, b, c\}^* : |w|_a = |w|_b = |w|_c\}$. The rules needed to derive all words are the following:

$$S \to aX, \quad X \to bY, \quad Y \to c \mid cS, \quad S \to \lambda,$$

$$Sa \leftrightarrow aS, \quad Sb \leftrightarrow bS, \quad Sc \leftrightarrow cS,$$

$$Xa \leftrightarrow aX, \quad Xb \leftrightarrow bX, \quad Xc \leftrightarrow cX,$$

$$Ya \leftrightarrow aY, \quad Yb \leftrightarrow bY, \quad Yc \leftrightarrow cY.$$

A word acb is derived in four steps:

$$S \Rightarrow aX \Rightarrow abY \Rightarrow aYb \Rightarrow acb.$$

Finally, let us clarify some basic concepts of parameterized complexity.

Let A be an alphabet. A language $L \subseteq A^* \times \mathbb{N}$ is a parameterized language. If $(x, k) \in L$, then k is called a parameter.

Having defined a parameterized language, we are ready to define a notion of reducibility. We will only take into account a standard parameterized reduction (or shortly FPT-reduction). In order to read more on this topic and compare with other definitions please refer to [3].

Definition 1. *We say that $L \subseteq A^* \times \mathbb{N}$ reduces to $K \subseteq B^* \times \mathbb{N}$ by a standard parameterized reduction if there are computable functions $f, g : \mathbb{N} \to \mathbb{N}$, a polynomial $p : \mathbb{N} \to \mathbb{N}$ and a function $F : A^* \times \mathbb{N} \to B^*$ such that:*

1. *$F : (x, k) \mapsto x'$ is computable in time $g(k)p(|x|)$,*
2. *$(x, k) \in L \Leftrightarrow (x', f(k)) \in K$.*

We say that a parameterized language $L \subseteq A^* \times \mathbb{N}$ is fixed-parameter tractable if there exists a Turing machine that decides L and for every input $(x, k) \in L$ it runs in time $f(k)p(|x|)$, where $f : \mathbb{N} \to \mathbb{N}$ is a computable function and $p : \mathbb{N} \to \mathbb{N}$ is a polynomial.

Fixed-parameter tractable language class (FPT) is closed under standard parameterized reduction and, as was mentioned in the previous section, has some good properties. On the other hand, there is a bigger family of parameterized languages called $W[1]$, i.e. $FPT \subseteq W[1]$ (the equality of classes is highly unlikely, but it was not proved). These are languages for which the input reduces by a standard parameterized reduction to a combinatorial circuit that has weft at most 1. For more detailed definitions, see [3].

3   The Non-uniform Membership Problem for *PermLin*

Let $A = \{a, b, c, d, e, f\}$ be an alphabet. The language $L = \{ab^k cde^k f : k \geq 0\}^\otimes$ over A is known to be NP-complete [13]. We show that it reduces to the language $L_\$ \in PermLin$. Before we go into technical details, let us have a quick motivational preface.

It is hard to show that $L \in PermLin$. The major problem in constructing a linear grammar with permutations generating this language is that the interchange rules can move the symbols too freely, resulting in such words as $abc\mathbf{bb}acdee\mathbf{f}def$, which is not in L. In other words, while shuffling the subword $abbcdeef$ into what was already generated ($abcdef$), there is a risk that the letters from this word change their order and that is not allowed.

In order to prevent that from happening, we add a special, guarding symbol \$ to the alphabet. Each time a letter from A is generated, it will come with the symbol \$ as a neighbor and only together it will be possible for this pair of symbols to be shuffled into already generated string of symbols. Then, after the shuffling of the new subword is done, the guarding symbols can be pushed to the sides of the configuration to prevent them from hindering the next steps of derivation process. Let us illustrate these steps on an example.

We want to shuffle $\boldsymbol{abbcdeef}$ into $abcdef$ to get $aabbbccddeeeff$.

Generate the first subword.	$a\$b\$c\$\$d\$e\$f$
Push the \$-signs to the left.	$\$^6abcdef$
Generate the first letter of the new subword.	$\$^6abcdef\boldsymbol{a}\$$
Move it left, but do not pass any \$-signs.	$\$^6\boldsymbol{aa}\$bcdef$
Generate the last letter of the new subword.	$\$^6\boldsymbol{aa}\$\$\boldsymbol{f}bcdef$
Move it right, but do not pass any \$-signs.	$\$^6\boldsymbol{aa}\$bcde\$\boldsymbol{f}f$
Generate the second letter of the new subword.	$\$^6\boldsymbol{aa}\$bcde\boldsymbol{b}\$\$\boldsymbol{f}f$
Move it left, but do not pass any \$-signs.	$\$^6\boldsymbol{aa}\$\boldsymbol{b}\$bcde\$\boldsymbol{f}f$

We repeat these steps for all letters of the new subword:

$$\$^6\boldsymbol{aa}\$b\$bb\$c\$c\$dd\$ee\$e\$ff$$

and after that push the \$-signs to the left: $\$^{14}aabbbccddeeeff$.

There are a few important observations to be made. The letters are generated in an alternating manner: the first one, the last one, the second one, the one before last and so on. The letters from the first half of the word (a, b, c) are always pushed left and they have their guarding symbol \$ on the right. The other letters (d, e, f) are pushed right and have their \$ on the left side. Such positions of the \$-signs are important: a letter b is moved left, but it cannot pass the letter a from the same subword. Therefore, a is protected with the \$ from the right. Imagine a worker sitting in a middle of the word, pushing letters left and right without being able to move past the letters he already placed in their position. This worker is actually a variable from the grammar that we now define.

The language with the guarding symbols over the alphabet $A \cup \{\$\}$ is $L_\$ = \bigcup_{w \in L, |w|=n} w \odot \$^{n+1}$. Observe that when we use an erasing homomorphism $h_\$(t) = t$ for $t \in A$ and $h_\$(\$) = \lambda$ (empty word) we acquire $h(L_\$) = L$. Now we construct a linear permutation grammar $G_\$ = (\{S, X_1, X_2, X_3, X_4, X_5\}, A \cup \{\$\}, P, S)$ generating $L_\$$. The set P contains the following linear rules:

$$S \xrightarrow{0} \$, \quad S \xrightarrow{1} a\$X_1, \quad X_1 \xrightarrow{2} X_2\$f, \quad X_2 \xrightarrow{3} b\$X_3,$$

$$X_3 \xrightarrow{4} X_2\$e, \quad X_2 \xrightarrow{5} c\$X_4, \quad X_4 \xrightarrow{6} X_5\$d, \quad X_5 \xrightarrow{7} S,$$

and for arbitrary $t \in A$ the following permutation rules:

$$St \xleftrightarrow{8} tS, \quad S\$ \xleftrightarrow{9} \$S, \quad tS\$ \xleftrightarrow{10} \$St,$$

$$ta\$X_1 \xrightarrow{11} a\$X_1t, \quad X_2\$ft \xrightarrow{12} tX_2\$f, \quad tb\$X_3 \xrightarrow{13} b\$X_3t,$$

$$X_2\$et \xrightarrow{14} tX_2\$e, \quad tc\$X_4 \xrightarrow{15} c\$X_4t, \quad X_5\$dt \xrightarrow{16} tX_5\$d.$$

The numbers above the arrows are just a notation to simplify the clarification of the derivation process. Let us look briefly at how the word $w = \$^{14}aabbbccddeeeff$ from previous example is derived:

$$S \stackrel{1-6}{\Longrightarrow} ... \stackrel{1-6}{\Longrightarrow} a\$b\$c\$X_5\$d\$e\$f \stackrel{7}{\Longrightarrow} a\$b\$c\$S\$d\$e\$f \stackrel{8-10}{\Longrightarrow} ...$$

$$\stackrel{8-10}{\Longrightarrow} \$^6 abcdefS \stackrel{1}{\Longrightarrow} \$^6 abcdefa\$X_1 \stackrel{11}{\Longrightarrow} ... \stackrel{11}{\Longrightarrow} \$^6 aa\$X_1 bcdef \stackrel{2}{\Longrightarrow}$$

$$\stackrel{2}{\Longrightarrow} \$^6 aa\$X_2\$fbcdef \stackrel{12}{\Longrightarrow} ... \stackrel{12}{\Longrightarrow} \$^6 aa\$bcdeX_2\$ff \stackrel{3}{\Longrightarrow}$$

$$\stackrel{3}{\Longrightarrow} \$^6 aa\$bcdeb\$X_3\$ff \stackrel{13}{\Longrightarrow} ... \stackrel{13}{\Longrightarrow} \$^6 aa\$b\$X_3 bcde\$ff \Longrightarrow$$

$$... \Longrightarrow \$^6 aa\$b\$bb\$c\$cS\$dd\$ee\$e\$ff \stackrel{8-10}{\Longrightarrow} ... \stackrel{0}{\Longrightarrow} w.$$

We show now that the grammar $G_\$$ indeed generates the language $L_\$$.

Lemma 1. $L_\$ = \mathcal{L}(G_\$)$

Proof. We start with proving $\subseteq$. Let $w \in L_\$$. We prove that $w \in \mathcal{L}(G_\$)$. By definition of the $L_\$$ language $w \in u \odot \$^{|u|+1}$, where $u \in L$. We may additionally assume that:

$$u \in (((\underbrace{ab^{n_1} cde^{n_1} f}_{u_1} \odot \underbrace{ab^{n_2} cde^{n_2} f}_{u_2}) \odot \underbrace{ab^{n_3} cde^{n_3} f}_{u_3}) \odot ...) \odot \underbrace{ab^{n_p} cde^{n_p} f}_{u_p}.$$

The idea of the proof is similar to the one presented in the above mentioned examples.

Using the rules 1–6, we generate u_1 with dollar signs $a\$(b\$)^{n_1} c\$S\$d(\$e)^{n_1}\$f$. Then we use permutations 8–10 to move all $-signs left: $\$^{2n_1+4} ab^{n_1} cde^{n_1} fS$. Next, we want to shuffle u_2 into u_1. We use the rule 1 and permutation 11 at most $2n_1 + 4$ times to put the letter a in the desired place. We then use the rule 2 and permutation 12 to put f in an arbitrary position. With rule 3 we can add a letter b, but it cannot go before the letter a of u_2, since permutations do not go past $-signs. We continue do so until all letters of u_2 are placed. Then we use permutations 8–10 to clean up the dollars getting a string of the form: $\$^{2n_1+2n_2+8}(ab^{n_1} cde^{n_1} f \odot ab^{n_2} cde^{n_2} f)S$. The procedure continues until we reach $\$^{|u|} uS$. We use the permutation 8–10 to move $-signs wherever we need and place the final $ with the rule 0, obtaining w.

Let us now prove $\supseteq$. It suffices to prove that if $w \in \mathcal{L}(G_\$)$, then in the one before last step of the derivation the string is in the set $S \odot w \odot \$^k$, where $w \in L$ and $|w| = k$. Then, one uses the rule 0 to end the derivation and the outcome is, by definition, a word of $L_\$$.

Observe that each derivation in $G_\$$ contains a loop, which starts with rule 1, contains rules 2–6 and possibly permutations 11–16. The loop ends with rule 7 and possibly permutations 8–10 afterwards. We call this an S-loop. Below, there is a derivation with strings γ_1, γ_2 after the two first S-loops.

$$S \Rightarrow ... \Rightarrow \underbrace{\gamma_1}_{\in S \odot u_1 \odot \$^{|u_1|}} \Rightarrow ... \Rightarrow \underbrace{\gamma_2}_{\in S \odot u_1 \odot u_2 \odot \$^{|u_1 \odot u_2|}} \Rightarrow ...$$

We show inductively over the number of S-loops that the derived string after each S-loop is of the form $S \odot w \odot \$^k$, where $w \in L$ and $|w| = k$.

In the base case, for $n = 0$ we omit all the productions and we are left with symbol S only. The derived string $S \odot \lambda$ is of the desired form.

Suppose that the statement is true for i and we acquired some string from the set $S \odot w_i \odot \$^k$, where $w_i \in L_1, |w_i| = k$. Having symbol S in the current derivation step, we can use permutation rules 8–10. As we already know, these rules are used to move S in the configuration and move the $-symbols to arbitrary positions. However afterwards the configuration is still of the form $S \odot w_i \odot \$^k$.

Let S be in the following position in the derived string $\alpha \$ y_0 S y_1 \$ \beta$, where $\alpha, \beta \in (A \cup \$)^*$, $y_0, y_1 \in A^*$. Other cases of strings ($\alpha \$ y_0 S y_1$ and $y_0 S y_1 \$ \beta$) are treated analogously. After using rule 1 we get $\alpha \$ y_0 a \$ X_1 y_1 \$ \beta$. Now we can use permutation 11 maximum $|y_0|$ times, because we cannot surpass symbol $. We get the following string of symbols $\alpha \$ z_1 a \$ X_1 y_2 \$ \beta$, where $y_0 y_1 = z_1 y_2$. In analogy to the step before, we use rule 2 and permutation 12 (maximum $|y_2|$ times) and get the following string $\alpha \$ z_1 a \$ y_3 X_2 \$ f z_2 \$ \beta$, where $y_0 y_1 = z_1 y_2 = z_1 y_3 z_2$.

For an arbitrary $m \geq 0$ we continue to do so for $2m+4$ steps (linear productions with their following permutations) to finally get:

$$\alpha \$ z_1 a \$ z_3 b \$ ... z_{2m+1} b \$ z_{2m+3} c \$ y_{2m+5} X_5 \$ d z_{2m+4} \$ e z_{2m+2} ... \$ e z_4 \$ f z_2 \$ \beta.$$

where $y_0 y_1 = z_1 y_2 = z_1 y_3 z_2 = ... = z_1 z_3 z_5 ... z_{2m+3} y_{2m+5} z_{2m+4} ... z_6 z_4 z_2$.

We see that the word $(a\$)(b\$)^m(c\$)(\$d)(\$e)^m(\$f)$ was shuffled into $y_0 y_1$-part of w_i. We can use rule 7 to switch to S. Then the permutations 8–10 move the $-signs to arbitrary positions. The string in derivation is of the form $S \odot w_i \odot ab^m cde^m f \odot \$^{k+2m+4}$. We clearly see that $w_{i+1} \in w_i \odot ab^m cde^m f$ and $w_{i+1} \in L$. The length $|w_{i+1}| = k + 2m + 4$ is the number of $-signs in the derivation step.

The induction proof is complete.

Since we proved that $L_\$ \in PermLin$, we are ready to show its NP-completeness.

Theorem 1. $L_\$ \in PermLin$ *is an NP-complete language.*

Proof. Let $f : A^* \to (A \cup \{\$\})^*$ be a polynomial-time reduction from L to $L_\$$ given with the formula $f(w) = \$^{k+1} w$, where $|w| = k$. Because L is NP-hard [13], then so is $L_\$$.

It is easy to construct a non-deterministic Turing machine so that it mimics $G_\$$ and accepts $L_\$$ in polynomial time. On the input tape we keep the word, on the second tape we guess and write the derivation of the word, for example, the numbers of the rules. On the third tape we derive a word using the encoded derivation. Finally we check whether the words from tape 1 and tape 3 are the same.

It is important to observe that the derivation of a word of length n requires $polynomial(n)$ steps. After each linear rule 1–6 we use the permutations 11–16 at most n times. After rule 7 is applied, the symbol S moves the $-signs in an arbitrary order using permutation rules 8–10. This can be done in $O(n^3)$ time. We find a $-sign in $O(n)$ time and move it to a desired place in $O(n)$ steps. This procedure is done for $O(n)$ $-signs.

Thus, we have proved the NP-completeness.

4   The Uniform Membership Problem for *PermReg*

In this section, we show that the uniform membership problem for *PermReg* is NP-hard.

Input: a grammar G, a word w,
Question: $w \in \mathcal{L}(G)$?

We do this by a reduction from the k-Clique problem, which is known to be NP-complete.

Input: a graph H, a number k
Question: Does graph H have a k-clique as subgraph?

The graph $H = (V_H, E_H)$ (where $|V_H| = n$, $|E_H| = m$) is undirected and we will not encode it as a standard adjacency list, but as a string $V'_H \# E'_H$, where $V'_H = v_1 v_2 \cdots v_n$ is a string containing all vertices sorted in an ascending manner in respect to their indices, $E'_H = \bigotimes_{v_i v_j \in E_H, i<j} v_i v_j \$$ is a string of edges also sorted in this way with respect to both indices of the vertices in the edge and separated with the \$-signs.

We assume that the alphabet is proportional to the number of vertices. One could encode each vertex as a binary number enabling the alphabet to be of constant size. This has a slight impact on the complexity. However, for the sake of clarity, we consider each vertex symbol to be a single sign.

The k-clique cannot consist of more edges than the graph, so we additionally assume that $m \geq \frac{k(k-1)}{2}$.

We construct a grammar $G_H = (N, A, P, S)$, where $N = \{S, X_1, X_2, ..., X_k, Y, T_1, T_2, ..., T_{m - \frac{k(k-1)}{2}}\}$, $A = \{v_1, v_2, ..., v_n, \$\}$ and the set P contains the following rules:

– the starting rule,

$$S \to \$^{\frac{k(k-1)}{2}} X_1$$

– the X-rules, which guess the vertices of the k-clique and generate $k-1$ of copies for each vertex:

$$X_1 \to v_1^{k-1} X_2 \mid v_2^{k-1} X_2 \mid ... \mid v_n^{k-1} X_2,$$
$$X_2 \to v_1^{k-1} X_3 \mid v_2^{k-1} X_3 \mid ... \mid v_n^{k-1} X_3,$$

$$...$$

$$X_k \to v_1^{k-1} Y \mid v_2^{k-1} Y \mid ... \mid v_n^{k-1} Y,$$

– the Y-permutations, which reorder all symbols in an arbitrary manner:

$$a_1 a_2 Y \to a_2 a_1 Y, \quad a_1 Y \leftrightarrow Y a_1, \text{ where } a_1, a_2 \in A, a_1 \neq a_2,$$

– the T-rules, which generate all edges outside of k-clique:

$$Y \to \lambda \mid T_1,$$

$$T_1 \to v_i v_j \$ T_2, \quad T_2 \to v_i v_j \$ T_3, \quad ..., \quad T_{m - \frac{k(k-1)}{2}} \to v_i v_j \$$$

where $1 \leq i < j \leq n$ and $v_i v_j \in E_H$.

– the T-permutations, which allow to put the missing edges in correct places

$$T_l v_i v_j \$ \to v_i v_j \$ T_l, \text{ where } l \in \{1, 2, ..., m - \frac{k(k-1)}{2}\}, 1 \le i < j \le n, v_i v_j \in E_H$$

Before we go into details of the proof let us present an example. Suppose we have the following graph and we ask whether it contains a 4-clique.

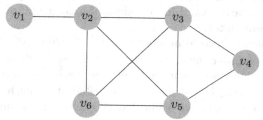

The derivation consists of three major steps:

1. We can nondeterministically guess that the 4-clique consists of vertices v_2, v_3, v_5, v_6. Using the X-rules we generate a string $\$^6 v_2^3 v_3^3 v_5^3 v_6^3 Y$.
2. Using the Y-permutations, we reorder the string to contain 6 edges that form the 4-clique: $Y v_2 v_3 \$ v_2 v_5 \$ v_2 v_6 \$ v_3 v_5 \$ v_3 v_6 \$ v_5 v_6 \$$.
3. We can generate all the 3 missing edges (bold font) which are not part of the k-clique. This is done with T-rules and T-permutations:
   $\boldsymbol{T_1} v_2 v_3 \$ v_2 v_5 \$ v_2 v_6 \$ v_3 v_5 \$ v_3 v_6 \$ v_5 v_6 \$ \Rightarrow$
   $\boldsymbol{v_1 v_2 \$} \boldsymbol{T_2} v_2 v_3 \$ v_2 v_5 \$ v_2 v_6 \$ v_3 v_5 \$ v_3 v_6 \$ v_5 v_6 \$ \Rightarrow ... \Rightarrow$
   $\boldsymbol{v_1 v_2 \$} v_2 v_3 \$ v_2 v_5 \$ v_2 v_6 \$ \boldsymbol{T_2} v_3 v_5 \$ v_3 v_6 \$ v_5 v_6 \$ \Rightarrow$
   $\boldsymbol{v_1 v_2 \$} v_2 v_3 \$ v_2 v_5 \$ v_2 v_6 \$ \boldsymbol{v_3 v_4 \$} \boldsymbol{T_3} v_3 v_5 \$ v_3 v_6 \$ v_5 v_6 \$ \Rightarrow ... \Rightarrow$
   $\boldsymbol{v_1 v_2 \$} v_2 v_3 \$ v_2 v_5 \$ v_2 v_6 \$ \boldsymbol{v_3 v_4 \$} v_3 v_5 \$ v_3 v_6 \$ \boldsymbol{T_3} v_5 v_6 \$ \Rightarrow$
   $\boldsymbol{v_1 v_2 \$} v_2 v_3 \$ v_2 v_5 \$ v_2 v_6 \$ \boldsymbol{v_3 v_4 \$} v_3 v_5 \$ v_3 v_6 \$ \boldsymbol{v_4 v_5 \$} v_5 v_6 \$$

With some intuition gained, we are ready to prove the following lemma.

Lemma 2. *Let $H = (V_H, E_H)$ be an undirected graph and $k \ge 1$. Then: H contains a k-clique $\Leftrightarrow E'_H \in \mathcal{L}(G_H)$.*

Proof. The right implication is easy to follow, once the above example is understood, so we only sketch the proof. Let C be the k-clique of H. By analogy to the graph H, we define V_C, V'_C, E_C and E'_C.

1. We start the derivation and using k X-rules we generate a string: $S \Rightarrow ... \Rightarrow \$^{\frac{k(k-1)}{2}} v_{i_1}^{k-1} v_{i_2}^{k-1} \cdots v_{i_r}^{k-1} Y$, where $V_C = \{v_{i_1}, ..., v_{i_r}\}$.
2. We use the Y-permutations to mix the vertices and \$-signs to get a string $Y E'_C$. It is important to move the symbol Y to the left. Thus, in the next step edges can be put inbetween any of the edges of C.
3. Using $m - \frac{k(k-1)}{2}$ T-rules and at most $\frac{k(k-1)}{2}$ T-permutations we add all the edges of H that are not in E_C. We derive the desired word: E'_H.

Notice that due to the freedom to move symbol by Y- and T-permutations, we are able to order the edges in the correct order.

We prove the implication in the other direction. We show that if $E'_H \in \mathcal{L}(G_H)$ is a valid string of edges, then H has a k-clique. A few key observations should be made:

- The starting rule, X-rules and Y-permutations, up to the point where $Y \to T_1$ is used, must generate a valid list of edges of a subgraph of H. The symbols cannot be reordered later.
- The X-rules generate a total number of $k(k-1)$ copies of vertices. Using Y-permutations we reorder these symbols to form $\frac{k(k-1)}{2}$ edges.
- Since the X-rules generate at most k vertices (each has many copies) and we have a total number of $\frac{k(k-1)}{2}$ edges, we conclude that each X-rule chooses a different vertex. Otherwise we would have copies of the same edge or loops of the form $v_i v_i$, which is not allowed.
- Each vertex must form an edge with the other $k-1$ vertices and therefore, it is within $k-1$ edges. Together they form a clique of size k.

The fact above should give us a clear image that the string generated is either a proper set of edges of a k-clique or a word that is not valid and cannot be translated to a graph.

After that, T-rules generate $m - \frac{k(k-1)}{2}$ missing edges from H to complete the list E'_H.

Theorem 2. *The uniform membership problem for PermReg is NP-hard.*

Proof. NP-hardness follows from Lemma 2.

5   Parameterized Complexity of the Uniform Membership Problem

In the previous section NP-hardness of the uniform membership problem was proved for the class *PermReg*. As was already said in the introductory section of this paper, such result does not give much information, because the time complexity is only considered with respect to the length of the input. Therefore, we investigate a parameterized version of this problem, which also takes into consideration a parameter k. Let us rewrite the problem from the previous section:

Input: a grammar G, a word w,
Parameter: k,
Question: Can w be derived in grammar G using permutation rules at most k times?

From practical point of view, it would be good, if this problem were in FPT class, as that would perhaps allow a certain degree of scalability. This is however not true, unless $FPT = W[1]$. We show that the problem is $W[1]$-hard.

We prove this using an FPT-reduction from the parameterized k-clique problem, which is known to be $W[1]$-hard.

Input: a graph H,
Parameter: k,
Question: Does graph H have a k-clique as subgraph?

We prove the following lemma:

Lemma 3. *The parameterized k-clique problem reduces by a standard parameterized reduction to the parameterized uniform membership problem for PermReg.*

Proof. We need to check if both conditions of the Definition 1 are fulfilled.

1. The function F translates the input of one problem to another. The graph H (given as a string $V'_H \# E'_H$) with parameter k should be transformed into a word with a grammar $w \# G_H$. The part with w is a straightforward copy of E'_H and can be written in linear time.

   It is easy to estimate that encoding G_H is polynomial over the length of the input. We roughly asses the number of steps we need to make.

   We add $k \cdot n$ X-rules, each with k symbols, which makes around $k^2 \cdot n$ steps. There are $n+1$ symbols in the alphabet (vertices and the symbol \$), so the total number of Y-permutations is $\frac{n(n+1)}{2} + 2(n+1)$, each written in constant time. Written in constant time are also all $m - \frac{k(k-1)}{2}$ T-rules and $\left(m - \frac{k(k-1)}{2}\right) \cdot m$ T-permutations, because all the edges are encoded in E'_H. When we sum the numbers, we get

$$k^2 \cdot n + \frac{n(n+1)}{2} + 2(n+1) + m - \frac{k(k-1)}{2} + \left(m - \frac{k(k-1)}{2}\right) \cdot m.$$

Since $n, k, m \leq |input|$, the time needed to encode everything is $O(|input|^3)$.

2. The equivalence follows directly from Lemma 2. It suffices to show that the function f properly translates the parameter k between problems, i.e., f : (size of the clique) $\mapsto$ (number of permutations used in derivation). It is important to show that f is only dependent on k and not on other parameters such as n or m.

   Having a clique of size k, the starting rule and X-rules generate a string of length $\frac{k(k-1)}{2} + k(k-1) = \frac{3k(k-1)}{2}$. To reorder symbols in this string in an arbitrary manner, for every of the $\frac{3k(k-1)}{2}$ symbols the variable Y must move itself to the desired place within $\frac{3k(k-1)}{2}$ steps and move this symbol to the right place using Y-permutation at most $2 \cdot \frac{3k(k-1)}{2}$ times. This results in $\frac{27k^3(k-1)^3}{4}$ applications of Y-permutations. We move Y to the beginning of the string ($\frac{3k(k-1)}{2}$ steps) and use T-permutations at most $\frac{k(k-1)}{2}$ times, since that is the maximal number of generated edges of the clique. Thus, in the end we get the formula:
   $f(k) = \frac{k^6(k-1)^3}{4} + \frac{3k(k-1)}{2} + \frac{k(k-1)}{2}$.
   That is of course a rough estimation but is sufficient to show that $f(k)$ is a polynomial over k and is not dependent on other parameters.

From the $W[1]$-hardness of k-clique the theorem follows immediately.

Theorem 3. *The parameterized uniform membership problem for PermReg with the number of applications of permutation rules set as parameter is $W[1]$-hard.*

6   Conclusions

We have proved that the non-uniform membership problem for a certain language in $PermLin$ is NP-complete. We have also shown NP-hardness of the uniform membership problem for $PermReg$. These results shed some light on the problems relating to recognizing permutation languages. It seems that imposing restrictions on context-free rules does not make the problem tractable. Therefore, it would be interesting to know if there existed a restriction on permutation rules that made the problem solvable in polynomial time.

We have also proved that uniform membership problem for $PermReg$ is $W[1]$-hard with the number of occurrences of permutation rules in derivation set as parameter. Due to this result, we know that these problems do not lie in FPT class, unless $FPT = W[1]$. There are, however, other possibilities of setting the parameter k to be studied, such as the number of permutation rules in grammar or the number of applications of context-free rules in derivation.

Notice also, that since the language $L_\$ $ from Sect. 3 is NP-complete, the uniform membership problem for $PermLin$ is NP-hard, but the completeness is not proved. It was shown that $L_\$ \in NP$, but the complexity of the whole class $PermLin$ or $PermReg$ was not analyzed. It would be interesting to show that $PermLin, PermReg \subseteq NP$ or perhaps prove an even stronger fact, that $PermReg \subseteq P$. $PermReg$ bears some resemblance to the shuffle languages which lie in P. Both rely on regular languages (either regular grammar or regular expression) with an interleaving mechanism (permutation rules or shuffle operation).

Finally, other decision problems could be analyzed in terms of decidability and complexity.

References

1. Berglund, M., Björklund, H., Björklund, J.: Shuffled languages - representation and recognition. Theor. Comput. Sci. **489–490**, 1–20 (2013)
2. Czerwinski, W., Lasota, S.: Partially-commutative context-free languages. EXPRESS/SOS, pp. 35–48 (2012)
3. Downey, R.G., Fellows, M.R.: Fundamentals of Parameterized Complexity. Springer, London (2013)
4. Esparza, J.: Petri nets, commutative context-free languages, and basic parallel processes. Fundam. Inform. **30**, 23–41 (1997)
5. Garg, V., Ragunath, M.: Concurrent regular expressions and their relationship to petri nets. Theor. Comput. Sci. **96**(2), 285–304 (1992)
6. Gischer, J.: Shuffle languages, petri nets, and context-sensitive grammars. Commun. ACM **24**(9), 597–605 (1981)
7. Hopcroft, J.E., Ullmann, J.D.: Introduction to Automata Theory, Languages, and Computation. Addison-Wesley, Cambridge (1979)
8. Jędrzejowicz, J., Szepietowski, A.: Shuffle languages are in P. Theor. Comput. Sci. **250**(1–2), 31–53 (2001)
9. Karttunen, L., Kay, M.: Parsing in a free word order language. In: Dowty, D.R., et al. (eds.) Natural Language Parsing: Psychological, Computational, and Theoretical Perspectives, pp. 279–306. Cambridge University Press, Cambridge (2005)

10. Madejski, G.: Infinite hierarchy of permutation languages. Fundam. Inform. **130**(3), 263–274 (2014)
11. Nagy, B.: Languages generated by context-free grammars extended by type AB → BA rules. J. Autom. Lang. Comb. **14**(2), 175–186 (2009)
12. Nagy, B.: On a hierarchy of permutation languages. In: Ito, M., Kobayashi, Y., Shoji, K., (eds.) Automata, Formal Languages and Algebraic Systems, pp. 163–178. World Scientific, Singapore (2010)
13. Warmuth, M.K., Haussler, D.: On the complexity of iterated shuffle. J. Comput. Syst. Sci. **28**(3), 345–358 (1984)

Classical and Quantum Counter Automata
on Promise Problems

Masaki Nakanishi[1]($\boxtimes$) and Abuzer Yakaryılmaz[2]

[1] Faculty of Education, Art and Science, Yamagata University,
Yamagata 990-8560, Japan
`masaki@cs.e.yamagata-u.ac.jp`
[2] National Laboratory for Scientific Computing, Petrópolis, Rj 25651-075, Brazil
`abuzer@lncc.br`

Abstract. In this paper, we show that one-way quantum one-counter automaton with zero-error is more powerful than its probabilistic counterpart on promise problems. Then, we obtain a similar separation result between Las Vegas one-way probabilistic one-counter automaton and one-way deterministic one-counter automaton. Lastly, it was conjectured that one-way probabilistic one blind-counter automata cannot recognize Kleene closure of equality language [A. Yakaryilmaz: Superiority of one-way and realtime quantum machines. RAIRO - Theor. Inf. and Applic. 46(4): 615–641 (2012)]. We show that this conjecture is false.

Keywords: Quantum automata · Counter automata · Promise problems · Blind counter · Exact Probabilistic and quantum computation

1 Introduction

Quantum computation is a generalization of probabilistic computation which is a generalization of deterministic computation. It is natural to ask whether a quantum model is more powerful than its probabilistic counterpart and similarly whether a probabilistic model is more powerful than its deterministic counterpart. For a fair comparison between these three types of models, bounded-error models of quantum and probabilistic should be considered, as we do in this paper.

Quantum automata models can be regarded as restricted models of quantum Turing machines; usually we restrict its memory and/or the direction of head movement. By analyzing such restricted models, we can find where the power of quantum computation comes from and what kinds of restrictions spoil advantages of quantum computation. In our case, we investigate how counter resources benefit from quantum computation.

We have a more complete picture for constant-space models (finite state automata) when compared to models using memories (finite state automata

---

M. Nakanishi—Partially supported by JSPS KAKENHI Grant Numbers 24500003 and 24106009, and also by the Asahi Glass Foundation.
A. Yakaryılmaz—Partially supported by CAPES with grant 88881.030338/2013-01.

© Springer International Publishing Switzerland 2015
F. Drewes (Ed.): CIAA 2015, LNCS 9223, pp. 224–237, 2015.
DOI: 10.1007/978-3-319-22360-5_19

augmented with counter(s), stack(s), tape(s), etc.). For example, one-way[1] deterministic finite automata (1DFAs) are equivalent to one-way probabilistic finite automata (1PFAs) and one-way quantum finite automata (1QFAs) – they define the class of regular languages [15, 20, 25]. On the other hand, in the case of two-way models[2] abbreviated respectively 2DFA, 2PFA, and 2QFA, 2DFAs are equivalent to 1DFAs, 2PFAs are more powerful than 2DFAs, and 2QFAs are more powerful than 2PFAs. [2, 7, 13, 24]. As a special case, one-way with ε-moves[3] (1_ε) quantum finite automata (1_εQFAs) can recognize some non-regular languages if the head is allowed to be in a superposition [1]. Note that ε-moves can be easily removed for the classical finite automata without increasing the number of states.

When considering finite automata using memory, there are more unanswered cases. The most challenging ones seem to be between quantum and probabilistic models. For example, 1PFAs with a counter (1P1CAs) are more powerful than 1DFAs with a counter (1D1CAs) [6] but we do not know whether 1QFAs with a counter (1Q1CAs) are more powerful than 1P1CAs – we have only an affirmative answer for one-sided bounded-error [22]. For two-way models, abbreviated respectively 2D1CAs, 2P1CAs, and 2Q1CAs, only 2Q1CAs were shown to be more powerful than 2D1CAs [28] and the other cases are still open. For one-way pushdown automata models, abbreviated respectively 1DPDAs, 1PPDAs, and 1QPDAs, 1DPDAs were shown to be weaker than even Las Vegas restriction of 1PPDAs [12] and the question is open between quantum and probabilistic models [29].

All mentioned results above are regarding language recognition. When considering solving promise problems, a generalization of language recognition such that the aim is to separate two disjoint languages that do not necessarily form the set of all strings, the picture can dramatically change [9, 17, 18, 21]. The separation results can be obtained even for one-way models or the case of zero-error – a very restricted case is that unary QFAs are more powerful than unary PFAs [8]. Also as pointed out in [9], the effects of randomness and quantumness can be more easily shown with promise problems and some open problems defined on language recognition can be answered in the case of solving promise problems. In [17, 18], exact[4] 1_εQPDAs are shown to be more powerful than exact 1_εPPDAs, which are 1_εDPDAs. In this paper, we obtain the same result between 1Q1CAs and 1P1CAs. That is, we show that exact 1Q1CAs can solve a certain promise problem that cannot be solved by exact 1P1CAs, which are 1D1CAs. As mentioned above, Las Vegas 1_εPPDAs are more powerful than 1_εDPDAs on language recognition. As the second separation, we obtain the same result for Las Vegas 1P1CAs and 1D1CAs on

---

[1] The input is read as a stream from left to right and a single symbol is fed to the machine in each step. We also use two end-markers to allow the machine making some pre- and post-processing.

[2] The input is written on a single-head read-only tape between two end-markers and the head can move in both directions or stay in the same tape square in each step.

[3] It is a restricted version of two-wayness such that the head cannot move to the left.

[4] A single answer is given with probability 1.

promise problems. In each separation, we define a new promise problem and give an algorithm for the more powerful model, and then, we show the impossibility result for the weaker model.

Additionally, we disprove the conjecture defined by Yakaryılmaz [27], in which the author separated 1QFAs with a blind counter from 1DFAs with a blind counter by using the language EQ*, the Kleene closure of EQ $= \{a^n b^n \mid n > 0\}$. Then he conjectured that the same language cannot be recognized by 1PFAs with a blind counter. However, we provide an algorithm for 1PFAs with a blind counter that recognizes EQ*.

In the next section, we provide the required background and then we present our main results under three subsections.

2   Definitions

We use the following notations: Σ, not containing ¢ and $ (the left and the right end-markers, respectively), denotes the input alphabet; $\tilde{\Sigma} = \Sigma \cup \{$¢, $\$\}$; Q is the set of (internal) states; $Q_a \subseteq Q$ (resp. $Q_r \subseteq Q$) is the set of accepting (resp. rejecting) states; q_0 is the initial state. For any $w \in \tilde{\Sigma}^*$, $w(i)$ is the i-th symbol of w.

For all models, the input $w \in \Sigma^*$ is placed on a read-only one-way infinite tape as $\tilde{w} = $ ¢$w\$$ between the cells indexed by 1 to $|\tilde{w}|$. At the beginning, the head is initially placed on the cell indexed by 1 and the value of the counter is set to zero. Also, in the following definitions, m denotes the maximum value by which the counter may be increased or decreased at each step.

A one-way probabilistic one-counter automaton (1P1CA) is a 5-tuple $M = (Q, \Sigma, \delta, q_0, Q_a)$, where $\delta : Q \times \tilde{\Sigma} \times \{Z, NZ\} \times Q \times \{-m, ..., m\} \longrightarrow [0, 1]$ is a transition function such that $\delta(q, \sigma, z, q', c) = p$ means that the transition from $q \in Q$ to $q' \in Q$ increasing the counter value by $c \in \{-m, ..., m\}$ occurs with probability $p \in [0, 1]$ if the scanned symbol is $\sigma \in \tilde{\Sigma}$ and the status of the counter value is z, where Z (resp. NZ) means zero (resp. non-zero). The transition function must satisfy the following condition since the overall probabilities must be 1 during the computation: $\forall (q, \sigma, z)$,

$$\sum_{q' \in Q, c \in \{-m, ..., m\}} \delta(q, \sigma, z, q', c) = 1.$$

The computation is terminated after reading ¢$w\$$ and the automaton accepts (resp. rejects) the input if the final state is in Q_a (resp. $Q \setminus Q_a$). Then, for each input, the acceptance (resp. rejection) probability can be calculated by summing up the probabilities of all the accepting (resp. rejecting) paths.

A one-way probabilistic blind one-counter automaton (1P1BCA) is a 1P1CA such that it cannot see the status of the counter during the computation and the input is automatically rejected if the value of the counter is non-zero [10]. A 1P1BCA is a 5-tuple $M = (Q, \Sigma, \delta, q_0, Q_a)$, where $\delta : Q \times \Sigma \times Q \times \{-m, ..., m\} \longrightarrow [0, 1]$ is a transition function such that $\delta(q, \sigma, q', c) = p$ means that the transition from $q \in Q$ to $q' \in Q$ increasing the counter value by $c \in \{-m, ..., m\}$ occurs with probability $p \in [0,1]$ if the scanned symbol is $\sigma \in \Sigma$. As described above, the transition function must satisfy the following condition: $\forall (q, \sigma)$,

$$\sum_{q' \in Q, c \in \{-m,...,m\}} \delta(q, \sigma, q', c) = 1.$$

The computation is terminated after reading $\mathcal{c}w\$$ and the automaton accepts the input if the counter value is zero and the state is in Q_a, otherwise it rejects the input.

A configuration of a counter automaton (regardless of whether blind or not) is a pair (q, v) of the current state and the current counter value. Here we do not consider the head position. In our proofs, this will not lead to any confusion.

For each of the above two models, we can define its deterministic version where the range of the transition function is restricted to $\{0, 1\}$. We abbreviate them respectively as 1D1CA and 1D1BCA.

Moreover, a one-way nondeterministic blind one-counter automaton (1N-1BCA) can be defined as a 1P1BCA with a special acceptance mode such that it accepts an input if the accepting probability is non-zero and it rejects the input if the accepting probability is zero. Here, each probabilistic choice (the probabilities are insignificant and can be removed) is called as a nondeterministic choice. Then, an input is accepted if and only if there is a path reaching an accepting condition.

Similarly, we can define a one-way universal blind one-counter automaton (1U1BCA) where the automaton accepts the input if the accepting probability is 1 and it rejects the input if the accepting probability is less than 1. In this case, each probabilistic choice (the probabilities are insignificant and can be removed) is called as a universal choice. Then, an input is accepted if and only if each path reaches an accepting condition.

A Las Vegas probabilistic machine is a probabilistic machine that (i) can also give the decision of "don't know" besides "accepting" and "rejection" and (ii) gives only one of decisions "accepting" and "rejection" on any input. For one-way Las Vegas automaton model, we split the set of states into three disjoint sets: the accepting, the rejecting, and neutral states. The automaton says "don't know" when it finishes its computation in a neutral state.

Since quantum computation is a generalization of probabilistic computation [26], any quantum model is expected to simulate its classical counterpart exactly. However, the earlier quantum finite automata (QFAs) models (e.g. [13,16]) were defined in a restrictive way and they do not reflect the full power of quantum computation. Even though they were shown to be more powerful than their classical counterparts in some special cases, these QFAs models cannot simulate classical finite automata. The first quantum counter automata model was defined based on these restricted models [14], and so, they were also shown not to be able to simulate its classical counterpart [31]. Nowadays, we know how to define general quantum automata models that generalize probabilistic automata [11,30].

Therefore, even a superiority result of a restricted model, as given in this paper, serves as a separation between the quantum and probabilistic model. Due to its simplicity, we give the definition of a restricted model that allows to represent our algorithm and we refer the reader to [22] for the definition of general quantum model. We assume the reader familiar with basics of quantum computation (see [23] for a short introduction and [19] for complete references).

A one-way quantum one-counter automaton (1Q1CA) is a 5-tuple $M = (Q, \Sigma, \delta, q_0, Q_a)$, where $\delta : Q \times \Sigma \times \{Z, NZ\} \times Q \times \{-m, ..., m\} \longrightarrow \mathbb{C}$ is a transition function; $\delta(q, \sigma, z, q', c) = p$ means that the transition from q to q' increasing the counter value by c occurs with probability amplitude p if the scanned symbol is σ and the status of the counter value is z.

$|q, v\rangle$ (resp. $\langle q, v|$), called a ket (resp. bra), denotes the column (resp. row) vector where the entry corresponding to (q, v) is one and the remaining entries are zeros. That is, $\{|q, v\rangle\}$ is an orthonormal basis of $l_2(Q \times \mathbb{Z})$. For each $\sigma \in \tilde{\Sigma}$, we define a time evolution operator U_σ as follows:

$$U_\sigma |q, v\rangle = \sum_{(q', c) \in Q \times \{-m, ..., m\}} \delta(q, \sigma, z(v), q', c)|q', v + c\rangle,$$

where $z(v) = Z$ (resp. $z(v) = NZ$) if $v = 0$ (resp. $v \neq 0$). In order to be a well-formed automaton, U_σ's must be unitary. The computation of an 1Q1CA is described by $|\Psi\rangle = U_{\tilde{w}(|\tilde{w}|)} U_{\tilde{w}(|\tilde{w}|-1)} \cdots U_{\tilde{w}(1)} |q_0, 0\rangle$. The following projective measurement P is applied to $|\Psi\rangle$ at the end of the computation:

$$P = \{P_a = \Sigma_{q \in Q_a, v \in \mathbb{Z}} |q, v\rangle\langle q, v|, P_r = \Sigma_{q \notin Q_a, v \in \mathbb{Z}} |q, v\rangle\langle q, v|\}.$$

Then, we have "a" (resp. "r") with probability $\langle \Psi | P_a | \Psi \rangle$ (resp. $\langle \Psi | P_r | \Psi \rangle$). The automaton accepts (resp. rejects) the input if we have "a" (resp. "r") as the outcome.

A promise problem $P = (P_{yes}, P_{no})$ defined on an alphabet Σ is composed by two disjoint languages $P_{yes} \subseteq \Sigma^*$ and $P_{no} \subseteq \Sigma^*$, called respectively the set of yes-instances and the set of no-instances.

A promise problem $P = (P_{yes}, P_{no})$ is said to be solved by a (probabilistic or quantum) machine M with error bound $\epsilon < \frac{1}{2}$ if any yes-instance is accepted with probability at least $1 - \epsilon$ and any no-instance is rejected with probability at least $1 - \epsilon$. It is also said that P is solved by M with bounded-error. If yes-instances (resp. no-instances) are accepted (resp. rejected) exactly, then it is said that P is solved by M with negative (resp. positive) one-sided error bound ϵ. If $\epsilon = 0$, then it is said that the promise problem is solved exactly.

A promise problem $P = (P_{yes}, P_{no})$ is said to be solved by a Las Vegas machine with success probability $p > 0$ if

- any yes-instance is accepted with probability at least p and rejected with 0 probability, and,
- any no-instance is rejected with probability at least p and accepted with 0 probability.

Remark that all non-accepting or non-rejecting probabilities go to the decision of "don't know".

3   Main Results

We start with the separation of exact quantum models from deterministic one and then we give the separation of Las Vegas probabilistic model from deterministic one. Lastly, we give our algorithm for 1P1BCA and also discuss other classical models.

3.1   Separation of Exact 1Q1CAs and 1D1CAs

We show that there exists a promise problem that can be solved by 1Q1CAs exactly but not by any 1D1CAs. For our purpose, we calculate XOR value of two comparisons. Let a, b, c, and d be four even positive numbers. Our first comparison is whether $a = c$ and the second one is whether $b = d$, and, our aim is to decide whether

$$((a = c)\,\mathsf{XOR}\,(b = d))$$

is true or false. Remark that this expression takes the value of true if and only if exactly one of the comparisons fails.

In order to implement this decision procedure by 1Q1CAs, we give the numbers as $0^a\#0^b\#0^c\#0^d$. However, due to some technical difficulties, we also append four more numbers as $\#0^{k_1}\#0^{k_2}\#0^{l_1}\#0^{l_2}$, which will help the automaton to set the counter to zero at the end of the computation so that an appropriate quantum interference can be done between the different configurations, i.e. two configurations having different counter values do not interfere.

Formally, we define our promise problem as follows. Let XOR-EQ be the set of strings of the form $0^a\#0^b\#0^c\#0^d\#0^{k_1}\#0^{k_2}\#0^{l_1}\#0^{l_2}$ such that a, b, c, and d are even and satisfy the following:

$$a - c - (-1)^{\delta_{a,c}}(k_1 - k_2) = b - d - (-1)^{\delta_{b,d}}(l_1 - l_2),$$

where $\delta_{u,v} = 1$ if $u = v$, and $\delta_{u,v} = 0$ otherwise. Then, the set XOR-EQ is our promise. We define yes-instances (XOR-EQ$_\mathsf{yes}$) as the set of strings in XOR-EQ such that $((a = c)$ xor $(b = d))$ takes the value of true. Then, no-instances (XOR-EQ$_\mathsf{no}$) are the ones taking the value of false, or equivalently XOR-EQ \ XOR-EQ$_\mathsf{yes}$.

Theorem 1. *The promise problem XOR-EQ can be solved by 1Q1CAs exactly.*

Proof. We can construct a one-way deterministic reversible one-counter automaton M_1, which is a special case of the 1Q1CA model,[5] that decides whether $a = c$ as follows.

1. M_1 reads the first block 0^a and increases the counter by one in each transition.
2. M_1 skips the second block 0^b.

---

[5] A classical reversible operation defined on the set of configurations is a unitary operator containing only 0 s and 1s.

3. M_1 reads the third block 0^c and decreases the counter by one in each transition. At the end of this block, M_1 decides whether $a = c$ or not.
4. M_1 skips the fourth block 0^d.
5. M_1 reads the fifth block 0^{k_1} and increases the counter by one if $a \neq c$ (decreases the counter by one if $a = c$) in each transition.
6. M_1 reads the sixth block 0^{k_2} and decreases the counter by one if $a \neq c$ (increases the counter by one if $a = c$) in each transition.
7. M_1 skips the seventh and the eighth blocks.

Similarly, we can construct a 1Q1CA M_2 that decides whether $b = d$ by comparing b with d using the counter and then the counter is set to zero after reading 0^{l_1} and 0^{l_2}. We illustrate M_1 and M_2 in Fig. 1.

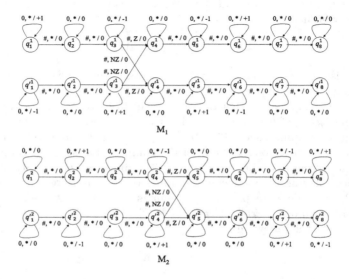

Fig. 1. Subautomata M_1 and M_2

In the figure, each label of the edges is of the form $(\sigma, z/c)$, where $\sigma \in \Sigma$, $z \in \{Z, NZ\}$, and $c \in \{-1, 0, +1\}$. A label $(\sigma, z/c)$ means that the transition occurs when the input symbol is σ and the status of the counter value is z ($*$ denotes a wild card which matches any of Z and NZ), and the counter value is updated by $c \in \{-1, 0, +1\}$. The initial state is q_1^1/q_1^2 for M_1/M_2, respectively. The set of accepting states is $\{q_8^1\}/\{q_8^2\}$ for M_1/M_2, respectively. Also the set of rejecting states is $\{q_8'^1\}/\{q_8'^2\}$ for M_1/M_2, respectively. It is easy to see that if we set the initial state to $q_1'^1$ for M_1 ($q_1'^2$ for M_2), the output is inverted.

We use the algorithm in [3] (the improved Deutsch-Jozsa algorithm [5]) to compute the exclusive-or exactly using the two sub-automata as the oracle for Deutsch's problem [4]. Note that the counter values are the same between M_1 and M_2 at the moment of reading the last input symbol. Thus, we can construct a

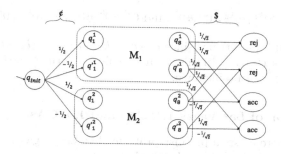

Fig. 2. Simulation of the Deutsch-Jozsa algorithm

1Q1CA that solves XOR-EQ by simulating the improved Deutsch-Jozsa algorithm [3] on it by running M_1 and M_2 in a superposition, which is illustrated in Fig. 2. In the figure, the value on each edge represents the amplitude associated with the transition. The first and the last transitions occur when it reads the left and the right end-markers, respectively. It is straightforward to see that the time evolution operators can be extended to unitary operators by adding dummy states and/or transitions.   □

Theorem 2. *No 1D1CA can solve* XOR-EQ.

Proof. We assume that there exists a 1D1CA M that solves XOR-EQ. Note that M can have at most $O(n)$ possible configurations for a string whose length is less than n; a constant number of possible states with $O(n)$ possible counter values. Also note that there are $\Theta(n^2)$ possible partial inputs of the form $0^a\#0^b\#$ whose length is less than n. Thus, there exist two distinct partial inputs $0^a\#0^b\#$ and $0^{a'}\#0^{b'}\#$ such that the configurations after reading them are the same. We will show that there exists a postfix string, $0^c\#0^d\#0^{k_1}\#0^{k_2}\#0^{l_1}\#0^{l_2}$, such that either (i) $u_1 = 0^a\#0^b\#0^c\#0^d\#0^{k_1}\#0^{k_2}\#0^{l_1}\#0^{l_2}$ is a yes-instance and $u_2 = 0^{a'}\#0^{b'}\#0^c\#0^d\#0^{k_1}\#0^{k_2}\#0^{l_1}\#0^{l_2}$ is a no-instance, or, (ii) vice versa. However, M cannot distinguish u_1 and u_2 since the two configurations after reading $0^a\#0^b\#$ and $0^{a'}\#0^{b'}\#$, respectively, are the same. This is a contradiction.

Now, we show how to obtain the required u_1 and u_2. We start with the case of $a \neq a'$. We set l_1 and l_2 so that $b \neq d$, $b' \neq d$, and d is even for $d = \frac{b+b'+a-a'}{2} + (l_1 - l_2)$. Note that this is possible since a, b, a', and b' are even. We also set k_1 and k_2 so that $-(k_1 - k_2) = b - d + (l_1 - l_2)$. Thus, both u_1 and u_2, respectively, $0^a\#0^b\#0^a\#0^d\#0^{k_1}\#0^{k_2}\#0^{l_1}\#0^{l_2}$ and $0^{a'}\#0^{b'}\#0^a\#0^d\#0^{k_1}\#0^{k_2}\#0^{l_1}\#0^{l_2}$, become promised input strings since $-(k_1 - k_2) = b - d + (l_1 - l_2)$ and $a' - a + (k_1 - k_2) = b' - d + (l_1 - l_2)$. In this setting, the former one is a yes-instance and the latter one is a no-instance.

In the following, we show how to obtain the required u_1 and u_2 when $a = a'$. Note that, in this case, $b \neq b'$.

We set k_1 and k_2 so that $a \neq c$, $a' \neq c$, and c is even for $c = \frac{a+a'+b-b'}{2} + (k_1 - k_2)$. Note that this is possible since a, b, a', and b' are even. We also set l_1 and l_2 so that $a - c + (k_1 - k_2) = -(l_1 - l_2)$. Thus, both u_1 and u_2, respectively,

$$0^a \#0^b \#0^c \#0^b \#0^{k_1} \#0^{k_2} \#0^{l_1} \#0^{l_2} \text{ and } 0^{a'} \#0^{b'} \#0^c \#0^b \#0^{k_1} \#0^{k_2} \#0^{l_1} \#0^{l_2},$$

become promised input strings since $a - c + (k_1 - k_2) = -(l_1 - l_2)$ and $a' - c + (k_1 - k_2) = b' - b + (l_1 - l_2)$. In this setting, again the former one is a yes-instance and the latter one is a no-instance. □

3.2 Separation of Las Vegas 1P1CAs and 1D1CAs

We show that there exists a promise problem that Las Vegas 1P1CAs can solve but 1D1CAs cannot. Our idea is inspired from [21].

Let ONE $\subset \{a, b, c\}^* d$ be the set of strings such that the numbers of symbols are equal for exactly one pair: (a, b), (b, c), or (c, a). Let NONE $\subset \{a, b, c\}^* d$ be the set of strings such that the numbers of symbols are equal for none of the pairs.

We define a promise problem ONE-NONE where ONE-NONE$_{yes}$, composed by yes instances, is formed by the concatenation ONE $\cdot$ NONE and ONE-NONE$_{no}$, composed by no instances, is formed by the concatenation NONE $\cdot$ ONE.

Theorem 3. *Promise problem ONE-NONE can be solved by a Las Vegas 1P1CA with success probability $\frac{1}{6}$.*

Proof. Let $udvd \subseteq \{a, b, c\}^* d \{a, b, c\}^* d$ be a promised input. The details of the automaton are as follows. At the beginning, the computation splits into 6 different paths with equal probabilities. The first three paths operate on ud and the last three paths operate on vd. Each of the first three paths compares the numbers of symbols for one of the three pairs, (a, b), (b, c) or (c, a), on u. If it finds equal number of symbols, then the automaton accepts the input in this path. Each of the last three paths makes a similar comparison on v, but, if it finds equal number of symbols, then the automaton rejects the input in this path. In all the other cases, the automaton says "don't know".

If the input is a yes instance, then the numbers of symbols are equal only for a single pair of u. Then, the input is accepted with probability $\frac{1}{6}$ in one of the first three paths, and the computation ends in a neutral state in all the other cases. Similarly, if the input is a no instance, then it is rejected with probability $\frac{1}{6}$ and the automaton says "don't know" with probability $\frac{5}{6}$. □

Theorem 4. *No 1D1CAs can solve ONE-NONE.*

Proof. We assume that there exists a 1D1CA M that solves ONE-NONE. Let $c(w)$ and $v(w)$ be the configuration and the value of the counter of M after reading the partial input string $\textcent w$, respectively. We consider the following two cases.

Case 1: There exists a 1D1CA M that solves ONE-NONE such that for at least one of $\{a, b, c\}$, say u, $|v(u^n)| \in \omega(1)$.

In this case, we set $m = f(n)$ such that $|v(u^{f(n)})| \in \omega(n)$. Without loss of generality, we pick $u = a$. We consider the input string $a^m b^i c^j d a^k b^l d$ $(i, j, k, l \in O(n))$. Note that M can increase or decrease the counter value by at most a constant amount at a single transition. Thus, the counter value cannot be zero during reading $b^i c^j da^k b^l d$ since $|v(a^m)| \in \omega(n)$. This implies the counter is useless during reading $b^i c^j da^k b^l d$, i.e. the status of the counter is always the same. Thus, by omitting the value of the counter, we define $c_{fa}(w)$ as the current state after reading the partial input string $\mathcal{c}w$.

For $t > |Q|$, we consider the sequence of states $c_{fa}(a^m b), c_{fa}(a^m b^2), \ldots,$ $c_{fa}(a^m b^t)$. Then, $c_{fa}(a^m b^{n_1}) = c_{fa}(a^m b^{n_2})$ for some distinct $n_1, n_2 < t$. Also for $t' > |Q|$, we consider the sequence of states $c_{fa}(a^m b^{n_1} c^{n_1} da), c_{fa}(a^m b^{n_1} c^{n_1} da^2),$ $\ldots, c_{fa}(a^m b^{n_1} c^{n_1} da^{t'})$. Then, $c_{fa}(a^m b^{n_1} c^{n_1} da^{n'_1}) = c_{fa}(a^m b^{n_1} c^{n_1} da^{n'_2})$ for some distinct $n'_1, n'_2 < t'$. Note that $c_{fa}(a^m b^{n_2} c^{n_1} da^{n'_1}) = c_{fa}(a^m b^{n_1} c^{n_1} da^{n'_1})$, and thus, $c_{fa}(a^m b^{n_2} c^{n_1} da^{n'_1}) = c_{fa}(a^m b^{n_1} c^{n_1} da^{n'_2})$. Therefore, $c_{fa}(a^m b^{n_2} c^{n_1} da^{n'_1} b^{n'_1}) = c_{fa}(a^m b^{n_1} c^{n_1} da^{n'_2} b^{n'_1})$. However, the former is a no-instance and the latter is a yes-instance. This is a contradiction.

Case 2: For any 1D1CA M that solves ONE-NONE, M satisfies that for all $u \in \{a, b, c\}$, $|v(u^n)| \in O(1)$.

In this case, there exist n_1 and n_2 ($n_1 < n_2$) such that $c(a^{n_1}) = c(a^{n_2})$ since the number of possible configurations is constant when the counter value is bounded by $O(1)$. Thus, $c(a^{n_1} b^{n_1} d) = c(a^{n_2} b^{n_1} d)$.

Now, we define another promise problem ONEORNONE such that ONE forms yes-instances and NONE forms no-instances.

The fact that $c(a^{n_1} b^{n_1} d) = c(a^{n_2} b^{n_1} d)$ implies that the decision of acceptance depends only on the second half of the input of ONE-NONE. Thus, based on M, we can build a 1D1CA M' that solves ONEORNONE; the subautomaton of M that starts in the configuration $c(a^{n_1} b^{n_1} d)$ can be regarded as the automaton M' that solves ONEORNONE. Obviously M' can be extended to solve ONE-NONE, say M'', as follows: it executes M' on the first half of the input of ONE-NONE and ignores the second half. To sum up, if M satisfying Case 2 solves ONE-NONE, then M'' satisfying Case 2 solves ONE-NONE by reading the input only until the first d. Due to Case 2, M'' also satisfies that $|v(a^n)| \in O(1)$. By using the same reasoning above, we can follow that there exist n'_1 and n'_2 for M'' such that $c(a^{n'_1} b^{n'_1} d) = c(a^{n'_2} b^{n'_1} d)$ and so M'' gives the same decisions to the strings $a^{n'_1} b^{n'_1} da^{n_2} b^{n'_1} d$ and $a^{n'_2} b^{n'_1} da^{n'_1} b^{n'_1} d$. This is a contradiction and so M cannot solve ONE-NONE. □

To get a better error bound, we can use the promise problem ONE-NONE(t) where yes-instances (ONE-NONE$_{yes}$(t)) are formed by (ONE-NONE$_{yes}$)^t and no-instances (ONE-NONE$_{no}$(t)) are formed by (ONE-NONE$_{no}$)^t. That is, the error bound can be reduced to $\frac{1}{6^t}$ for 1P1CAs, where $t > 1$. We leave as a future work whether 1D1CA can solve the promise problem ONE-NONE(t).

3.3   A New Result on Blind Counter Automata

In this section, we present a 1P1BCA algorithm for the Kleene closure of unary equality language:

$$\text{EQ}^* = \{\varepsilon\} \cup \{a^{n_1} b^{n_1} \cdots a^{n_k} b^{n_k} | n_i > 0 (1 \leq i \leq k), k \geq 1\},$$

which was shown not to be recognized by any one-way deterministic finite automaton with multi blind counters [10]. Recently, Yakaryılmaz presented a negative one-sided error 1Q1BCA algorithm for this language and he conjectured that it cannot be recognized by 1P1BCA [27]. Now, we show that this conjecture is false. It is also surprising that our new algorithm is kind of a probabilistic adaptation of the quantum algorithm given by Yakaryılmaz.

Theorem 5. *The language* EQ* *can be recognized by a 1P1BCA* M *with negative one-sided error bound* $\frac{1}{3}$.

Proof. We assume that the input is of the form $a^{n_1} b^{m_1} \cdots a^{n_k} b^{m_k}$. Otherwise, M rejects the input deterministically (exactly). At the beginning of each block $a^{n_l} b^{m_l}$ $(1 \le l \le k)$, M selects one of the following three paths ($Path_i$'s) with equal probability:

$$Path_i(1 \le i \le 3) : M \text{ increases (resp. decreases) the counter by } i$$
$$\text{each time reading an } a \text{ (resp. a } b) \text{ of the block.}$$

The computation always ends in an accepting state (except the deterministic check mentioned at the beginning). Thus, the input is accepted if and only if the value of counter is zero. It is obvious that M accepts any member of EQ* with certainty. We consider the case that the input $w \notin$ EQ*. Let i_{max} be the greatest index satisfying $n_{i_{max}} \ne m_{i_{max}}$, i.e., $a^{n_{i_{max}}} b^{m_{i_{max}}}$ is the last block satisfying $n_{i_{max}} \ne m_{i_{max}}$. Let $path'$ be a probabilistic path before reading the i_{max}-th block having the counter value c. This path will split into three sub-paths $subpath'_1$, $subpath'_2$, and $subpath'_3$ and each subpath reads the block as described above. Let c_1, c_2, and c_3 be the counter values of these sub-paths, respectively, after reading the block. Any computation starts from $subpath'_i$ will have the same counter value of c_i at the end of the computation since the remaining blocks have the same numbers of a's and b's, where $1 \le i \le 3$. Assume that $subpath'_i$ leads to a decision of acceptance. This is possible only if $c_i = 0$. Let $d = n_{i_{max}} - m_{i_{max}} \ne 0$. Then the values of c_1, c_2, and c_3 are $c + d$, $c + 2d$, and $c + 3d$, respectively. Therefore, only one of them can be zero. That is, the maximum accepting probability that $path'$ can contribute is $\frac{1}{3}$. This is the case also for all other probabilistic paths that exist just before reading the i_{max}-th block. Therefore the overall accepting path can be bounded by $\frac{1}{3}$.   □

It is clear from the analysis given in the proof that the error bound can be reduced to $\frac{1}{k}$ for any k by spiting into k probabilistic paths on each block instead of 3.

Corollary 1. *The language* EQ* *can be recognized by a 1P1BCA* M *with any negative one-sided error bound* $\epsilon \le \frac{1}{2}$.

Even though any number of blind counters is useless for a 1DFA, a single non-blind counter is enough to recognize EQ*, i.e. 1D1CA can recognize EQ*. Another related result is that Freivalds [6] proved that $EQ^3 = \{a^n b^n c^n \mid n \ge 0\}$ can be recognized by a 1P1BCAs with arbitrary small one-sided error bound and this non-context free language, of course, cannot be recognized by a 1D1CA.

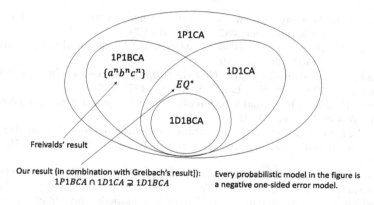

Our result (in combination with Greibach's result]):   Every probabilistic model in the figure is
$1P1BCA \cap 1D1CA \supsetneq 1D1BCA$ a negative one-sided error model.

Fig. 3. Hierarchy of various models of counter automata

Our above result implies that $\mathcal{L}(1D1BCA) \subsetneq \mathcal{L}(1P1BCA) \cap \mathcal{L}(1D1CA)$, where $\mathcal{L}(Model)$ is the class of languages recognized by $Model$. We represent our result with known facts in Fig. 3, from which it is seen that the whole picture is still incomplete. Moreover, we still do not know whether bounded-error 1Q1BCAs are more powerful than bounded-error 1P1BCAs.

We close the section with some discussions on nondeterministic and universal models. The language EQ* can be recognized by a 1U1BCA: it universally picks each block of a^+b^+ and then deterministically determines whether the numbers of a's and b's are equal. The input is accepted if and only if each block has equal number of a's and b's. One may ask whether we can recognize the complement of EQ*. Here any input that is not of the form $(a^+b^+)^+$ can be deterministically detected. The difficult task is to detect a block of a^+b^+ having different number of a's and b's. Currently, we know only a 1_εN1BCA to catch such a block: It nondeterministically picks such a block and then increases (resp. decreases) the counter for each a (resp. b). Let the counter value after reading the block be c. Then, the automaton switches to an accepting state after nondeterministically setting the value of counter to one of $\ldots, c-2, c-1, c+1, c+2, \ldots$. One of these values must be zero if c is not zero. (Note that, none of these values is zero if c is zero.) Thus, our one-way with ε-move machine can recognize the language.

Here we find interesting to identify whether there is an alternation hierarchy for one-way blind-counter automata with and without ε-moves. We also leave this as a future work.

Acknowledgement. We thank Klaus Reinhardt for answering our question regarding the subject matter of this paper.

References

1. Amano, M., Iwama, K.: Undecidability on quantum finite automata. In: STOC 1999: Proceedings of the Thirty-First Annual ACM Symposium on Theory of Computing, pp. 368–375 (1999)

2. Ambainis, A., Watrous, J.: Two-way finite automata with quantum and classical states. Theoret. Comput. Sci. **287**(1), 299–311 (2002)
3. Cleve, R., Ekert, A., Macchiavello, C., Mosca, M.: Quantum algorithms revisited. Proc. R. Soc. A **454**, 339–354 (1998)
4. Deutsch, D.: Quantum theory, the Church-Turing principle and the universal quantum computer. Proc. R. Soc. Lond. A **400**, 97–117 (1985)
5. Deutsch, D., Jozsa, R.: Rapid solution of problem by quantum computation. Proc. R. Soc. A **439**, 553–558 (1992)
6. Freivalds, R.: Fast probabilistic algorithms. In: Bečvář, J. (ed.) Mathematical Foundations of Computer Science 1979. LNCS, vol. 74, pp. 57–69. Springer, Heidelberg (1979)
7. Freivalds, R.: Probabilistic two-way machines. In: Proceedings of the International Symposium on Mathematical Foundations of Computer Science, pp. 33–45 (1981)
8. Gainutdinova, A., Yakaryılmaz, A.: Unary probabilistic and quantum automata on promise problems. In: Developments in Language Theory, LNCS, vol. 9168, pp. 252–263. Springer International Publishing (2015). (arXiv:1502.01462)
9. Geffert, V., Yakaryılmaz, A.: Classical automata on promise problems. In: Jürgensen, H., Karhumäki, J., Okhotin, A. (eds.) DCFS 2014. LNCS, vol. 8614, pp. 126–137. Springer, Heidelberg (2014). (ECCC:TR14-136)
10. Greibach, S.A.: Remarks on blind and partially blind one-way multicounter machines. Theoret. Comput. Sci. **7**, 311–324 (1978)
11. Hirvensalo, M.: Quantum automata with open time evolution. Int. J. Nat. Comput. **1**(1), 70–85 (2010)
12. Hromkovič, J., Schnitger, G.: On probabilistic pushdown automata. Inf. Comput. **208**(8), 982–995 (2010)
13. Kondacs, A., Watrous, J.: On the power of quantum finite state automata. In: FOCS 1997, pp. 66–75 (1997)
14. Kravtsev, M.: Quantum finite one-counter automata. In: Bartosek, M., Tel, G., Pavelka, J. (eds.) SOFSEM 1999. LNCS, vol. 1725, p. 431. Springer, Heidelberg (1999)
15. Li, L., Qiu, D., Zou, X., Li, L., Wu, L., Mateus, P.: Characterizations of one-way general quantum finite automata. Theoret. Comput. Sci. **419**, 73–91 (2012)
16. Moore, C., Crutchfield, J.P.: Quantum automata and quantum grammars. Theoret. Comput. Sci. **237**(1–2), 275–306 (2000)
17. Murakami, Y., Nakanishi, M., Yamashita, S., Watanabe, K.: Quantum versus classical pushdown automata in exact computation. IPSJ Digit. Cour. **1**, 426–435 (2005)
18. Nakanishi, M.: Quantum pushdown automata with a garbage tape. In: Italiano, G.F., Margaria-Steffen, T., Pokorný, J., Quisquater, J.-J., Wattenhofer, R. (eds.) SOFSEM 2015. LNCS, vol. 8939, pp. 352–363. Springer, Heidelberg (2015). arXiv:1402.3449
19. Nielsen, M.A., Chuang, I.L.: Quantum Computation and Quantum Information. Cambridge University Press, Cambridge (2000)
20. Rabin, M.O.: Probabilistic automata. Inf. Control **6**, 230–243 (1963)
21. Rashid, J., Yakaryılmaz, A.: Implications of quantum automata for contextuality. In: Holzer, M., Kutrib, M. (eds.) CIAA 2014. LNCS, vol. 8587, pp. 318–331. Springer, Heidelberg (2014). arXiv:1404.2761
22. Say, A.C.C., Yakaryılmaz, A.: Quantum counter automata. Int. J. Found. Comput. Sci. **23**(5), 1099–1116 (2012)

23. Say, A.C.C., Yakaryılmaz, A.: Quantum finite automata: A modern introduction. In: Calude, C.S., Freivalds, R., Kazuo, I. (eds.) Gruska Festschrift. LNCS, vol. 8808, pp. 208–222. Springer, Heidelberg (2014)
24. Shepherdson, J.C.: The reduction of two-way automata to one-way automata. IBM J. Res. Dev. **3**, 198–200 (1959)
25. Sipser, M.: Introduction to the Theory of Computation, 2nd edn. Thomson Course Technology, USA (2006)
26. Watrous, J.: Quantum computational complexity. In: Meyers, R.A. (ed.) Encyclopedia of Complexity and System Science. Springer, New york (2009). arXiv:0804.3401
27. Yakaryılmaz, A.: Superiority of one-way and realtime quantum machines. RAIRO Theoret. Inf. Appl. **46**(4), 615–641 (2012)
28. Yakaryılmaz, A.: One-counter verifiers for decidable languages. In: Bulatov, A.A., Shur, A.M. (eds.) CSR 2013. LNCS, vol. 7913, pp. 366–377. Springer, Heidelberg (2013)
29. Yakaryılmaz, A., Freivalds, R., Say, A.C.C., Agadzanyan, R.: Quantum computation with write-only memory. Nat. Comput. **11**(1), 81–94 (2012)
30. Yakaryılmaz, A., Say, A.C.C.: Unbounded-error quantum computation with small space bounds. Inf. Comput. **279**(6), 873–892 (2011)
31. Yamasaki, T., Kobayashi, H., Imai, H.: Quantum versus deterministic counter automata. Theoret. Comput. Sci. **334**(1–3), 275–297 (2005)

State Complexity of Prefix Distance

Timothy Ng, David Rappaport, and Kai Salomaa[✉]

School of Computing, Queen's University, Kingston, ON K7L 3N6, Canada
{ng,daver,ksalomaa}@cs.queensu.ca

Abstract. The prefix distance between strings x and y is the number of symbol occurrences in the strings that do not belong to the longest common prefix of x and y. The suffix and the substring distance are defined analogously in terms of the longest common suffix and longest common substring, respectively, of two strings. We show that the set of strings within prefix distance k from an n state DFA (deterministic finite automaton) language can be recognized by a DFA with $(k+1) \cdot n - \frac{k(k+1)}{2}$ states and this number of states is needed in the worst case. Also we give tight bounds for the nondeterministic state complexity of the set of strings within prefix, suffix or substring distance k from a regular language.

1 Introduction

Various similarity measures between strings and languages have been considered for information transmission applications. The edit distance counts the number of substitution, insertion and deletion operations that are needed to transform one string to another. The Hamming distance counts the number of positions in which two equal length strings differ. A distance measure between words can be extended in various ways as a distance between sets of strings (or languages) [3,4] and algorithms for computing the distance between languages are important for error-detection and error-correction applications [4,9,10]. The descriptive complexity of error/edit systems has been considered by Kari and Konstantinidis [8]. Other types of sequence similarity measures have been considered e.g. by Apostolico [1].

Instead of counting the number of edit operations, the similarity of strings can be defined by way of their longest common prefix, suffix, or substring, respectively [4]. For example, the prefix distance of strings x and y is the sum of the length of the suffix of x and the suffix of y that occurs after their longest common prefix. A parameterized prefix distance between regular languages has been considered by Kutrib et al. [11] for estimating the fault tolerance of information transmission applications.

The neighbourhood of radius k of a language L consists of all strings that are within distance k from some string in L. Calude et al. [3] have shown that the neighbourhood of a regular language with respect to an additive distance is regular. A distance is said to be additive if it, in a certain sense, respects string concatenation. This gives rise to the question how large is the (non)deterministic

© Springer International Publishing Switzerland 2015
F. Drewes (Ed.): CIAA 2015, LNCS 9223, pp. 238–249, 2015.
DOI: 10.1007/978-3-319-22360-5_20

finite automaton (DFA, respectively, NFA) needed to recognize the neighbourhood of a regular language, that is, what is the state complexity of neighbourhoods of regular languages.

Povarov [15] has given an improved upper bound and a closely matching lower bound for the state complexity of Hamming neighbourhoods of radius one. Upper bounds for the state complexity of neighbourhoods with respect to an additive distance or quasi-distance have been obtained by the authors [14,16] using a construction based on weighted finite automata.

It follows from Choffrut and Pighizzini [4] that the prefix, suffix and substring distance preserve regularity, that is, the neighbourhood of a regular language of finite radius remains regular. Here we study the state complexity of these neighbourhoods. We show that if L is recognized by a deterministic finite automaton (DFA) of size n, the prefix neighbourhood of L of radius $k < n$ has a DFA of size $(k+1) \cdot n - \frac{k(k+1)}{2}$ and that this bound cannot be improved in the worst case. Our lower bound construction uses an alphabet of size $n+1$ and we show that the general upper bound cannot be reached using languages defined over a fixed alphabet.

We consider also the nondeterministic state complexity of prefix, suffix and substring neighbourhoods. If L has a nondeterministic finite automaton (NFA) of size n, the neighbourhood of L of radius k can be recognized by an NFA of size $n+k$. The upper bound for the substring neighbourhood of L of radius k is $(k+1) \cdot n + 2k$. In all cases we give matching lower bounds for nondeterministic state complexity, and in the lower bound constructions L has, in fact, a DFA of size n.

2   Preliminaries

Here we briefly recall some definitions and notation used in the paper. For all unexplained notions on finite automata and regular languages the reader may consult the textbook by Shallit [17] or the survey by Yu [18]. A survey of distances is given by Deza and Deza [5]. Recent surveys on descriptional complexity of regular languages include [6,7,12].

In the following Σ is always a finite alphabet, the set of strings of Σ is Σ^* and ε is the empty string. The reversal of a string $x \in \Sigma^*$ is x^R. The set of nonnegative integers is $\mathbb{N}_0$. The cardinality of a finite set S is denoted $|S|$ and the powerset of S is 2^S. A string $w \in \Sigma^*$ is a *substring* or *factor* of x if there exist strings $u, v \in \Sigma^*$ such that $x = uwv$. If $u = \varepsilon$, then w is a *prefix* of x. If $v = \varepsilon$, then w is a *suffix* of x.

A *nondeterministic finite automaton* (NFA) is a 5-tuple $A = (Q, \Sigma, \delta, Q_0, F)$ where Q is a finite set of states, Σ is an alphabet, δ is a multi-valued transition function $\delta : Q \times \Sigma \to 2^Q$, $Q_0 \subseteq Q$ is a set of initial states, and $F \subseteq Q$ is a set of final states. We extend the transition function δ to $Q \times \Sigma^* \to 2^Q$ in the usual way. A string $w \in \Sigma^*$ is *accepted* by A if, for some $q_0 \in Q_0$, $\delta(q_0, w) \cap F \neq \emptyset$ and the language recognized by A consists of all strings accepted by A. An ε-NFA is an extension of an NFA where transitions can be labeled by the empty string

ε [17,18], i.e., δ is a function $Q \times (\Sigma \cup \{\varepsilon\}) \to 2^Q$. It is known that every ε-NFA has an equivalent NFA without ε-transitions and with the same number of states. An NFA $A = (Q, \Sigma, \delta, Q_0, F)$ is a *deterministic finite automaton* (DFA) if $|Q_0| = 1$ and, for all $q \in Q$ and $a \in \Sigma$, $\delta(q, a)$ either consists of one state or is undefined. Two states p and q of a DFA A are equivalent if $\delta(p, w) \in F$ if and only if $\delta(q, w) \in F$ for every string $w \in \Sigma^*$. A DFA A is *minimal* if each state $q \in Q$ is reachable from the initial state and no two states are equivalent.

Note that our definition of a DFA allows some transitions to be undefined, that is, by a DFA we mean an incomplete DFA. It is well known that, for a regular language L, the sizes of the minimal incomplete and complete DFAs differ by at most one. The constructions in Sect. 3 are more convenient to formulate using incomplete DFAs but our results would not change in any significant way if we were to require that all DFAs are complete.

The (incomplete deterministic) *state complexity* of a regular language L, $sc(L)$, is the size of the minimal DFA recognizing L. The *nondeterministic state complexity* of L, $nsc(L)$, is the size of the minimal NFA recognizing L. The minimal NFA recognizing a regular language need not be unique. A common way of establishing lower bounds for nondeterministic state complexity relies on fooling sets.

Definition 1. *A set of pair of strings* $S = \{(x_1, y_1), \ldots, (x_m, y_m)\}$, $x_i, y_i \in \Sigma^*$, $i = 1, \ldots, m$, *is a* fooling set *for a language* L *if* $x_i y_i \in L$, $i = 1, \ldots, m$ *and, for all* $1 \le i < j \le m$, $x_i y_j \notin L$ *or* $x_j y_i \notin L$.

Proposition 1 ([2,7]). *If* L *has a fooling set* S *then* $nsc(L) \ge |S|$.

To conclude this section, we recall definitions of the distance measures used in the following. Generally, a function $d : \Sigma^* \times \Sigma^* \to [0, \infty)$ is a *distance* if it satisfies for all $x, y, z \in \Sigma^*$, the conditions $d(x, y) = 0$ if and only if $x = y$, $d(x, y) = d(y, x)$, and $d(x, z) \le d(x, y) + d(y, z)$. The *neighbourhood* of a language L of radius k with respect to a distance d is the set

$$E(L, d, k) = \{w \in \Sigma^* \mid (\exists x \in L)d(w, x) \le k\}.$$

Let $x, y \in \Sigma^*$. The *prefix distance* of x and y counts the number of symbols which do not belong to the longest common prefix of x and y [4]. It is defined by

$$d_p(x, y) = |x| + |y| - 2 \cdot \max_{z \in \Sigma^*} \{|z| \mid x, y \in z\Sigma^*\}.$$

Similarly, the *suffix distance* of x and y counts the number of symbols which do not belong to the longest common suffix of x and y and is defined

$$d_s(x, y) = |x| + |y| - 2 \cdot \max_{z \in \Sigma^*} \{|z| \mid x, y \in \Sigma^* z\}.$$

The *substring distance* measures the similarity of x and y based on their longest common continuous substring (or factor) and is defined

$$d_f(x, y) = |x| + |y| - 2 \cdot \max_{z \in \Sigma^*} \{|z| \mid x, y \in \Sigma^* z\Sigma^*\}.$$

The paper [4] refers to d_f as the *subword distance*. The term "subword distance" has been used also for a distance defined in terms of the longest common noncontinuous subword [13].

3   State Complexity of Prefix Neighbourhoods

In this section we consider the deterministic state complexity of prefix neighbourhoods. We construct a DFA for the neighbourhood of radius k with respect to the prefix distance d_p. After that we show that the construction is optimal by giving a matching lower bound. The lower bound construction uses an alphabet of size $n + 1$ where n is the number of states of the DFA. We show that the upper bound cannot be reached by languages defined over a constant size alphabet.

Proposition 2. *Let $n > k \geq 0$ and L be a regular language recognized by a DFA with n states. Then there is a DFA recognizing $E(L, d_p, k)$ with at most $n \cdot (k + 1) - \frac{k(k+1)}{2}$ states.*

Proof. Let $A = (Q, \Sigma, \delta, q_0, F)$ be the DFA that recognizes L. We define the function $\varphi : Q \to \mathbb{N}_0$ by

$$\varphi(q) = \min_{w \in \Sigma^*} \{|w| \mid \delta(q, w) \in F\}.$$

The function $\varphi(q)$ gives the length of the shortest path from a state q to the closest, or next, reachable final state. Note that under this definition, if $q \in F$, then $\varphi(q) = 0$.

We construct a DFA $A' = (Q', \Sigma, \delta', q_0', F')$ that recognizes the neighbourhood $E(L, d_p, k)$. We define the state set

$$Q' = ((Q - F) \times \{1, \ldots, k + 1\}) \cup F \cup \{p_1, \ldots, p_k\}.$$

Note that some states of Q' are always unreachable and at the end of the proof we calculate an upper bound for the number of reachable states. The initial state q_0' is defined

$$q_0' = \begin{cases} q_0, & \text{if } q_0 \in F; \\ (q_0, \varphi(q_0)) & \text{if } q_0 \notin F \text{ and } \varphi(q_0) \leq k; \\ (q_0, k + 1) & \text{if } q_0 \notin F \text{ and } \varphi(q_0) > k. \end{cases}$$

The set of final states is given by

$$F' = ((Q - F) \times \{1, \ldots, k\}) \cup F \cup \{p_1, \ldots, p_k\}.$$

Let $q_{i,a} = \delta(i, a)$ for $i \in Q$ and $a \in \Sigma$, if $\delta(i, a)$ is defined. Then for all $a \in \Sigma$, the transition function δ' is defined for states $i \in F$ by

$$\delta'(i, a) = \begin{cases} (q_{i,a}, 1), & \text{if } q_{i,a} \in Q - F; \\ q_{i,a}, & \text{if } q_{i,a} \in F; \\ p_1, & \text{if } \delta(i, a) \text{ is undefined.} \end{cases}$$

For states $(i, j) \in Q - F \times \{1, \ldots, k+1\}$, δ' is defined

$$\delta'((i,j), a) = \begin{cases} q_{i,a}, & \text{if } q_{i,a} \in F; \\ (q_{i,a}, \min\{j+1, \varphi(q_{i,a})\}), & \text{if } \varphi(q_{i,a}) \text{ or } j+1 \le k; \\ (q_{i,a}, k+1), & \text{if } \varphi(q_{i,a}) \text{ and } j+1 > k; \\ p_{j+1}, & \text{if } \delta(i, a) \text{ is undefined.} \end{cases}$$

Finally, we define δ' for states p_ℓ for $\ell = 1, \ldots, k-1$ by $\delta'(p_\ell, a) = p_{\ell+1}$. The machine A' has three types of states. The first type consists of final states of A. The second type are new states p_ℓ, which form a chain of error states. When a transition that was undefined in A is encountered during some computation, A' is taken to the chain of error states p_i. The third type of states consists of states of A which are not final states and are paired with a counter. For a state (i, j), the counter component j keeps track of the distance of the current computation to the closest final state of A.

On input $w \in \Sigma^*$, there are three cases to consider. Let $x \in L$ be a closest string to w according to the prefix distance d_p.

1. First, suppose that $x = wx'$ for some $x' \in \Sigma^*$. Then $w \in E(L, d_p, k)$ if and only if $|x'| \le k$. Consider the computation on w, which must end in some state (i, j). Otherwise, the computation either ends in a final state, in which case $x = w$, or it ends in some state p_ℓ, which cannot be the case as w is a proper prefix of a word in L. Since x is the closest word in L to w, there must be a shortest path of length $|x'|$ in the original DFA A from state i to a final state of A. By definition, (i, j) is a final state if $j = \varphi(i) \le k$. Thus, $j = \varphi(i) = |x'|$ and (i, j) is a final state if and only if $j = |x'| \le k$.

2. Next, suppose that $w = xw'$ for some $w' \in \Sigma^*$. In this case, $w \in E(L, d_p, k)$ if $|w'| \le k$. The machine reaches some final state f of A once it reads all of x. Then the machine continues reading w' until it reaches some state $q \in Q'$. The state q is either a state (i, j) or a state p_ℓ, since otherwise, $q \in F$ and $w' = \varepsilon$.

   (a) Consider $q = (i, j)$. By definition, (i, j) is a final state if $j \le k$. Since x is a closest word in L to w, $j = |w'|$ must be the distance of the current computation from the closest final state f unless $|w'| > k$, in which case $j = k + 1$. Otherwise, there was some state (i', j') that was encountered during the computation of w' with a final state f' that was closer than f. Thus, if $|w'| > k$, then $j = k+1$ and (i, j) is not a final state. Otherwise, $j = |w'| \le k$ and (i, j) is a final state.

   (b) Now consider when $q \ne (i, j)$ and let $w' = w'_1 w'_2$. The computation from f on w'_1 reaches some state $q' = (i', j')$ for which there is no transition in A defined for the first symbol of w'_2. By the same reasoning as above, $j' = |w'_1| < k$. Since an undefined transition was encountered on the first symbol of w'_2, the machine goes to state $p_{|w'_1|+1}$. From state $p_{|w'_1|+1}$, the machine reads the rest of w'_2. Now, if $|w'| > k$, then $|w'_2| > k - |w'_1|$ and the computation on the rest of w'_2 fails when it reaches p_k and there are no further transitions. Otherwise, $|w'| \leq k$ and the computation of w'_2 ends in a state $p_{|k'_1|+|k'_2|}$, which is a final state since $|w'| = |w'_1|+|w'_2| \leq k$.

3. Finally, suppose that $w = pw'$ and $x = px'$ with $p, w', x' \in \Sigma^*$ such that p is the longest common prefix of w and x. Note that if $w' = \varepsilon$, then it becomes Case 1, and if $x' = \varepsilon$, then Case 2 applies. Thus $w \in E(L, d_p, k)$ if and only if $|w'| + |x'| \leq k$. In this case, A' reads w until it reaches a state (i_p, j_p) on the prefix p. At this point, reading x' from (i_p, j_p) will take the machine to some final state f, while reading w' from (i_p, j_p) takes the machine to some other state $q \in Q'$. Note that $|x'| \leq k$, since otherwise $|x'| + |w'| > k$, and $j_p = \varphi(i_p) = |x'|$, since otherwise x would not be a closest word to w. Now, q is either of the form (i, j) or a state p_ℓ.

   (a) Suppose q is of the form (i, j). Then j is either $|w'| + |x'|$ or $k + 1$. If $|w'| > k - |x'|$, then $j = k + 1$ and (i, j) is not a final state. If $j \leq |w'| + |x'|$, then there must be some final state f' closer to a state on the computation path of w' from (i_p, j_p) which cannot be the case if x is a closest word to w. Thus, $j = |w'| + |x'| \leq k$ and (i, j) is a final state.

   (b) Now, suppose $q \neq (i, j)$ and let $w' = w'_1 w'_2$. The computation from (i_p, j_p) on w'_1 reaches some state $q' = (i', j')$ for which there is no transition in A on the first symbol of w'_2. By the same reasoning as above, $j' = |w'_1| < k - |x'|$. Since an undefined transition was encountered on the first symbol of w'_2, the machine goes to state $p_{|w'_1|+|x'|+1}$. From state $p_{|w'_1|+|x'|+1}$, the rest of w'_2 is read. If $|w'_2| > k - (|w'_1| + |x'|)$, then the computation of w'_2 falls off at p_k. Otherwise, the computation ends in state $p_{|w'_2|+|w'_1|+|x'|}$. We have

$$|w'_2| + |w'_1| + |x'| = |w'| + |x'| \leq k$$

   and thus, $p_{|w'_2|+|w'_1|+|x'|}$ is a final state.

The set of states Q' has $(n - f) \cdot (k + 1) + k + f$ elements but they cannot all be reachable. Based on the definition of the transitions of δ' we observe that if there is a transition entering a state (q, j), $q \in Q - F$, $1 \leq j \leq k + 1$, then $\varphi(q)$ must be at least j. Thus, all elements of the set

$$S_{ur} = \{(q, j) \mid q \in Q - F, 1 \leq j \leq k + 1, \; j > \varphi(q)\}$$

are unreachable as states of A'. Since increasing the number of final states of A by one decreases the cardinality of Q' by k and decreases the cardinality of S_{ur} by at most k, it is clear that an upper bound for the cardinality of the set of potentially reachable states $Q' - S_{ur}$ is obtained by choosing $f = 1$. Using the observation that all useful states must reach a final state, in the case when $F = \{q_f\}$ is a singleton set, the cardinality of S_{ur} is minimized when, in the DFA A for each $1 \leq i \leq k$, exactly one non-final state q_i has a shortest path of length i that reaches q_f. In this case $S_{ur} = \{(q_i, j) \mid i < j \leq k + 1, i = 1, \ldots, k\}$ and $|S_{ur}| = \frac{k(k+1)}{2}$.

We have verified that at most $n \cdot (k+1) - \frac{k(k+1)}{2}$ states of A' can be reachable. $\square$

The lower bound construction that we present uses an alphabet with variable size. We will show later that it is impossible to reach the upper bound (for all n) with an alphabet of fixed size.

Lemma 1. *For $n > k \in \mathbb{N}$, there exists a DFA A_n with n states over an alphabet of size $n + 1$ such that*

$$\mathrm{sc}(E(L(A), d_p, k)) \geq n \cdot (k + 1) - \frac{k(k + 1)}{2}.$$

Proof. We define a DFA $A_n = (Q_n, \Sigma_n, \delta_n, q_0, F)$ (Fig. 1) by choosing

$$Q_n = \{0, \ldots, n - 1\}, \quad \Sigma_n = \{a, b, c_1, \ldots, c_{n-1}\},$$

$q_0 = 0$, $F = \{0\}$, and the transition function is given by

- $\delta_n(q, a) = q$ for all $q \in Q_n$,
- $\delta_n(q, b) = q + 1 \mod n$ for $q = 1, \ldots, n - 1$,
- $\delta_n(0, c_i) = i$ for $i = 1, \ldots, n - 1$,

Note that for every state $q \in Q_n$, we have $\varphi(q) = n - q$.

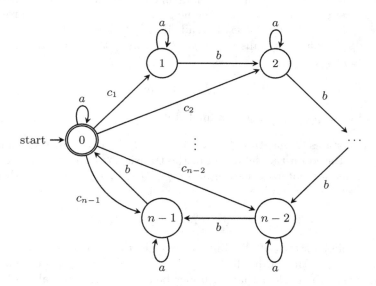

Fig. 1. The DFA A_n.

We transform A_n into the DFA $A'_n = (Q'_n, \Sigma_n, \delta'_n, q'_0, F')$ by following the construction from Proposition 2. To determine the reachable states of Q'_n, we first consider states of the form $(i, j) \in (Q_n - \{q_0\}) \times \{1, \ldots, k + 1\}$. For states $i \in Q_n - \{q_0\}$ with $\varphi(i) > k$, we can reach state (i, j) via the word $c_i a^j$ for

$j = 1, \ldots, k + 1$. For states $i \in Q_n - \{q_0\}$ with $\varphi(i) \leq k$, we can reach state (i, j) via the word $c_i a^j$ for $j = 1, \ldots, \varphi(i)$. However, states (i, j) with $j > \varphi(i)$ are unreachable by definition of A'_n. Thus the number of unreachable states in $(Q_n - \{q_0\}) \times \{1, \ldots, k + 1\}$ is

$$\sum_{i=n-k}^{n-1} |\{i\} \times \{\varphi(i) + 1, \ldots, k + 1\}| = \sum_{i=1}^{k} |\{i + 1, \ldots, k + 1\}| = \sum_{i=1}^{k} i = \frac{k(k + 1)}{2}.$$

Now consider states $p_1, \ldots, p_k$. The state p_ℓ is reachable on the word b^ℓ. Finally, 0 is reachable since it is the initial state. Thus, the number of reachable states is

$$(n - 1) \cdot (k + 1) - \frac{k(k + 1)}{2} + k + 1 = n \cdot (k + 1) - \frac{k(k + 1)}{2}.$$

Now, we show that all reachable states are pairwise inequivalent. First, note that 0 can be distinguished from any other state by the word ε. Next, we distinguish states of the form (i, j) from states of the form p_ℓ via the word $a^k b^{n-i}$. From state (i, j), reading a^k takes the machine to state $(i, \min\{\varphi(i), k + 1\})$. Subsequently reading b^{n-i} takes the machine to the final state 0. However, for every state p_ℓ, reading a^k forces the machine beyond state p_k, after which there are no transitions defined.

Next, without loss of generality, we let $\ell < \ell'$ and consider states p_ℓ and $p_{\ell'}$. From above, the state p_ℓ can be reached by a word b^ℓ and $p_{\ell'}$ is reached by a word $b^{\ell'}$. Choose $z = a^{k-\ell}$. The string z takes state p_ℓ to the state p_k, where it is accepted. However, the computation on string z from state $p_{\ell'}$ is undefined since $\ell' + k - \ell > k$.

Finally, we consider states of the form (i, j). Let $i < i'$ and consider states (i, j) and (i', j'). Recall that (i, j) can be reached by a word $c_i a^j$ and (i', j') is reached by a word $c_{i'} a^{j'}$. Let $z = b^{n-i+k}$. From state (i, j), the word z goes to state 0 on b^{n-i}. Then by reading b^k from state 0, we reach state p_k and thus, $c_i a^j \cdot z \in E(L(A_n), d_p, k)$. However, when reading z from state (i', j'), we reach state 0 on $b^{n-i'}$, since $i' > i$. We are then left with $b^{i'-i+k}$. Reading b^k takes us to state p_k, where we still have $b^{i'-i}$ and no further defined transitions. Thus, $c_{i'} a^{j'} \cdot z \notin E(L(A_n), d_p, k)$.

Next, we fix i and let $j < j'$. The state (i, j) is reachable by the word $c_i a^j$ and (i, j') is reachable by $c_i a^{j'}$. First, consider the case when $\varphi(i) > k$. Then let $z = a^k - j$. Reading z from (i, j) takes us to state (i, k), which is a final state, so we have $c_i a^j \cdot z \in E(L(A_n), d_p, k)$. However, from (i, j'), reading z brings us to state $(i, k + 1)$ and we have $c_i a^{j'} \cdot z \notin E(L(A_n), d_p, k)$.

Now, consider the case when $\varphi(i) \leq k$. Let $z = c_i a^{k-j-1}$. From state (i, j), reading c_i takes the machine to state p_{j+1} and reading a^{k-j-1} puts the machine in state p_k. Thus, $c_i a^j \cdot z \in E(L(A_n), d_p, k)$. From (i, j'), reading z takes us to state p_k with $a^{j'-j}$ still unread since $j' + k - j - 1 > k$ and thus with no further transitions available, we have $c_i a^{j'} \cdot z \notin E(L(A_n), d_p, k)$.

Thus, we have shown that there are $n \cdot (k + 1) - \frac{k(k+1)}{2}$ reachable states and that all reachable states are pairwise inequivalent. □

Taking Proposition 2 together with Lemma 1, we get the following theorem.

Theorem 1. *For $n > k \geq 0$, if $\mathrm{sc}(L) = n$ then*

$$\mathrm{sc}(E(L, d_p, k)) \leq n \cdot (k+1) - \frac{k(k+1)}{2}$$

and this bound can be reached in the worst case.

The proof of Lemma 1 uses an alphabet of size $n+1$. To conclude this section we observe that the general upper bound cannot be reached by languages defined over a fixed alphabet.

Proposition 3. *Let A be a DFA with n states. If the state complexity of $E(L(A), d_p, n)$ equals $n \cdot (k+1) - \frac{k(k+1)}{2}$, then the alphabet of A needs at least $n-1$ letters.*

4   Nondeterministic State Complexity

We consider the nondeterministic state complexity of neighbourhoods of a regular language with respect to the prefix-, the suffix- and the substring distance, respectively.

4.1   Prefix and Suffix Distance

We consider first neighbourhoods with respect to the prefix distance, and the results for the suffix distance are obtained as a consequence of the fact that the nondeterministic state complexity of a regular language L is the same as the nondeterministic state complexity of the reversal of L and using the observation $d_s(x, y) = d_p(x^R, y^R)$ for all strings x and y.

We give an upper bound for the nondeterministic state complexity of the neighbourhood of radius k with respect to the prefix distance d_p and give a matching lower bound construction.

Proposition 4. *Let $k \geq 0$ and L be a regular language recognized by an NFA with n states. Then there is an NFA recognizing $E(L, d_p, k)$ with at most $n + k$ states.*

Proof. Let $A = (Q, \Sigma, \delta, Q_0, F)$ be the NFA recognizing L. We define an NFA $A' = (Q', \Sigma, \delta', I, F)$ for the language $E(L, d_p, k)$ by

- $Q' = Q \cup \{p_1, \ldots, p_k\}$, $I = Q_0$,
- $F' = F \cup \{p_1, \ldots, p_k\} \cup \{q \in Q \mid \varphi(q) \leq k\}$.

Recall that for $q \in Q$, $\varphi(q)$ denotes the length of the shortest string that takes q to a final state. The transition function is defined for all $a \in \Sigma$ by

- $\delta'(q, a) = \delta(q, a) \cup \{p_1\}$ for all $q \in F$,
- $\delta'(q, a) = \delta(q, a) \cup \{p_{\varphi(q)+1}\}$ for all $q \in Q$ with $\varphi(q) < k$,
- $\delta'(p_i, a) = p_{i+1}$ for $i = 1, \ldots, k-1$.

$\square$

Using the fooling sets of Proposition 1 we get a matching lower bound.

Lemma 2. *For* $n, k \in \mathbb{N}$, *there exists a DFA* A *with* n *states over* $\Sigma = \{a, b\}$ *such that any NFA for* $E(L(A), d_p, k)$ *requires* $n + k$ *states.*

Theorem 2. *For a regular language* $L \subseteq \Sigma^*$ *recognized by an NFA with* n *states and an integer* $k \geq 0$,
$$\mathrm{nsc}(E(L, d_p, k)) \leq n + k.$$
There exists a DFA A *with* n *states such that for all* $k \geq 0$,
$$\mathrm{nsc}(E(L(A), d_p, k)) = n + k.$$

We get the results for the suffix distance neighbourhoods as a corollary of Theorem 2 and the observation that, for all strings x and y, $d_s(x, y) = d_p(x^R, y^R)$.

Corollary 1. *Let* $k \geq 0$ *and* L *be a regular language recognized by a DFA with* n *states. Then there is an NFA recognizing* $E(L, d_s, k)$ *with at most* $n + k$ *states.*

The following lemma is a symmetric variant of the lower bound construction for prefix distance neighbourhoods. As a consequence of Corollary 1 and Lemma 3 we then get a tight bound for the nondeterministic state complexity of suffix neighbourhoods.

Lemma 3. *For* $n, k \in \mathbb{N}$, *there exists a DFA* A *with* n *states over* $\Sigma = \{a, b\}$ *such that any NFA for* $E(L(A), d_s, k)$ *requires* $n + k$ *states.*

Theorem 3. *For a regular language* $L \subseteq \Sigma^*$ *recognized by an NFA with* n *states and an integer* $k \geq 0$,
$$\mathrm{nsc}(E(L, d_s, k)) \leq n + k.$$
There exists a DFA A *with* n *states such that for all* $k \geq 0$,
$$\mathrm{nsc}(E(L(A), d_s, k)) = n + k.$$

4.2 Substring Distance Neighbourhoods

A neighbourhood with respect to the substring distance can be recognized by an NFA that, roughly speaking, makes $k + 1$ copies of the NFA A recognizing the original language. Later we will show that the construction is optimal.

Lemma 4. *If* A *is an* n-*state NFA and* $k \in \mathbb{N}_0$, *the neighbourhood* $E(L(A), d_f, k)$ *can be recognized by an NFA with* $(k+1) \cdot n + 2k$ *states.*

Proof sketch. Combining the constructions used for Proposition 4 and Corollary 1, an NFA B for the language $E(L(A), d_f, k)$ uses a chain of k states both at the beginning and at the end of the computation to keep track of the length of the nonmatching prefixes (respectively, suffixes) of the input and a word of $L(A)$. After processing the prefixes, the NFA B has to "remember" the sum of the lengths of the nonmatching prefixes (which can be up to k), and for this reason B is equipped with $k + 1$ copies of the original NFA A. □

Although both the upper bound and the construction used in the proof of Lemma 4 differ significantly from the corresponding bound and construction for prefix distance (or suffix distance) neighbourhoods, it turns out that for the lower bound, we can use the same cyclic languages.

Lemma 5. *There exists a DFA A with n states such that, for all $k \geq 0$,*

$$\mathrm{nsc}(E(L(A), d_f, k)) \geq (k + 1) \cdot n + 2k.$$

Proof sketch. By choosing $\Sigma = \{a, b\}$ and $L = (a^n)^*$, the minimal incomplete DFA for L has n states. Define

$$S_1 = \{(b^\ell a^i, \ a^{n-i} b^{k-\ell}) \mid 0 \leq i \leq n - 1, \ 0 \leq \ell \leq k\},$$

$$S_2 = \{(a^n b^j, \ b^{k-j}) \mid 1 \leq j \leq k\}, \quad S_3 = \{(b^j, \ b^{k-j} a^n) \mid 1 \leq j \leq k\}.$$

When $k \leq n$, it can be verified that $S_1 \cup S_2 \cup S_3$ is a fooling set for $E(L, d_f, k)$. When $n < k$, we can modify the definition of S_2 and S_3 to construct a fooling set of cardinality $(k + 1) \cdot n + 2k$ for $E(L, d_f, k)$. □

As a consequence of Lemmas 4 and 5 we have an exact bound for the nondeterministic state complexity of neighbourhoods with respect to the substring distance:

Theorem 4. *If L has an NFA with n states and $k \in \mathbb{N}_0$,*

$$\mathrm{nsc}(E(L, d_f, k)) \leq (k + 1) \cdot n + 2k.$$

For every $n \in \mathbb{N}$ there exists a DFA A with n states such that for all $k \in \mathbb{N}_0$, $\mathrm{nsc}(E(L(A), d_f, k)) = (k + 1) \cdot n + 2k$.

5   Conclusion

We have given a tight bound for the deterministic state complexity of neighbourhoods with respect to the prefix distance and tight bounds for the nondeterministic state complexity of the prefix, suffix and substring distance neighbourhoods.

Due to the fact that the reversal of a regular language L can be recognized by an NFA having the same size as an NFA for L, the bounds for the nondeterministic state complexity of suffix neighbourhoods were obtained as a corollary of the corresponding bounds for prefix neighbourhoods. The situation is essentially different for DFAs since, for a DFA A with n states, the incomplete DFA recognizing $L(A)^R$ needs in the worst case $2^n - 1$ states. Obtaining tight bounds for the deterministic state complexity of neighbourhoods with respect to the suffix distance, or the substring distance, remains an open problem.

References

1. Apostolico, A.: Maximal words in sequence comparisons based on subword composition. In: Elomaa, T., Mannila, H., Orponen, P. (eds.) Ukkonen Festschrift 2010. LNCS, vol. 6060, pp. 34–44. Springer, Heidelberg (2010)
2. Birget, J.C.: Intersection and union of regular languages and state complexity. Inf. Process. Lett. **43**, 185–190 (1992)
3. Calude, C.S., Salomaa, K., Yu, S.: Additive distances and quasi-distances between words. J. Univ. Comput. Sci. **8**(2), 141–152 (2002)
4. Choffrut, C., Pighizzini, G.: Distances between languages and reflexivity of relations. Theor. Comput. Sci. **286**(1), 117–138 (2002)
5. Deza, M.M., Deza, E.: Encyclopedia of Distances. Springer, Berlin Heidelberg (2009)
6. Gao, Y., Moreira, N., Reis, R., Yu, S.: A review on state complexity of individual operations. Faculdade de Ciencias, Universidade do Porto, Technical report DCC-2011-8 www.dcc.fc.up.pt/dcc/Pubs/TReports/TR11/dcc-2011-08.pdf to appear in Computer Science Review
7. Holzer, M., Kutrib, M.: Descriptional and computational complexity of finite automata – a survey. Inf. Comput. **209**, 456–470 (2011)
8. Kari, L., Konstantinidis, S.: Descriptional complexity of error/edit systems. J. Automata Lang. Comb. **9**, 293–309 (2004)
9. Kari, L., Konstantinidis, S., Kopecki, S., Yang, M.: An efficient algorithm for computing the edit distance of a regular language via input-altering transducers. CoRR abs/1406.1041 (2014)
10. Konstantinidis, S.: Computing the edit distance of a regular language. Inf. Comput. **205**, 1307–1316 (2007)
11. Kutrib, M., Meckel, K., Wendlandt, M.: Parameterized prefix distance between regular languages. In: Geffert, V., Preneel, B., Rovan, B., Štuller, J., Tjoa, A.M. (eds.) SOFSEM 2014. LNCS, vol. 8327, pp. 419–430. Springer, Heidelberg (2014)
12. Kutrib, M., Pighizzini, G.: Recent trends in descriptional complexity of formal languages. Bull. EATCS **111**, 70–86 (2013)
13. Lothaire, M.: Applied Combinatorics on Words, Ch. 1 Algorithms on Words. Encyclopedia of Mathematics and It's Applications 105. Cambridge University Press, New York (2005)
14. Ng, T., Rappaport, D., Salomaa, K.: Quasi-distances and weighted finite automata. In: Shallit, J., Okhotin, A. (eds.) DCFS 2015. LNCS, vol. 9118, pp. 209–219. Springer, Heidelberg (2015)
15. Povarov, G.: Descriptive complexity of the hamming neighborhood of a regular language. In: Language and Automata Theory and Applications, pp. 509–520 (2007)
16. Salomaa, K., Schofield, P.: State complexity of additive weighted finite automata. Int. J. Found. Comput. Sci. **18**(06), 1407–1416 (2007)
17. Shallit, J.: A Second Course in Formal Languages and Automata Theory. Cambridge University Press, Cambridge (2009)
18. Yu, S.: Regular languages. In: Rozenberg, G., Salomaa, A. (eds.) Handbook of Formal Languages, pp. 41–110. Springer-Verlag, Berlin (1997)

(Un)decidability of the Emptiness Problem for Multi-dimensional Context-Free Grammars

Daniel Průša[✉]

Czech Institute of Informatics, Robotics and Cybernetics, Czech Technical
University, Zikova 1903/4, 166 36 Prague 6, Czech Republic
prusapa1@cmp.felk.cvut.cz

Abstract. We study how dimensionality and form of context-free productions affect the power of multi-dimensional context-free grammars
over unary alphabets. Attention is paid to the emptiness decision problem. It is an open question whether or not it is decidable for two-
dimensional Kolam type context-free grammars of Siromoney. We show
that the undecidability can be proved in the three-dimensional setting.
For the two-dimensional variant, we present several results revealing that
the process of generating is still much more complex than that one of
the classical one-dimensional context-free grammar.

Keywords: Picture languages · Multi-dimensional context-free
grammars · Emptiness problem · Undecidability

1 Introduction

The theory of two-dimensional languages generalizes notions from the theory of
formal languages. The basic entity, which is the *string*, is replaced by a rectangular array of symbols, called a *picture*. A motivation for such a generalization
comes from the area of image processing, image recognition and two-dimensional
pattern matching.

Several models of two-dimensional automata and grammars have been
proposed to recognize/generate pictures. The early models of context-free
picture grammars include matrix and Kolam type grammars of Siromoney
et al. [17,18]. Kolam grammars were independently proposed by Matz [8] and by
Schlesinger [15,16] who designed them as a tool for structural pattern recognition. Two extensions of the grammars are known – two-dimensional context-free
grammars of Průša [13] and regional tile grammars of Pradella et al. [12]. The
grammars are also related to the grid grammars of Drewes et al. [3].

It is a well known phenomenon that the two-dimensional topology changes
a lot of properties of accepted/generated languages. For example, the *four-way
finite automaton* of Blum and Hewitt [1], which is the straightforward generalization of the two-way finite automaton, is more powerful with nondeterminism
than without it. Questions concerning decidability are another example. Several

© Springer International Publishing Switzerland 2015
F. Drewes (Ed.): CIAA 2015, LNCS 9223, pp. 250–262, 2015.
DOI: 10.1007/978-3-319-22360-5_21

problems decidable in the one-dimensional setting become undecidable. This is the case of the emptiness or finiteness problems for finite automata.

Multi-dimensional arrays are the natural extension of pictures. Especially three- and four-dimensional structures arising in application areas such as computer animation, virtual reality systems or motion image processing have practical importance. Three- and four-dimensional automata were studied e.g. in [6, 19].

At first sight, increasing the dimensionality may seem less appealing as the two-dimensional case already includes the core complexity of the multi-dimensional topology. However, we can find problems whose solution required a greater effort in the two-dimensional setting. The problem of whether or not the language of connected pictures over $\{0, 1\}$ (a connected picture has at most one connected component of 1's) is accepted by a four-way finite automaton was open for a long time. Nakamura answered this negatively first in the three-dimensional setting [9]. He showed later that the negative result is also valid in the two-dimensional setting [10].

In this paper, we face a similar situation. We study complexity of context-free grammars with respect to their dimensionality and the form of their productions. The considered criterion is decidability of the emptiness problem. While it is an open question, whether the problem is decidable for Kolam grammars, it is undecidable for the more general grammars of Průša [14]. We extend this result by showing its decidability for matrix grammars and undecidability for three-dimensional Kolam grammars. We also present results indicating that two-dimensional Kolam grammars generate quite complex unary languages.

We give the basic notions and notations on picture languages and definitions of two-dimensional context-free grammars in Sect. 2. Results related to the emptiness problem for two- and three- dimensional grammars are presented in Sects. 3 and 4, respectively. In Sect. 5, we show which functions and equations can be represented by the two-dimensional Kolam grammar. Finally, we conclude with a summary and discussion in Sect. 6.

2   Two-Dimensional Context-Free Grammars

We use the common notation and terms on pictures and picture languages (see, e.g., [4]). If Σ is a finite alphabet, then $\Sigma^{*,*}$ is used to denote the set of all rectangular pictures over Σ, that is, if $P \in \Sigma^{*,*}$, then P is a two-dimensional array of symbols from Σ. If P has m rows and n columns, we say it is of size $m \times n$, and we write $P \in \Sigma^{m,n}$, $\ell_1(P) = m$ and $\ell_2(P) = n$. If P is a square picture of size $n \times n$, we shortly say P is of size n. We also write $a^{m,n}$ to denote the picture over $\{a\}$ of size $m \times n$. The empty picture Λ is defined as the only picture of size 0×0. Moreover, $\Sigma^{+,+}$ is the set of non-empty pictures, i.e., $\Sigma^{+,+} = \Sigma^{*,*} \setminus \{\Lambda\}$. Each $a \in \Sigma$ is also treated as a picture of size 1×1.

Two (partial) binary operations are introduced to concatenate pictures. Let A be a picture of size $k \times \ell$ such that a_{ij} is the symbol in the i-th row and j-th column. Similarly, let B be a picture of size $m \times n$ with symbols b_{ij}. The *column concatenation* $A \oplus B$ is defined iff $k = m$, and the *row concatenation* $A \ominus B$ is defined iff $\ell = n$. The products are specified by the following schemes:

$$A \oplus B = \begin{bmatrix} a_{11} \dots a_{1\ell} & b_{11} \dots b_{1n} \\ \vdots \ddots \vdots & \vdots \ddots \vdots \\ a_{k1} \dots a_{k\ell} & b_{m1} \dots b_{mn} \end{bmatrix} \quad \text{and} \quad A \ominus B = \begin{bmatrix} a_{11} \dots a_{1\ell} \\ \vdots \ddots \vdots \\ a_{k1} \dots a_{k\ell} \\ b_{11} \dots b_{1n} \\ \vdots \ddots \vdots \\ b_{m1} \dots b_{mn} \end{bmatrix}.$$

Beside that, both operations are always defined when at least one of the operands is Λ. In this case, Λ is the neutral element, so $\Lambda \ominus P = P \ominus \Lambda = \Lambda \oplus P = P \oplus \Lambda = P$ for any picture P.

The operations extend to picture languages. For $L_1, L_2 \in \Sigma^{*,*}$, we define

$$L_1 \oplus L_2 = \{P \mid P = P_1 \oplus P_2 \wedge P_1 \in L_1 \wedge P_2 \in L_2\},$$
$$L_1 \ominus L_2 = \{P \mid P = P_1 \ominus P_2 \wedge P_1 \in L_1 \wedge P_2 \in L_2\}.$$

Definition 1. *A two-dimensional Kolam grammar (2KG) is a tuple* $\mathcal{G} = (V_N, V_T, \mathcal{P}, S_0)$, *where* V_N *is a finite set of nonterminals,* V_T *is a finite set of terminals,* $S_0 \in V_N$ *is the initial nonterminal and* $\mathcal{P}$ *is a finite set of productions in one of the following forms:*

$$N \to a \quad (1) \qquad\qquad S_0 \to \Lambda \quad (2)$$

$$N \to A B \quad (3) \qquad\qquad N \to \begin{matrix} A \\ B \end{matrix} \quad (4)$$

where $N, A, B \in V_N$ *and* $a \in V_T$.

Definition 2. *Let* $\mathcal{G} = (V_N, V_T, \mathcal{P}, S_0)$ *be a 2KG. For each* $N \in V_N$, $L(\mathcal{G}, N)$ *is the set of pictures generated by* $\mathcal{G}$ *from* N. *All these sets are the smallest sets fulfilling the following rules.*

1. *If* $N \to a$ *is a production in* $\mathcal{P}$ *then* $a \in L(\mathcal{G}, N)$,
2. *if* $S_0 \to \Lambda$ *is in* $\mathcal{P}$ *then* $\Lambda \in L(\mathcal{G}, S_0)$,
3. *if* $N \to A B$ *is in* $\mathcal{P}$, $P = P_1 \oplus P_2$, $P_1 \in L(\mathcal{G}, A)$ *and* $P_2 \in L(\mathcal{G}, B)$, *then* $P \in L(\mathcal{G}, N)$, *and*
4. *if* $N \to \begin{matrix} A \\ B \end{matrix}$ *is in* $\mathcal{P}$, $P = P_1 \ominus P_2$, $P_1 \in L(\mathcal{G}, A)$ *and* $P_2 \in L(\mathcal{G}, B)$, *then* $P \in L(\mathcal{G}, N)$.

The picture language generated by $\mathcal{G}$ *is defined as* $L(\mathcal{G}) = L(\mathcal{G}, S_0)$.

Example 3 (Square pictures). Let $\mathcal{G} = (V_N, V_T, \mathcal{P}, Q)$ be a 2KG where $V_N = \{R, C, U, Q\}$, $V_T = \{a\}$ and $\mathcal{P}$ is the set of productions

$$R \to a, \qquad R \to R R, \qquad C \to a, \qquad C \to \begin{matrix} C \\ C \end{matrix},$$

$$Q \to a, \qquad Q \to \frac{U}{R}, \qquad U \to QC.$$

Then, $L(\mathcal{G}, R)$ consists of all one-row pictures of a's, $L(\mathcal{G}, C)$ consists of all one-column pictures of a's, $L(\mathcal{G}, U)$ consists of pictures of size $n \times (n+1)$, $n \in \mathbb{N}^+$, and $L(\mathcal{G}, Q) = L(\mathcal{G})$ is the picture language of non-empty square pictures.

Example 4 (Exponentially sized pictures). Let $\mathcal{G} = (V_N, V_T, \mathcal{P}, E)$ be a 2KG where $V_N = \{A, R, D, E\}$, $V_T = \{a\}$ and $\mathcal{P}$ is the set of productions

$$R \to a, \qquad R \to RR, \qquad A \to a, \qquad E \to AA, \qquad E \to \frac{D}{R}, \qquad D \to EE.$$

Again, $L(\mathcal{G}, R)$ consists of all one-row pictures of a's. The picture languages $L(\mathcal{G}, D)$ and $L(\mathcal{G}, E) = L(\mathcal{G})$ consist of all pictures over $\{a\}$ of size $n \times 2^{n+1}$ and $n \times 2^n$, respectively $(n \in \mathbb{N}^+)$. Recursive patterns applied in both examples are depicted in Fig. 1.

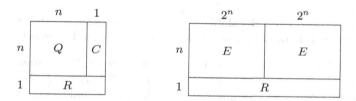

Fig. 1. Schemes showing how pictures $(n+1) \times (n+1)$ and $(n+1) \times 2^{n+1}$ in Examples 3 and 4, respectively, are assembled from smaller parts.

A matrix grammar can be seen as a special type of Kolam grammar with the usage of productions restricted in the following way. Productions of type (3) generate a row of nonterminals from S_0, then productions of type (4) generate columns of terminals of the same length from the nonterminals. A formal definition follows.

Definition 5. *A two-dimensional matrix grammar (2MG) is a tuple* $\mathcal{G} = (V_1, V_2, V_T, \mathcal{P}, S_0)$ *where*

- $(V_1 \cup V_2, V_T, \mathcal{P}, S_0)$ *is a 2KG,*
- $S_0 \in V_1$,
- *if* $N \to AB$ *is a production in* $\mathcal{P}$ *then* $N \in V_1$,
- *if* $N \to \frac{A}{B}$ *is a production in* $\mathcal{P}$ *then* $N, A, B \in V_2$*, and*
- *if* $N \to a$ *is a production in* $\mathcal{P}$ *then* $N \in V_2$.

The two-dimensional context-free grammar from [13] is a generalization of 2KG. Productions have general matrices of terminals and nonterminals on their right-hand sides. It is known, that the generative power increases when increasing size of the matrices. However, we will consider here only the basic productions (1)-(4) of 2KG and productions of the form

$$N \to \begin{matrix} A_1 \ A_2 \\ A_3 \ A_4 \end{matrix} \qquad (5)$$

where all A_i and N are nonterminals. A picture P is generated by a grammar $\mathcal{G}$ from N using production (5) iff there are pictures $P_i \in L(\mathcal{G}, A_i)$ such that

$$P = (P_1 \oplus P_2) \ominus (P_3 \oplus P_4) = (P_1 \ominus P_3) \oplus (P_2 \ominus P_4).$$

This extension is sufficient for a Turing machine simulation presented in [14] (all the productions used there can be turned into 2×2 form, however, not into 1×2, 2×1 form). We denote such a grammar as 2CFG. The ability to simulate Turing machines implies the undecidability of the emptiness problem for 2CFG.

3   Emptiness Problem

The emptiness problem is decidable for one-dimensional context-free grammars thanks to the well known *pumping lemma* (a.k.a. *uvwxy* Theorem).

Theorem 6 ([5]). *Let* $\mathcal{G} = (V_N, V_T, \mathcal{P}, S)$ *be a context-free grammar in the Chomsky normal form. Let* $p = 2^{|V_N|-1}$ *and* $q = 2^{|V_N|}$. *If* $z \in L(\mathcal{G})$ *and* $|z| > p$, *then* z *can be written as* $z = uvwxy$, *where* $|vwx| \le q$ *and* $|vx| > 0$, *such that for each* $i \in \mathbb{N}$, $uv^i wx^i y \in L(\mathcal{G})$.

Let CFG denote the one-dimensional context-free grammar in the Chomsky normal form. It can be treated as a two-dimensional grammar generating one-row pictures. Let X be a class of two-dimensional grammars and X_n its subset of grammars with exactly n nonterminals. Define $\sigma : X \to \mathbb{Z}$, $\delta_X : \mathbb{N}^+ \to \mathbb{N}^+$ as follows:

$$\sigma(\mathcal{G}) = -1 \quad \text{if } L(\mathcal{G}) = \emptyset,$$
$$\sigma(\mathcal{G}) = \min_{P \in L(\mathcal{G})} \max\{\ell_1(P), \ell_2(P)\} \quad \text{if } L(\mathcal{G}) \ne \emptyset,$$
$$\delta_X(n) = \max_{\mathcal{G} \in X_n} \sigma(\mathcal{G}).$$

This means $\delta_X(n)$ is the maximum among sizes of the smallest objects generated by grammars from X_n. Sizes of pictures are compared by their largest dimension. Theorem 6 implies that $\delta_{\mathsf{CFG}} = \mathcal{O}(2^n)$. Moreover, context-free grammars with n nonterminals generating the only string of length 2^n can be constructed, thus $\delta_{\mathsf{CFG}} = \Theta(2^n)$. Since there is a parsing algorithm for each 2CFG [13], showing that δ_X is recursive proves decidability of the emptiness problem for $X \in \{2\mathsf{KG}, 2\mathsf{MG}\}$.

Theorem 7. $\delta_{2\mathsf{MG}}(n) = \mathcal{O}(2^{n^3})$.

Proof. Let $\mathcal{G} = (V_1, V_2, V_T, \mathcal{P}, S)$ be a 2MG. Denote $n = |V_N|$. Assume there is $P \in L(\mathcal{G})$. By inspecting how P is generated, we show it is always possible to generate a picture whose dimensions do not exceed $\mathcal{O}(2^{n^3})$. To obtain P, productions of type (3) generate a string $w \in (V_1)^+$ from S. Then, nonterminals of w are substituted by one-column pictures of the same length. It thus holds

$$L = \bigcap_{N \in V(w)} L(\mathcal{G}, N) \neq \emptyset$$

where $V(w)$ is the set of all nonterminals appearing in w.

If $|w| > 2^n$, it is possible to generate w' over $V(w)$ of length $\mathcal{O}(2^n)$ instead of w (Theorem 6). The language L is the intersection of unary context-free languages where each context-free grammar has at most n nonterminals. Results in [11] show that each such a grammar has an equivalent deterministic finite automaton (DFA) with $\mathcal{O}(2^{n^2})$ states. Thus, L is accepted by the product automaton of at most n DFAs. It has $\mathcal{O}(2^{n^3})$ states. If it accepts a nonempty language, it accepts a string of length $\mathcal{O}(2^{n^3})$. Hence, $L(\mathcal{G})$ contains a picture with $\mathcal{O}(2^n)$ rows and $\mathcal{O}(2^{n^3})$ columns.   □

We use Knuth's up-arrow notation to denote the power tower operation. For $a \in \mathbb{N}^+$ and $b \in \mathbb{Z}$, we define

$$a \uparrow\uparrow b = 1 \quad \text{if } b \leq 0, \qquad a \uparrow\uparrow b = \underbrace{a^{a^{\cdot^{\cdot^{\cdot^a}}}}}_{b} \quad \text{if } b \geq 1.$$

Theorem 8. $\delta_{2\mathsf{KG}}(n) = \Omega(2 \uparrow\uparrow \frac{n-8}{2})$.

Proof. For $n \in \mathbb{N}^+$, define 2KG $\mathcal{G}_n = (V_n, \{a\}, \mathcal{P}_n, F_n)$ so that $V_n = \{R, C, U, Q, A, D, E\} \cup \{F_1, \ldots, F_n\} \cup \{H_1, \ldots, H_{n-1}\}$ and $\mathcal{P}_n$ is the union of the set of productions from Example 3, Example 4 and the following productions:

$$F_1 \rightarrow a, \qquad F_{i+1} \rightarrow \frac{H_i}{Q}, \qquad H_i \rightarrow F_i\, E, \qquad \forall i \in \{1, \ldots, n-1\}.$$

By induction on i, observe that $|L(\mathcal{G}_n, H_i)| = 1$ and $|L(\mathcal{G}_n, F_i)| = 1$ for all admissible i. Let P_i be the only picture in $L(\mathcal{G}_n, F_i)$. Denote $r_i = \ell_1(P_i)$ and $c_i = \ell_2(P_i)$. It holds

$$c_1 = 1, \quad r_1 = 1,$$
$$c_{i+1} = c_i + 2^{r_i}, \quad r_{i+1} = r_i + c_{i+1},$$

which implies

$$r_i \geq c_i, \quad c_{i+1} \geq c_i + 2^{c_i} \geq 2^{c_i} \quad \forall i \in \mathbb{N}^+.$$

The number of rows as well as columns of P_n is thus at least $2 \uparrow\uparrow (n-1)$. Since $|V_n| = 2n+6$ and thus $n - 1 = (|V_N| - 8)/2$, we derive $\delta_{2\mathsf{KG}}(k) = \Omega(2 \uparrow\uparrow \frac{k-8}{2})$.   □

The following theorem is a kind of pumping lemma for very wide or very high pictures. It also gives a constraint on the relation between the number of rows and columns of the smallest picture generated by a 2KG.

Theorem 9. *Let L be a picture language over $\{a\}$ generated by a 2KG with a set of nonterminals V_N. Let $a^{m,n}$ be a picture in L. It holds that $n \geq 2^{m|V_N|}$ implies $a^{m,n+i\cdot n!} \in L$ and $m \geq 2^{n|V_N|}$ implies $a^{m+i\cdot m!,n} \in L$ for all $i \in \mathbb{N}$.*

Proof. W.l.o.g, we prove the theorem for wide pictures. To simplify the notation within the proof, we write (m,n) to denote the picture $a^{m,n}$. Let $\mathcal{G} = (V_N, \{a\}, \mathcal{P}, S_0)$ be a 2KG. Define one-dimensional context-free grammar $\mathcal{G}' = (V_N, \{a\}, \mathcal{P}', S_0)$ where $\mathcal{P}'$ consists of those productions in $\mathcal{P}$ which are in the form (1), (2) and (3). For every $N \in V_N$ and $m \in \mathbb{N}^+$, define picture language $L(N, m)$ as follows:

$$L(N, m) = \{P \mid P \in L(\mathcal{G}, N) \wedge \ell_1(P) = m \wedge \ell_2(P) \geq 2^{m|V_N|}\}.$$

Proceed by induction on m. Let P be a picture in $L(N, 1)$. It is a one-row picture of length $n = \ell_2(P) \geq 2^{|V_N|}$. Theorem 6 is applicable and it yields $(1, n + j \cdot k) \in L(N, 1)$ for every $j \in \mathbb{N}$ and some $1 \leq k \leq n$, thus $(1, n + i \cdot n!) \in L(N, 1)$ for every $i \in \mathbb{N}$ by choosing $j = i \cdot (n!/k)$.

Let $m > 1$. A picture $P \in L(N, m)$ can be written as

1. $P = P_1 \ominus P_2$ where $P_1 \in L(\mathcal{G}, A_1)$, $P_2 \in L(\mathcal{G}, A_2)$, $N \to \begin{smallmatrix} A_1 \\ A_2 \end{smallmatrix} \in \mathcal{P}$, or
2. $P = P_1 \oplus P_2$ where $P_1 \in L(\mathcal{G}, A_1)$, $P_2 \in L(\mathcal{G}, A_2)$, $N \to A_1 A_2 \in \mathcal{P}$.

Assume, w.l.o.g, that the initial nonterminal S_0 is not a part of the right-hand side of any production, hence P_1, P_2 are nonempty. Denote again $n = \ell_2(P)$. In the first case it holds $P_1 = (m_1, n)$, $P_2 = (m_2, n)$ where $m_1, m_2 < m$, $P_1 \in L(A_1, m_1)$ and $P_2 \in L(A_2, m_2)$. The induction hypotheses yields $(m_1, n+i\cdot n!) \in L(A_1, m_1)$ and $(m_2, n + i\cdot n!) \in L(A_2, m_2)$ for every $i \in \mathbb{N}$. It is thus possible to generate any $(m, n + i \cdot n!)$ from N.

In the second case, consider a more extensive decomposition of P defined as follows. Take a picture P_i, $i \in \{1, 2\}$ with the maximal number of columns. There is again a production of type (2) or (3) and a decomposition of P_i into two parts proving that $P_i \in L(\mathcal{G}, A_i)$. The process decomposing a picture with the maximal number of columns can be repeated at most $2^{|V_N|} - 1$ times until one of two following states is reached:

1. $P = U_1 \oplus \ldots \oplus U_s$ where $s = 2^{|V_N|}$, or
2. $P = U_1 \oplus \ldots \oplus U_{j-1} \oplus (U_j \ominus U_{j+1}) \oplus U_{j+2} \oplus \ldots \oplus U_s$ where $s \leq 2^{|V_N|}$.

Let B_i be that nonterminal on the right-hand side of the production used during the decomposition process to produce U_i, so it holds $U_i \in L(\mathcal{G}, B_i)$. In the first case, only one-row productions of $\mathcal{P}'$ are used, we can thus write $N \Rightarrow^*_{\mathcal{G}'} B_1 \ldots B_s$, meaning that a sentential form of length $s = 2^{|V_N|}$ is generated from N in $\mathcal{G}'$. Theorem 6 applies to it. Substituting U_i's for B_i's in the pumped sentential forms proves that $(m, n + j \cdot k) \in L(N, m)$ for $k \leq n$ and all $j \in \mathbb{N}$. Again, choosing $j = i \cdot (n!/k)$ shows that $(m, n + i \cdot n!) \in L(N, m)$ for all $i \in \mathbb{N}$.

In the second case, let $U = U_j \ominus U_{j+1}$ denote the picture decomposed as the last one. Its number of columns is maximal when compared to the number of columns of pictures U_i, $i \in \{1,\dots,s\} \setminus \{j, j+1\}$, hence $\ell_2(U) \geq \ell_2(P)/s \geq 2^{m|V_N|}/2^{|V_N|} = 2^{(m-1)|V_N|}$. Since $\ell_2(U_j), \ell_2(U_{j+1}) \leq m - 1$, by the induction hypotheses, it is possible to pump U_j and U_{j+1} so that any $(m, n + i \cdot n!)$ is generated from N. □

4   Three-Dimensional Kolam Grammar

The three-dimensional Kolam Grammar (3KG) extends 2KG. It generates three-dimensional arrays called *cuboids*. Analogously to context-free productions of type (3) and (4), additional type performing concatenation in the third dimension (the depth) is added. For a cuboid P, its depth is denoted as $\ell_3(P)$.

The well known undecidable Post Correspondence Problem (PCP) is defined as follows. Let $\alpha_1,\dots,\alpha_n$ and $\beta_1,\dots,\beta_n$ be two finite lists of strings over $\{0,1\}$. The task is to decide whether there is a finite sequence of indices $(i_k)_{1 \leq k \leq K}$ with $K \geq 1$ and $1 \leq i_k \leq n$ for all k, such that $\alpha_{i_1} \dots \alpha_{i_K} = \beta_{i_1} \dots \beta_{i_K}$.

Theorem 10. *The emptiness problem is not decidable for* 3KG.

Proof. For a given instance of PCP $\alpha_1,\dots,\alpha_n$ and $\beta_1,\dots,\beta_n$, we show how to construct a 3KG $\mathcal{G} = (V_N, \{a\}, \mathcal{P}, S)$ such that $L(\mathcal{G}) \neq \emptyset$ iff the PCP instance has a solution. The grammar $\mathcal{G}$ will have two nonterminals A and B generating representatives of all strings $\alpha_{i_1} \dots \alpha_{i_K}$ and $\beta_{i_1} \dots \beta_{i_K}$, respectively. For convenience, we treat positive integers as binary strings and vice versa. Let $I_n = \{1,\dots,n\}$ and $\ell = \lceil \log_2(n+1) \rceil$. For $i \in I_n$, define code(i) as the binary string of length ℓ which represents i (i.e., the string is i written in binary, possibly supplemented by leading zeros to reach length ℓ). For a finite sequence of indices $\mathcal{I} = (i_k)_{1 \leq k \leq K}$ where every $i_k \in I_n$, define

$$\text{code}(\mathcal{I}) = 1\,\text{code}(i_1)\,\text{code}(i_2)\dots\text{code}(i_K),$$

$$\text{str}_\alpha(\mathcal{I}) = 1\alpha_{I_1}\alpha_{i_2}\dots\alpha_{i_K}, \qquad \text{str}_\beta(\mathcal{I}) = 1\beta_{I_1}\beta_{i_2}\dots\beta_{i_K}.$$

Moreover, define $P_\alpha = \text{cub}_\alpha(\mathcal{I})$ and $P_\beta = \text{cub}_\beta(\mathcal{I})$ as the cuboids over $\{A\}$ such that $\ell_1(P_\alpha) = \ell_2(P_\alpha) = \ell_1(P_\beta) = \ell_2(P_\beta) = \text{code}(\mathcal{I})$, $\ell_3(P_\alpha) = \text{str}_\alpha(\mathcal{I})$ and $\ell_3(P_\beta) = \text{str}_\beta(\mathcal{I})$. Let $\mathcal{I}'$ be $\mathcal{I}$ prolonged by one more element $j = i_{K+1} \in I_n$. Assume code$(j) = c_1 \dots c_\ell$ and $\alpha_j = a_1 \dots a_m$ where $c_i, a_i \in \{0,1\}$. Cuboid $\text{cub}_\alpha(\mathcal{I}')$ can be obtained from $\text{cub}_\alpha(\mathcal{I})$ by prolonging its size. We can observe that doubling $\text{cub}_\alpha(\mathcal{I})$ as depicted in Fig. 2 changes its number of columns (written in binary) from $\ell_2(\text{cub}_\alpha(\mathcal{I}))$ to $\ell_2(\text{cub}_\alpha(\mathcal{I}))0$. If the doubling is followed by appending a picture of width 1, the resulting number of columns equals $\ell_2(\text{cub}_\alpha(\mathcal{I}))1$. Repeating these operations, it is thus possible to append bits to reach length $\ell_2(\text{cub}_\alpha(\mathcal{I}))c_1 \dots c_\ell$. If A generates $\text{cub}_\alpha(\mathcal{I})$, then the described process is represented by productions

$$A_1^j \to \begin{cases} A\,A & \text{if } c_1 = 0 \\ A\,A\,C & \text{if } c_1 = 1 \end{cases}, \quad A_{i+1}^j \to \begin{cases} A_i^j\,A_i^j & \text{if } c_i = 0 \\ A_i^j\,A_i^j\,C & \text{if } c_i = 1 \end{cases}, \quad i = 1,\dots,m-1.$$

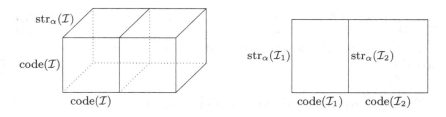

Fig. 2. Encoding of code($\mathcal{I}$) and str$_\alpha$($\mathcal{I}$) by cub$_\alpha$($\mathcal{I}$). Doubling the cuboid by applying a context-free production appends bit 0 to its width written in binary. It is not applicable to pictures – it would be possible to concatenate different representatives along the side representing str$_\alpha$($\mathcal{I}$) when there are $\mathcal{I}_1 \neq \mathcal{I}_2$ such that str$_\alpha$($\mathcal{I}_1$) = str$_\alpha$($\mathcal{I}_2$).

where C is a nonterminal generating all cuboids of size $s \times 1 \times t$ with $s, t \in \mathbb{N}^+$. Productions with three non-terminals on the right-hand side are used for brevity and can be easily turned into the Chomsky normal form. Similar productions can be added to change height and depth (nonterminal A_m^j is taken as the starting point instead of A). If also productions generating cub$_\alpha$($\mathcal{I}_0$) from A for all one-element sequences $\mathcal{I}_0$ are added (there are finitely many such cuboids, suitable productions thus exist), we obtain a complete set of productions generating cub$_\alpha$($\mathcal{I}$) from A for all $\mathcal{I}$. Analogously, there is a set of productions generating all cub$_\beta$($\mathcal{I}$) from B. Finally, for the initial nonterminal S, we add production $S \rightarrow A B$. A cuboid is generated from S only if the input PCP has a solution.

Note that, as explained in Fig. 2, the construction is not applicable to pictures. □

5   Representable Functions

In this section, we further illustrate the complexity of 2KG over unary alphabets by showing which functions and equations it can express. A characterization of representable functions is known e.g. for tiling systems [4].

Definition 11. *A function $f : \mathbb{N}^+ \rightarrow \mathbb{N}^+$ is called* representable *by* 2KG *if the picture language $L(f) = \{ a^{n, f(n)} \mid n \in \mathbb{N}^+ \}$ is generated by a* 2KG.

We will utilize the fact that the class of languages generated by 2KG is closed under the concatenation operations.

Lemma 12. *Let $\mathcal{G}_1 = (V_1, V_T, \mathcal{P}_1, S_1)$, $\mathcal{G}_2 = (V_2, V_T, \mathcal{P}_2, S_2)$ be* 2KG. *Then, $L(\mathcal{G}_1) \oplus L(\mathcal{G}_2)$ as well as $L(\mathcal{G}_1) \ominus L(\mathcal{G}_2)$ can be generated by a* 2KG.

Proof. W.l.o.g, assume $V_1 \cap V_2 = \emptyset$ and $S \notin V_1 \cup V_2$. Then, e.g., $L(\mathcal{G}_1) \oplus L(\mathcal{G}_2)$ is generated by $\mathcal{G} = (V_1 \cup V_2, V_T, \mathcal{P}_1 \cup \mathcal{P}_2 \cup \{S \rightarrow S_1 S_2\}, S)$. □

Lemma 13. *If f, g are two functions representable by* 2KG *and $c \in \mathbb{N}^+$, then cf and $f + g$ are also representable by* 2KG.

Proof. We can write $L(f+g) = L(f) \oplus L(g)$ and $L(cf) = L(\lfloor c/2 \rfloor f) \oplus L(\lceil c/2 \rceil f)$, which can be recursively applied to reduce the multiplier c to 1. By Lemma 12, the concatenation products can be generated by 2KG. □

Lemma 14. *For every $d \in \mathbb{N}$, function $f(n) = n^d$ is representable by 2KG.*

Proof. We prove the lemma by induction on d. If $d = 0$, the constant function $f(n) = 1$ is represented by a 2KG generating all one-column pictures. If $d = 1$, function $f(n) = n$ is represented by the picture language of squares from Example 3. Let $d > 1$. For $n > 1$, an application of the binomial theorem gives

$$n^d = ((n-1)+1)^d = (n-1)^d + \sum_{i=1}^{d} \binom{d}{i}(n-1)^{d-i} = (n-1)^d + h(n-1)$$

where h denotes a function. By the induction hypothesis and Lemma 13, there is a 2KG $\mathcal{G} = (V_N, \{a\}, \mathcal{P}, H)$ such that $L(\mathcal{G}) = L(h)$. Extend $\mathcal{G}$ to $\mathcal{G}' = (V_N \cup \{S, U, R\}, \{a\}, \mathcal{P}', S)$ where S, U, R are not contained in V_N and $\mathcal{P}'$ is $\mathcal{P}$ extended by productions

$$R \to a, \qquad R \to RR, \qquad S \to a, \qquad S \to \frac{U}{R}, \qquad U \to SH.$$

Then, $L(\mathcal{G}') = L(n^d)$. □

Lemma 15. *The exponential function $f(n) = 2^n$ is representable by 2KG.*

Proof. $L(2^n)$ is the picture language from Example 4. □

Taking into account Theorem 9, we can observe that functions which are of a greater than exponential growth cannot be represented by a 2KG.

Corollary 16. *If f is representable by 2KG then $f(n) = 2^{\mathcal{O}(n)}$.*

Note that all the presented results coincide with those known for functions representable by tiling systems [4].

Proposition 17. *Let $f_1, \ldots, f_m$ and $g_1, \ldots, g_n$ be functions representable by 2KG. There is a 2KG $\mathcal{G}$ computable uniformly in representatives of the functions such that*

$$\sum_{i=1}^{m} f_i(x_i) = \sum_{j=1}^{n} g_j(y_j)$$

has a solution $(\overline{x}_1, \ldots, \overline{x}_m, \overline{y}_1, \ldots, \overline{y}_n) \in (\mathbb{N}^+)^{m+n}$ iff $L(\mathcal{G}) \neq \emptyset$.

Proof. Define languages

$$L_1 = \left(L(f_1) \ominus \{a\}^{+,+}\right) \oplus \ldots \oplus \left(L(f_m) \ominus \{a\}^{+,+}\right),$$
$$L_2 = \left(L(g_1) \ominus \{a\}^{+,+}\right) \oplus \ldots \oplus \left(L(g_n) \ominus \{a\}^{+,+}\right),$$
$$L_3 = L_1 \ominus L_2.$$

The language of nonempty pictures $\{a\}^{+,+}$ can be easily generated by a 2KG. For any $x_1, \ldots, x_m \in \mathbb{N}^+$ and $y_1, \ldots, y_n \in \mathbb{N}^+$, there is a picture in L_1 and L_2 with $\sum_{i=1}^m f(x_i)$ and $\sum_{i=1}^n g(y_i)$ columns, respectively. The language L_3 is thus nonempty if and only if the equation has a solution.   □

Example 18 (Exponential Diophantine Equations). A Diophantine equation is a polynomial equation with integral coefficients and one or more unknowns. The existence of an integral solution is undecidable for it [7]. An exponential Diophantine equation is an extension with unknowns occurring also as exponents. We have proved that 2KG can represent a subclass of these equations, namely equations of the form

$$f(x_1, \ldots, x_m, y_1, \ldots, y_n) = c + \sum_{i=1}^m a_i x_i^{d_i} + \sum_{j=1}^n b_j 2^{y_j} = 0 \qquad (6)$$

where $c, a_i, b_j \in \mathbb{Z}$ and $d_i \in \mathbb{N}^+$ for all $i = 1, \ldots, m$, $j = 1, \ldots, n$. Note that Proposition 17 is applicable after rearranging the equation by moving summands with negative coefficients to the right-hand side.

It is unlikely that the solvability of (6) is undecidable. On the other hand, the smallest solution can be a vector of very large integers. For example, components of the smallest positive integral solution to

$$x^3 + y^3 = 4981z^3$$

have over 16 million digits [2]. This coincides with the smallest picture generated by a 2KG representing this equation. It is thus worth to give another important lower bound on δ_{2KG}. For all $n \in \mathbb{N}^+$, $\delta_{2KG}(n)$ equals or is greater than the largest integer among smallest solutions of equations (6) which can be represented by a 2KG with n nonterminals.

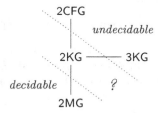

X	CFG	2MG	2KG	2CFG	3KG
δ_X	$\Theta\left(2^n\right)$	$\mathcal{O}\left(2^{n^3}\right)$	$\Omega\left(2 \uparrow\uparrow \frac{n-8}{2}\right)$	NR	NR

Fig. 3. A summary: (un)decidability of the emptiness problem and bounds on function δ for the studied multi-dimensional context-free grammars (NR stands for non-recursive).

6   Conclusion

In Fig. 3, we summarize our main findings, complemented by related known results. We have shown that the emptiness problem is undecidable for the three-dimensional Kolam grammar. The presented proof is easier than the rather technical proof for 2CFG in [14]. Increasing dimensionality therefore seems to have a stronger effect than strengthening the form of productions.

Decidability of the emptiness problem remains open for 2KG. This problem is challenging and important. Showing its decidability would result in revealing a significant difference between two- and three-dimensional world of languages. The presented results however indicate that proving the decidability (assuming it holds) can be very difficult. The derived exponentiation tower lower bound shows that the function δ_{2KG} has a very rapid growth. Moreover, the process of generating pictures by 2KG includes the complexity of some exponential Diophantine equations.

It is even possible that the emptiness problem is undecidable for 2KG. This would also be an interesting finding, saying that quite elementary context-free productions are very powerful in the two-dimensional setting.

Acknowledgement. The author would like to thank Markus Holzer for his suggestions that became the basis for this paper. This work was supported by the Czech Science Foundation under grant no. 15-04960S.

References

1. Blum, M., Hewitt, C.: Automata on a 2-dimensional tape. In: Proceedings of the 8th Annual Symposium on Switching and Automata Theory (SWAT 1967), FOCS 1967, pp. 155–160. IEEE Computer Society, Washington, DC (1967)
2. Bremner, A.: Positively prodigious powers or how Dudeney done it? Math. Mag. **84**(2), 120–125 (2011)
3. Drewes, F., Ewert, S., Klempien-Hinrichs, R., Kreowski, H.: Computing raster images from grid picture grammars. J. Automata, Lang. Comb. **8**(3), 499–519 (2003)
4. Giammarresi, D., Restivo, A.: Two-dimensional languages. In: Rozenberg, G., Salomaa, A. (eds.) Handbook of Formal Languages, vol. 3, pp. 215–267. Springer, New York (1997)
5. Hopcroft, J., Ullman, J.: Formal languages and their relation to automata. Addison-Wesley, Reading (1969)
6. Ito, T., Sakamoto, M., Okabe, H., Furutani, H., Kono, M., Ikeda, S.: Marker versus inkdot over three-dimensional patterns. Artif. Life Robot. **13**(1), 65–68 (2008)
7. Matiyasevich, Y.: Hilbert's tenth problem: Diophantine equations in the twentieth century. In: Bolibruch, A., Osipov, Y., Sinai, Y. (eds.) Mathematical Events of the Twentieth Century, pp. 185–213. Springer, Heidelberg (2006)
8. Matz, O.: Regular expressions and context-free grammars for picture languages. In: Reischuk, R., Morvan, M. (eds.) STACS 1997. LNCS, vol. 1200, pp. 283–294. Springer, Heidelberg (1997)
9. Nakamura, A.: Three-dimensional connected pictures are not recognizable by finite-state acceptors. Inf. Sci. **66**(3), 225–234 (1992)
10. Nakamura, A.: Two-dimensional connected pictures are not recognizable by finite-state acceptors. Inf. Sci. **69**(1–2), 55–64 (1993)

11. Pighizzini, G., Shallit, J., Wang, M.: Unary context-free grammars and pushdown automata, descriptional complexity and auxiliary space lower bounds. J. Comput. Syst. Sci. **65**(2), 393–414 (2002)
12. Pradella, M., Cherubini, A., Reghizzi, S.C.: A unifying approach to picture grammars. Inf. Comput. **209**(9), 1246–1267 (2011)
13. Průša, D.: Two-dimensional Languages. Ph.D. thesis, Faculty of Mathematics and Physics, Charles University, Prague, Czech Republic (2004)
14. Průša, D.: Non-recursive trade-offs between two-dimensional automata and grammars. In: Jürgensen, H., Karhumäki, J., Okhotin, A. (eds.) DCFS 2014. LNCS, vol. 8614, pp. 352–363. Springer, Heidelberg (2014)
15. Schlesinger, M.I.: Matematiceskie sredstva obrabotki izobrazenij (Mathematic tools for image processing). Naukova Dumka, Kiev (1989) (in Russian)
16. Schlesinger, M.I., Hlaváč, V.: Ten Lectures on Statistical and Structural Pattern Recognition (Computational Imaging and Vision). 1st edn. Springer, Heidelberg, May 2012
17. Siromoney, G., Siromoney, R., Krithivasan, K.: Abstract families of matrices and picture languages. Comput. Graph. Image Proces. **1**(3), 284–307 (1972)
18. Siromoney, G., Siromoney, R., Krithivasan, K.: Picture languages with array rewriting rules. Inf. Control **22**(5), 447–470 (1973)
19. Uchida, Y., Ito, T., Sakamoto, M., Uchida, K., Ide, T., Katamune, R., Furutani, H., Kono, M., Yoshinaga, T.: Cooperating systems of four-dimensional finite automata. Artif. Life Robot. **16**(4), 555–558 (2012)

On the Disambiguation of Weighted Automata

Mehryar Mohri[1,2] and Michael D. Riley[2]([⊠])

[1] Courant Institute of Mathematical Sciences, New York, NY, USA
[2] Google Research, New York, NY, USA
riley@google.com

Abstract. We present a disambiguation algorithm for weighted automata. The algorithm admits two main stages: a pre-disambiguation stage followed by a transition removal stage. We give a detailed description of the algorithm and the proof of its correctness. The algorithm is not applicable to all weighted automata but we prove sufficient conditions for its applicability in the case of the tropical semiring by introducing the *weak twins property*. In particular, the algorithm can be used with all acyclic weighted automata and more generally any determinizable weighted automata. While disambiguation can sometimes be achieved using determinization, our disambiguation algorithm in some cases can return a result that is exponentially smaller than any equivalent deterministic automaton. We also present some empirical evidence of the space benefits of disambiguation over determinization in speech recognition and machine translation applications.

1 Introduction

Weighted finite automata and transducers are widely used in applications. Most modern speech recognition systems used for hand-held devices or spoken-dialog applications use weighted automata and their corresponding algorithms for the representation of their models and their efficient combination and search [2,18]. Similarly, weighted automata are commonly used for a variety of tasks in machine translation [9] and other natural language processing applications [10], computational biology [6], image processing [1], optical character recognition [5], and many other areas.

A problem that arises in several applications is that of *disambiguation of weighted automata*: given an input weighted automaton, the problem consists of computing an equivalent weighted automaton that is *unambiguous*, that is one with no two accepting paths labeled with the same string. The need for disambiguation is often motivated by the computation of the marginals given a weighted transducer, or the common problem of determining the most probable string or more generally the n most likely strings, $n \geq 1$, of a *lattice*, an acyclic weighted automaton generated by a complex model, such as those used in machine translation, speech recognition, information extraction, and many other natural language processing and computational biology systems. A lattice compactly represents the model's most likely hypotheses. It defines a probability

© Springer International Publishing Switzerland 2015
F. Drewes (Ed.): CIAA 2015, LNCS 9223, pp. 263–278, 2015.
DOI: 10.1007/978-3-319-22360-5_22

distribution over the strings and is used as follows: the weight of an accepting path is obtained by multiplying the weights of its component transitions and the weight of a string obtained by summing up the weights of accepting paths labeled with that string. In general, there may be many accepting paths labeled with a given string. Clearly, if the lattice were unambiguous, a standard shortest-paths or n-shortest-paths algorithm [8] could be used to efficiently determine the n most likely strings. When the lattice is not unambiguous, the problem is more complex and can be solved using weighted determinization [19]. An alternative solution, which we will show has benefits, consists of first finding an unambiguous weighted automaton equivalent to the lattice and then running an n-shortest-paths algorithm on the resulting weighted automaton.

In general, one way to determine an equivalent unambiguous weighted automaton is to use the weighted determinization algorithm [16]. This, however, admits several drawbacks. First, weighted determinization cannot be applied to all weighted automata. This is both because not all weighted automata admit an equivalent deterministic weighted automaton but also because even for some that do, the weighted determinization algorithm may not halt. Sufficient conditions for the application of the algorithm have been given [3,16]. In particular the algorithm can be applied to all acyclic weighted automata. Nevertheless, a second issue is that in some cases where weighted determinization can be used, the size of the resulting deterministic automaton is prohibitively large.

This paper presents a new disambiguation algorithm for weighted automata extending to the weighted case the algorithm of [17] – the weighted case is significantly more complex and this extension non-trivial. As we shall see, our disambiguation algorithm applies to a broader family of weighted automata than determinization: we show that, for the tropical semiring, if a weighted automaton can be determinized using the algorithm of [16], then it can also be disambiguated using the algorithm presented in this paper. Furthermore, for some weighted automata, the size of the unambiguous weighted automaton returned by our algorithm is exponentially smaller than that of any equivalent deterministic weighted automata. In particular, our algorithm leaves the input unchanged if it is unambiguous, while the size of the automaton returned by determinization for some unambiguous weighted automata is exponentially larger. We also present empirical evidence that shows the benefits of weighted disambiguation over determinization in applications. Our algorithm applies in particular to unweighted finite automata. Note that it is known that for some non-deterministic finite automata of size n the size of an equivalent unambiguous automaton is at least $\Omega(2^{\sqrt{n}})$ [22], which gives a lower bound on the time and space complexity of any disambiguation algorithm for finite automata.

Our disambiguation algorithm for weighted automata is presented in a general way and for a broad class of semirings. Nevertheless, the algorithm is limited in several ways. First, not all weighted automata admit an equivalent unambiguous weighted automaton. But, even for some that do, our algorithm may not succeed. The situation is thus similar to that of weighted determinization. However, we present sufficient conditions based on a new notion of *weak twins property* under which our algorithm can be used. In particular, our algorithm applies to all acyclic weighted automata and more generally to all determinizable weighted automata. Our algorithm admits two stages. The first stage called *pre-disambiguation* constructs a weighted automaton with several key properties, including the property that paths leaving the initial state and labeled with the same string have the same weight. The second stage consists of removing some transitions to make the result unambiguous. Our disambiguation algorithm can be applied whenever pre-disambiguation terminates.

We refer to [17] for an extensive discussion of disambiguation algorithms for unweighted automata and finite-state transducers, in particular the algorithm of Schützenberger. In the weighted case, we already mentioned and discussed weighted determinization [16] as a possible disambiguation algorithm in some cases. A procedure was described by [14] for the special case of the disambiguation of finitely ambiguous min-plus automata, which is a straightforward application of Schützenberger's algorithm for the disambiguation of functional transducers. That procedure does not extend to the general case of weighted automata we are considering because in the general case, the removal of transitions causing ambiguity cannot be executed correctly in that way.[1] An alternative procedure was also described by [13][pp. 598–599] for constructing an unambiguous weighted automaton (when it exists) in the specific case of polynomially ambiguous min-plus weighted automata. The construction is rather intricate and further relies on the prior determination of a threshold value Y. The authors do not give an explicit algorithm for computing Y but state that it can be inferred from [13, Proposition 5.1]. However, the corresponding procedure seems intractable. In fact, as indicated by the authors, the cost of determining Y using that property is super-exponential. The authors of [13] do not give the running-time complexity of their procedure and do not detail various aspects, which makes a comparison difficult. But, our algorithm is much simpler and seems to be significantly more efficient. Our algorithm is also more general since it applies in particular to weighted automata over the tropical semirings that verify the weak twins property and that may be exponentially ambiguous. It is also given for a broader family of semirings. While we are not presenting guarantees for its applicability for semirings different from the tropical semiring, its applicability for at least acyclic weighted automata for those semirings is clear. One advantage of the procedures described by [13] is that the existence of an unambiguous weighted automaton is first tested, though that test procedure appears also to be very costly. Finally, let us mention that an algorithm of Eilenberg [7] bears the same name, disambiguation, but it is in fact designed for an entirely different problem.

---

[1] The removal of ambiguous transitions requires the following key property which is guaranteed by our R-pre-disambiguation algorithm: after removal of ambiguous transitions, the weight of a remaining path must be precisely the same as the weight assigned to the string labeling that path by the original automaton. Let us also emphasize that the procedure of [14] is not a special instance of our algorithm and in particular does not benefit from the crucial use of the relation R*.

The paper is organized as follows. In Sect. 2, we introduce some preliminary definitions and notation relevant to the description of our algorithm. Section 3 describes our pre-disambiguation algorithm and proves some key properties of its result. We describe in fact a family of pre-disambiguation algorithms parameterized by a relation R over the set of pairs of states. A simple instance of that relation is for two states to be equivalent when they admit a path labeled by the same string leading to a final state. In Sect. 4, we describe the second stage, which consists of transition removal, and prove the correctness of our disambiguation algorithm. In Sect. 5, we introduce the notion of *weak twins property* which we use to prove the sufficient conditions for the application of pre-disambiguation and thus the full disambiguation algorithm. The proofs for this section are given in the case of weighted automata over the tropical semiring. Finally, in Sect. 6, we present experiments that compare weighted disambiguation to determinization in speech recognition and machine translation applications. Our implementation of these algorithms used in these experiments is available through a freely available OpenFst library [4]. Detailed proofs for most of our results are given in the [20].

2 Preliminaries

Given an alphabet Σ, we will denote by $|x|$ the length of a string $x \in \Sigma^*$ and by ϵ the *empty string* for which $|\epsilon| = 0$.

The weighted automata we consider are defined over a broad class of *semirings*. A semiring is a system $(\mathbb{S}, \oplus, \otimes, \overline{0}, \overline{1})$ where $(\mathbb{S}, \oplus, \overline{0})$ is a commutative monoid with $\overline{0}$ as the identity element for $\oplus$, $(\mathbb{S}, \otimes, \overline{1})$ is a monoid with $\overline{1}$ as the identity element for $\otimes$, $\otimes$ distributes over $\oplus$, and $\overline{0}$ is an annihilator for $\otimes$.

A semiring is said to be *commutative* when $\otimes$ is commutative. Some familiar examples of (commutative) semirings are the tropical semiring $(\mathbb{R}_+ \cup \{+\infty\}, \min, +, +\infty, 0)$ or the semiring of non-negative integers $(\mathbb{N}, +, \times, 0, 1)$. The multiplicative operation of a semiring $(\mathbb{S}, \oplus, \otimes, \overline{0}, \overline{1})$ is said to be *cancellative* if for any x, x' and z in $\mathbb{S}$ with $z \neq \overline{0}$, $x \otimes z = x' \otimes z$ implies $x = x'$. When that property holds, the semiring $(\mathbb{S}, \oplus, \otimes, \overline{0}, \overline{1})$ is also said to be *cancellative*.

A semiring $(\mathbb{S}, \oplus, \otimes, \overline{0}, \overline{1})$ is said to be *left divisible* if any element $x \in \mathbb{S} - \{\overline{0}\}$ admits a left inverse $x' \in \mathbb{S}$, that is $x' \otimes x = \overline{1}$. $(\mathbb{S}, \oplus, \otimes, \overline{0}, \overline{1})$ is said to be *weakly left divisible* if for any x and x' in $\mathbb{S}$ such that $x \oplus x' \neq \overline{0}$, there exists at least one z such that $x = (x \oplus x') \otimes z$. When the $\otimes$ operation is cancellative, z is unique and we can then write: $z = (x \oplus x')^{-1} \otimes x$.

Weighted finite automata (WFAs) are automata in which the transitions are labeled with weights in addition to the usual alphabet symbols which are elements of a semiring [15]. A WFA $A = (\Sigma, Q, I, F, E, \lambda, \rho)$ over $\mathbb{S}$ is a 7-tuple where: Σ is the finite alphabet of the automaton, Q is a finite set of states, $I \subseteq Q$ the set of initial states, $F \subseteq Q$ the set of final states, E a finite multiset of transitions which are elements of $Q \times \Sigma \times \mathbb{S} \times Q$, $\lambda{:}I \to \mathbb{S}$ an initial weight function, and $\rho{:}F \to \mathbb{S}$ the final weight function mapping F to $\mathbb{S}$.

A path π of a WFA is an element of E^* with consecutive transitions. We denote by orig$[\pi]$ the origin state and by dest$[\pi]$ the destination state of the path. A path is said to be *accepting* or *successful* when orig$[\pi] \in I$ and dest$[\pi] \in F$.

We denote by $w[e]$ the weight of a transition e and similarly by $w[\pi]$ the weight of path $\pi = e_1 \cdots e_n$ obtained by $\otimes$-multiplying the weights of its constituent transitions: $w[\pi] = w[e_1] \otimes \cdots \otimes w[e_n]$. When orig$[\pi]$ is in I, we denote by $w_\mathcal{I}[\pi] = \lambda(\text{orig}[\pi]) \otimes w[\pi]$ the weight of the path including the initial weight of the origin state. For any two subsets $U, V \subseteq Q$ and any string $x \in \Sigma^*$, we denote by $P(U, x, V)$ the set of paths labeled with x from a state in U to a state in V and by $W(U, x, V)$ the $\oplus$-sum of their weights:

$$W(U, x, V) = \bigoplus_{\pi \in P(U,x,V)} w[\pi].$$

When U is reduced to a singleton, $U = \{p\}$, we will simply write $W(p, x, V)$ instead of $W(\{p\}, x, V)$ and similarly for V. To include initial weights, we denote:

$$W_\mathcal{I}(x, V) = \bigoplus_{\pi \in P(I,x,V)} w_\mathcal{I}[\pi].$$

We also denote by $\delta(U, x)$ the set of states reached by paths starting in U and labeled with $x \in \Sigma^*$. The weight associated by A to a string $x \in \Sigma^*$ is defined by

$$A(x) = \bigoplus_{\pi \in P(I,x,F)} w_\mathcal{I}[\pi] \otimes \rho(\text{dest}[\pi]), \tag{1}$$

when $P(I, x, F) \neq \emptyset$. $A(x)$ is defined to be $\overline{0}$ when $P(I, x, F) = \emptyset$.

A state q of a WFA A is said to be *accessible* if q can be reached by a path originating in I. It is *coaccessible* if a final state can be reached by a path from q. Two states q and q' are *co-reachable* if they each can be reached by a path from I labeled with a common string $x \in \Sigma^*$. A WFA A is *trim* if all states of A are both accessible and coaccessible. A is *unambiguous* if any string $x \in \Sigma^*$ labels at most one accepting path. The intersection of two WFAs is a WFA that satisfies $(A_1 \cap A_2)(x) = A_1(x) \otimes A_2(x)$.

In all that follows, we will consider weighted automata over a weakly left divisible cancellative semiring.[2]

3   R-Pre-disambiguation of Weighted Automata

3.1   Relation R over $Q \times Q$

Two states $q, q' \in Q$ are said to share a common future if there exists a string $x \in \Sigma^*$ such that $P(q, x, F)$ and $P(q', x, F)$ are not empty. Let R* be the relation defined over $Q \times Q$ by $q \, \mathsf{R}^* \, q'$ iff $q = q'$ or q and q' share a common future in

---

[2] Our algorithms can be straightforwardly extended to the case of weakly left divisible left semirings [3].

A. Clearly, R^* is reflexive and symmetric, but in general it is not transitive. Observe that R^* is *compatible with the inverse transition function*, that is, if $q\, R^*\, q'$, $q \in \delta(p,x)$ and $q' \in \delta(p',x)$ for some $x \in \Sigma^*$ with $(p,p') \in Q^2$, then $p\, R^*\, p'$. We will also denote by R_0 the complete relation defined by $q\, R_0\, q'$ for all $(q,q') \in Q^2$. Clearly, R_0 is also compatible with the inverse transition function.

The construction we will define holds for any relation R out of the set of admissible relations $\mathcal{R}$ defined as the reflexive relations over $Q \times Q$ that are compatible with the inverse transition function and coarser than R^*. Thus, $\mathcal{R}$ includes R^* and R_0, as well as any reflexive relation R compatible with the inverse transition function that is coarser than R^*, that is, for all $(q,q') \in Q^2$, $q\, R^*\, q' \implies q\, R\, q'$. Thus, for a relation R in $\mathcal{R}$, two states q and q' that share the same future are necessarily in relation, but they may also be in relation without sharing the same future. Note in particular that R is always reflexive.

3.2   Construction

Fix a relation $R \in \mathcal{R}$. For any $x \in \Sigma^*$, and $q \in \delta(U,x)$, we also denote by $\delta_q(U,x)$ the set of states in $\delta(U,x)$ that are in relation with q:

$$\delta_q(U,x) = \delta(U,x) \cap \{p{:}p\, R\, q\}.$$

Note that, since R is reflexive, by definition, $\delta_q(I,x)$ contains q. We will assume that $W_{\mathcal{I}}(x, \{p_1, \ldots, p_t\}) \neq \overline{0}$ for any $x \in \Sigma^*$, otherwise the subset corresponding to x needs not be constructed. For any $x \in \Sigma^*$ and $q \in \delta(I,x)$, we define the weighted subset $s(x,q)$ by

$$s(x,q) = \Big\{ (p_1, w_1), \ldots, (p_t, w_t){:}(\{p_1, \ldots, p_t\} = \delta_q(I,x))$$
$$\wedge \big(\forall i \in [1,t], w_i = W_{\mathcal{I}}(x, \{p_1, \ldots, p_t\})^{-1} \otimes W_{\mathcal{I}}(x, p_i)\big) \Big\}.$$

For a weighted subset s, define $\mathrm{set}(s) = \{p_1, \ldots, p_t\}$. For any automaton A define $A' = (\Sigma, Q', I', F', E', \lambda', \rho')$ as follows:

$$Q' = \{(q, s(x,q)){:}x \in \Sigma^*, q \in \delta(I,x)\}$$
$$I' = \{(q, s(\epsilon, q)){:}q \in I\} \quad \text{and} \quad F' = \{(q, s(x,q)){:}x \in \Sigma^*, q \in \delta(I,x) \cap F\}$$

$$E' = \Big\{ ((q,s), a, w, (q',s')){:}(q,s), (q',s') \in Q', a \in \Sigma,$$
$$\exists x \in \Sigma^* \mid s = s(x,q) = \{(p_1,w_1), \ldots, (p_t,w_t)\},$$
$$s' = s(xa,q') = \{(p'_1, w'_1), \ldots, (p'_{t'}, w'_{t'})\},$$
$$q' \in \delta(q,a), w = \bigoplus_{i=1}^{t} \big(w_i \otimes W(p_i, a, \mathrm{set}(s'))\big),$$
$$\forall j \in [1,t'], w'_j = w^{-1} \otimes \Big(\bigoplus_{i=1}^{t} w_i \otimes W(p_i, a, p'_j)\Big) \Big\}$$

and $\forall (q, s) \in I', s = \{(p_1, w_1), \ldots, (p_t, w_t)\}, \lambda'((q, s)) = \bigoplus_{i \in [1,t]} \lambda(p_i)$.

$$\forall (q, s) \in F', s = \{(p_1, w_1), \ldots, (p_t, w_t)\}, \rho'((q, s)) = \bigoplus_{\substack{p_i \in F \\ i \in [1,t]}} (w_i \otimes \rho(p_i)).$$

Note that in definition of the transition set E' above, the property set$(s') = \delta_{q'}(\text{set}(s), a)$ always holds. In particular, if p' is in $\delta_{q'}(\text{set}(s), a)$, then there is a path from I to some $p \in$ set(s) labeled x and a transition from p to p' labeled with a and $p' R q'$ so p' is in set(s'). Conversely, if p' is in set(s') then there exists p reachable by x with a transition labeled with a from p to p'. Since p' is in set(s'), p' is in $\delta_{q'}(I, xa)$, thus $p' R q'$. Since there exists a transition labeled with a from q to q' and from p to p', this implies that $p R q$. Since $p R q$ and p is reachable via x, p is $\delta_q(I, x)$.

When the set of states Q' is finite, A' is a WFA with a finite set of states and transitions and is defined as the result of the R-*pre-disambiguation of A*. In general, R-pre-disambiguation is thus defined only for a subset of weighted automata, which we will refer to as the set of R-*pre-disambiguable weighted automata*. We will show later sufficient conditions for an automaton A to be R-*pre-disambiguable* in the case of the tropical semiring. Figure 1 illustrates the R-pre-disambiguation construction.

3.3   Properties of the Resulting WFA

In this section, we assume that the input WFA $A = (\Sigma, Q, I, F, E, \lambda, \rho)$ is R-pre-disambiguable. In general, the WFA A' constructed by R-pre-disambiguation is not equivalent to A, but the weight of each path from an initial state equals the $\oplus$-sum of the weights of all paths with the same label in the input automaton starting at an initial state.

Proposition 1. *Let* $A' = (\Sigma, Q', I', F', E', \lambda', \rho')$ *be the finite automaton returned by the* R-*pre-disambiguation of the WFA* $A = (\Sigma, Q, I, F, E, \lambda, \rho)$. *Then, the following equalities hold for any path* $\pi \in P(I', x, (q, s))$ *in* A', *with* $x \in \Sigma^*$ *and* $s = \{(p_1, w_1), \ldots, (p_t, w_t)\}$:

$$w_{\mathcal{I}}[\pi] = W_{\mathcal{I}}(x, \text{set}(s)) \quad and \quad \forall i \in [1, t], \ w_{\mathcal{I}}[\pi] \otimes w_i = W_{\mathcal{I}}(x, p_i).$$

The proof of this proposition, as well as others not included here due to space limitations, can be found in the full version of this paper [20].

Proposition 2. *Let* $A' = (\Sigma, Q', I', F', E', \lambda', \rho')$ *be the finite automaton returned by the* R-*pre-disambiguation of the WFA* $A = (\Sigma, Q, I, F, E, \lambda, \rho)$. *Then, for any accepting path* $\pi \in P(I', x, (q, s))$ *in* A', *with* $x \in \Sigma^*$ *and* $(q, s) \in F'$, *the following equality holds:*

$$w_{\mathcal{I}}[\pi] \otimes \rho'((q, s)) = A(x).$$

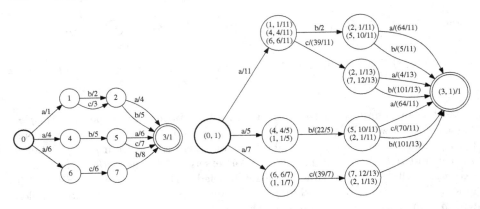

Fig. 1. Illustration of the R-pre-disambiguation construction in the semiring $(\mathbb{R}_+, +, \times, 0, 1)$. Initial states are depicted by a bold circle (always with initial weight $\bar{1}$ in figures here) and final states by double circles. For each state (q, s) of the result, the subset s is explicitly shown. q is the state of the first pair in s shown. The weights are rational numbers, for example $\frac{1}{11} \approx .091$.

Proof. Let $s = \{(p_1, w_1), \ldots, (p_t, w_t)\}$. By definition of ρ', we can write

$$w_\mathcal{I}[\pi] \otimes \rho'((q, s)) = w_\mathcal{I}[\pi] \otimes \bigoplus_{\substack{p_i \in F \\ i \in [1, t]}} (w_i \otimes \rho(p_i)) = \bigoplus_{\substack{p_i \in F \\ i \in [1, t]}} (w_\mathcal{I}[\pi] \otimes w_i \otimes \rho(p_i)).$$

Plugging in the expression of $(w_\mathcal{I}[\pi] \otimes w_i)$ given by Proposition 1 yields

$$w_\mathcal{I}[\pi] \otimes \rho'((q, s)) = \bigoplus_{\substack{p_i \in F \\ i \in [1, t]}} (W_\mathcal{I}(x, p_i) \otimes \rho(p_i)). \qquad (2)$$

By definition of R-pre-disambiguation, q is a final state. Any state $p \in \delta(I, x) \cap F$ shares a common future with q since both p and q are final states, thus we must have $p \, R \, q$, which implies $p \in \text{set}(s)$. Thus, the $\oplus$-sum in (2) is exactly over the set of states $\delta(I, x) \cap F$, which proves that $w_\mathcal{I}[\pi] \otimes \rho'((q, s)) = A(x)$. $\qquad \square$

Proposition 3. *Let* $A' = (\Sigma, Q', I', F', E', \lambda' \rho')$ *be the finite automaton returned by the R-pre-disambiguation of the WFA* $A = (\Sigma, Q, I, F, E, \lambda, \rho)$. *Then, any string* $x \in \Sigma^*$ *accepted by* A *is accepted by* A'.

Proof. Let $(q_0, a_1, w_1, q_1) \cdots (q_{n-1}, a_n, w_n, q_n)$ be an accepting path in A with $a_1 \cdots a_n = x$. By construction, $((q_0, s_0), a_1, w_1', (q_1, s_1)) \cdots ((q_{n-1}, s_{n-1}), a_n, w_n', (q_n, s_n))$ is a path in A' for some $w_i' \in \mathbb{S}$ and with $s_i = s(a_1 \cdots a_i, q_i)$ for all $i \in [1, n]$ and $s_0 = \epsilon$ and by definition of finality in R-pre-disambiguation, (q_n, s_n) is final. Thus, x is accepted by A'. $\qquad \square$

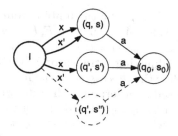

Fig. 2. Illustration of the proof of Lemma 1. The lemma proves the existence of the dashed transitions and the dashed state when $(q, s) \neq (q', s')$ and $x \neq x'$.

4 Disambiguation Algorithm

Propositions 1, 2 and 3 show that the strings accepted by A' are exactly those accepted by A and that the weight of any path in A' accepting $x \in \Sigma^*$ is $A(x)$. Thus, if for any x, we could eliminate from A' all but one of the paths labeled with x, the resulting WFA would be unambiguous and equivalent to A. Removing transitions to achieve this objective without changing the function represented by the WFA turns out not to be straightforward. The following two lemmas (Lemmas 1 and 2) and their proofs are the critical technical ingredients helping us define the transition removal and prove its correctness. This first lemma provides a useful tool for the proof of the second.

Lemma 1. *Let $A' = (\Sigma, Q', I', F', E', \lambda', \rho')$ be the finite automaton returned by the R-pre-disambiguation of the WFA $A = (\Sigma, Q, I, F, E, \lambda, \rho)$. Let (q, s) and (q', s') be two distinct states of A' both admitting a transition labeled with $a \in \Sigma$ to the same state (q_0, s_0) (or both final states), and such that $(q, s) \in \delta(I', x)$ and $(q', s') \in \delta(I', x)$ for some $x \in \Sigma^*$. Then, if $(q, s) \in \delta(I', x')$ for some $x' \neq x$, $x' \in \Sigma^*$, there exists a state $(q', s'') \in \delta(I', x')$ with $(q', s'') \neq (q, s)$ and such that (q', s'') admits a transition labeled with a to (q_0, s_0) (resp. is a final state).*

Proof. Figure 2 illustrates the proof of the lemma. First, note that since $s = s(q, x)$ and $s' = s(q', x)$, $q = q'$ implies $(q, s) = (q', s')$. By contraposition, since $(q, s) \neq (q', s')$, we must have $q \neq q'$. Since both $q_0 \in \delta(q, a)$ and $q_0 \in \delta(q', a)$ in A (or both q and q' are final states), q and q' share a common future, which implies $q \, \mathsf{R} \, q'$. Since (q', s') is reachable by x in A' from I', q' must be reachable by x from I in A. This, combined with $q \, \mathsf{R} \, q'$, implies that q' must be in set(s). Since $(q, s) \in \delta(I', x')$, all states in set$(s)$ must be reachable by x' from I in A, in particular q'. Thus, by definition of the R-pre-disambiguation construction, A' admits a state $(q', s(q', x'))$, which is distinct from (q, s) since $q \neq q'$. If (q, s) admits a transition labeled with a to (q_0, s_0), then we have $s_0 = s(q_0, x'a)$. If (q', s') also admits a transition labeled with a to (q_0, s_0), then q' admits a transition labeled with a to q_0 and by definition of the R-pre-disambiguation construction, $(q', s(q', x'))$ must admit a transition by a to $(q_0, s(q_0, x'a)) = (q_0, s_0)$. Finally, in the case where both (q, s) and (q', s') are final states, then q' is final in A and thus $(q', s(q', x'))$ is a final state in A'. $\qquad\square$

Let $A' = (\Sigma, Q', I', F', E', \lambda', \rho')$ be the finite automaton returned by the R-pre-disambiguation of the WFA $A = (\Sigma, Q, I, F, E, \lambda, \rho)$. For any state (q_0, s_0) of A' and label $a \in \Sigma$, let $\mathcal{L}(q_0, s_0, a) = ((q_1, s_1), \ldots, (q_n, s_n))$, $n \geq 1$, be the list of all distinct states of A' admitting a transition labeled with $a \in \Sigma$ to (q_0, s_0), with $q_1 \leq \cdots \leq q_n$. We define the *processing* of the list $\mathcal{L}(q_0, s_0, a)$ as follows: the states of the list are processed in order; for each state (q_j, s_j), $j \geq 2$, this consists of removing its a-transition to (q_0, s_0) if and only if there exists a co-reachable state (q_i, s_i) with $1 \leq i < j$ whose a-transition to (q_0, s_0) has not been removed.[3] Note that, by definition, the a-transition to (q_0, s_0) of the first state (q_1, s_1) is kept.

We define in a similar way the processing of the list $\mathcal{F} = ((q_1, s_1), \ldots, (q_n, s_n))$, $n \geq 1$, of all distinct final states of A', with an arbitrary order $q_1 \leq \cdots \leq q_n$ as follows: the states of the list are processed in order; for each state (q_j, s_j), $j \geq 1$, this consists of making it non-final if and only if there exists a co-reachable state (q_i, s_i) with $i < j$ whose finality has been maintained. By definition, the finality of state (q_1, s_1) is maintained.

Lemma 2. *Let* $A' = (\Sigma, Q', I', F', E', \lambda', \rho')$ *be the finite automaton returned by the R-pre-disambiguation of the WFA* $A = (\Sigma, Q, I, F, E, \lambda, \rho)$. *Let* (q_0, s_0) *be a state of* A' *and* $a \in \Sigma$, *then, the automaton* A'' *resulting from processing the list* $\mathcal{L}(q_0, s_0, a)$ *accepts the same strings as* A'. *Similarly, the processing of the list of final states* $\mathcal{F}$ *of* A' *does not affect the set of strings accepted by* A'.

Assume that A is R-pre-disambiguable. Then, this helps us define a disambiguation algorithm DISAMBIGUATION for A defined as follows:

1. construct A', the result of the R-pre-disambiguation of A;
2. for any state (q_0, s_0) of A' and label $a \in \Sigma$, process $\mathcal{L}(q_0, s_0, a)$; process the list of final states $\mathcal{F}$.

Theorem 1. *Let* $A = (\Sigma, Q, I, F, E, \lambda, \rho)$ *be a R-pre-disambiguable weighted automaton. Then, algorithm* DISAMBIGUATION *run on input* A *generates an unambiguous WFA* B *equivalent to* A.

Proof. Let $A' = (\Sigma, Q', I', F', E', \lambda', \rho')$ be the WFA returned by R-pre-disambiguation run with input A. By Lemma 2, the set of strings accepted after processing the lists $\mathcal{L}(q_0, s_0, a)$ and $\mathcal{F}$ remains the same[4]. Furthermore, in view of the Propositions 1, 2 and 3, the weight of the unique path labeled with an accepted string x in B $\otimes$-multiplied by its final weight is exactly $A(x)$. Finally, by definition of the processing operations, the resulting WFA is unambiguous, thus B is an unambiguous WFA equivalent to A.   □

------

[3] This condition can in fact be relaxed: it suffices that there exists a co-reachable state (q_i, s_i) with $i < j$ since it can be shown that in that case, there exists necessarily such a state with a a-transition to (q_0, s_0).

[4] The lemma is stated as processing one list, but from the proof it is clear it applies to multiple lists.

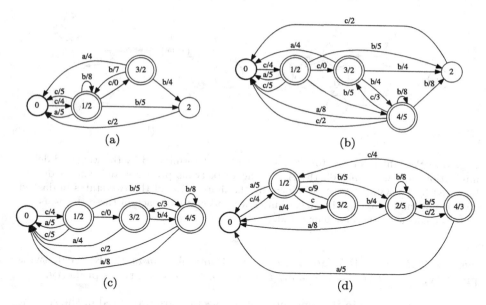

Fig. 3. Example illustrating the full disambiguation algorithm applied to a non-acyclic WFA. (a) WFA A over the tropical semiring. (b) WFA A' obtained from A by application of pre-disambiguation. (c) WFA A'' result of our disambiguation algorithm applied to A. A'' is obtained from A' by removal of the transition from state 2 labeled with $c/2$ and trimming. (d) WFA obtained from A by application of determinization.

Differing numberings of the states can lead to different orderings in each list and thus to different transition or finality removals, thereby resulting in different weighted automata, with potentially different sizes after trimming. Nevertheless, all such resulting weighted automata are equivalent.

Figure 3 gives an example illustrating the pre-disambiguation and transition-removal stages of our disambiguation algorithm and also shows the result of determinization.

5   Sufficient Conditions

The definition of siblings and that of twins property for weighted automata were previously given by [16] (see also [3]). We will use a weaker (sufficient) condition for R-pre-disambiguability.

Definition 1. *Two states p and q of a WFA A are said to be* siblings *if there exist two strings $x, y \in \Sigma^*$ such that both p and q can be reached from an initial state by paths labeled with x and there are cycles at both p and q labeled with y.*

Two sibling states p and q are said to be twins *if for any such x and y, $W(p, y, p) = W(q, y, q)$. A is said to have the* twins *property when any two siblings are twins. It is said to have the* R-weak twins *property when any two*

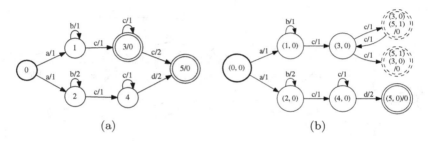

Fig. 4. (a) Weighted automaton A that cannot be determinized by the weighted determinization algorithm of [16]. (b) A has the weak twins property and can be disambiguated by DISAMBIGUATIONas shown by the figure. One of the two states in dashed
style is not made final by the algorithm. The head state for each of these states, is the
state appearing in the first pair listed.

siblings that are in R *relation are twins. When A admits the* R*-weak twins
property, we will also say in short that it admits the* weak twins property.

The results given in the remainder of this section are presented in the specific
case of the tropical semiring. To show the following theorem we partly use a proof
technique from [16] for showing that the twins property is a sufficient condition
for weighted determinizability.

Theorem 2. *Let A be a WFA over the tropical semiring that admits the* R-weak
twins property. *Then, A is* R-pre-disambiguable.

The theorem implies in particular that if A has the twins property then A
is R-pre-disambiguable. In particular, any acyclic weighted automaton is R-pre-
disambiguable.

A WFA A is said to be *determinizable* when the weighted determinization
algorithm of [16] terminates with input A (see also [3]). In that case, the output
of the algorithm is a deterministic automaton equivalent to A.

Theorem 3. *Let A be a determinizable WFA over the tropical semiring, then
A is* R-pre-disambiguable.

By the results of [11], this also implies that any polynomially ambiguous
WFA that has the *clones property* is R-pre-disambiguable and can be disambiguated using DISAMBIGUATION. There are however weighted automata that
are R-pre-disambiguable and thus can be disambiguated using DISAMBIGUA-
TION but that cannot be determinized using the algorithm of [16]. Figure 4 gives
an example of such a WFA. To see that the WFA A of Fig. 4 cannot be determinized, consider instead B obtained from A by removing the transition from
state 3 to 5. B is unambiguous and does not admit the twins property (cycles at
states 1 and 2 have distinct weights), thus it is not determinizable by theorem
12 of [16]. Weighted determinization creates infinitely many subsets of the form

$\{(1,0), (2,n)\}$, $n \in \mathbb{N}$, for paths from the initial state labeled with ab^n. Precisely the same subets are created when applying determinization to A.

On the tropical semiring, define $-A$ as the WFA in which each non-infinite weight in A is replaced by its negation. The following result can be proven in a way that is similar to the proof of the analogous result for the twins property given by [3].[5]

Theorem 4. *Let A be a trim polynomially ambiguous WFA over the tropical semiring. Then, A has the weak twins property iff the weight of any cycle in $B = \text{TRIM}(A \cap (-A))$ is 0.*

This leads to an algorithm for testing the weak twins property for polynomially ambiguous automata in time $O(|Q|^2 + |E|^2)$. It was recently shown that the twins property is a decidable property that is PSPACE-complete for WFAs over the tropical semiring [12]. It would be interesting to determine if the weak twins property we just introduced is also decidable.

6   Experiments

In order to experiment with weighted disambiguation, we implemented the algorithm (using the R^* relation) in the *OpenFst* C++ library [4]. For comparison, an implementation of weighted determinization is also available in that library [16].

For a first test corpus, we generated 500 speech *lattices* drawn from a randomized, anonymized utterance sampling of voice searches on the Google Android platform [21]. Each lattice is a weighted acyclic automaton over spoken words that contains many weighted paths. Each path represents a hypothesis of what was uttered along with the automatic speech recognizer's (ASR) estimate of the probability of that path. Such lattices are useful for passing compact hypothesis sets to subsequent processing without commitment to, say, just one solution at the current stage.

The size of a lattice is determined by a probability threshold with respect to the most likely estimated path in the lattice; hypotheses within the threshold are retained in the lattice. Using $|A| = |Q| + |E|$ to measure automata size, the mean size for these lattices was 2384 and the standard deviation was 3241.

The ASR lattices are typically non-deterministic and ambiguous due to both the models and the decoding strategies used. Determinization can be applied to reduce redundant computation in subsequent stages; disambiguation can be applied to determine the combined probability estimate of a string that may be distributed among several otherwise identically-labels paths.

Disambiguation has a mean expansion of 1.23 and a standard deviation of 0.59. Determinization has a mean expansion of 1.31 and a standard deviation of 1.35. For this data, disambiguation has a slightly less mean expansion compared to determinization but a very substantially less standard deviation.

---

[5] In [3], the authors use instead the terminology of *cycle-unambiguous* weighted automata, which coincides with that of polynomially ambiguous weighted automata.

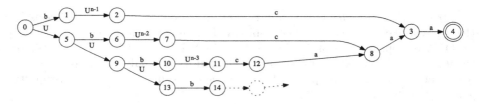

Fig. 5. Unambiguous automaton over the alphabet $\{a, b, c\}$ accepting the language $L = \{(a + b)^{k-1}b(a + b)^{n-k}ca^k : 1 \leq k \leq n\}$. For any $k \geq 0$, U^k serves as a shorthand for $(a + b)^k$.

As a second test corpus, we used 100 automata that are the compact representation of hypothesized Chinese-to-English translations from the DARPA Gale task [9]. These automata may contain cycles due to details of the particular translation system, which provides an interesting contrast to the acyclic speech case. Some fail to determinize within the allotted memory (1 GB) and about two-thirds of those also fail to disambiguate, possible when cycles are present.

Considering only those which are both determinizable and disambiguable, disambiguation has a mean expansion of 4.53 and a standard deviation of 6.0. Determinization has a mean expansion of 54.5 and a standard deviation of 90.5. For this data, disambiguation has a much smaller mean and standard deviation of expansion compared to determinization.

As a final example, Fig. 5 shows an acyclic unambiguous (unweighted) automaton whose size is in $O(n^2)$. No equivalent deterministic automaton can have less than 2^n states since such an automaton must have a distinct state for each of the prefixes of the strings $\{(a + b)^{k-1}b(a + b)^{n-k}ca^k : 1 \leq k \leq n\}$, which are prefixes of L. Thus, while our disambiguation algorithm leaves the automaton of Fig. 5 unchanged, determinization would result in this case in an automaton with more than 2^n states.

7   Conclusion

We presented an algorithm for the disambiguation of WFAs. The algorithm applies to a family of WFAs defined over the tropical semiring verifying a sufficient condition that we described, which includes all acyclic and, more generally, all determinizable WFAs. Our experiments showed the favorable properties of this algorithm in applications related to speech recognition and machine translation. The algorithm is likely to admit a large number of applications in areas such as natural language processing, speech processing, computational biology, and many other areas where WFAs are commonly used. The study of the theoretical properties we initiated raises a number of novel questions which include the following: the decidability of the weak twins property for arbitrary WFAs, the characterization of WFAs that admit an equivalent unambiguous WFA, the characterization of WFAs to which our algorithm can apply and perhaps an extension of our algorithm to a wider domain, and finally the proof and study of these questions for other semirings than the tropical semiring.

Acknowledgments. We thank Cyril Allauzen for discussions about the topic of this research. This work was partly funded by the NSF award IIS-1117591.

References

1. Albert, J., Kari, J.: Digital image compression. In: Handbook of Weighted Automata. Springer, Heidelberg (2009)
2. Allauzen, C., Benson, E., Chelba, C., Riley, M., Schalkwyk, J.: Voice query refinement. In: Interspeech (2012)
3. Allauzen, C., Mohri, M.: Efficient algorithms for testing the twins property. J. Automata, Lang. Comb. **8**(2), 117–144 (2003)
4. Allauzen, C., Riley, M., Schalkwyk, J., Skut, W., Mohri, M.: OpenFst Library (2007). http://www.openfst.org
5. Breuel, T.M.: The OCRopus open source OCR system. In: Proceedings of IS&T/SPIE 20th Annual Symposium (2008)
6. Durbin, R., Eddy, S.R., Krogh, A., Mitchison, G.J.: Biological Sequence Analysis: Probabilistic Models of Proteins and Nucleic Acids. Camb. Univ. Press, Cambridge (1998)
7. Eilenberg, S.: Automata, Languages and Machines. Academic Press, New York (1974)
8. Eppstein, D.: Finding the k shortest paths. SIAM J. Comp. **28**(2), 652–673 (1998)
9. Iglesias, G., Allauzen, C., Byrne, W., de Gispert, A., Riley, M.: Hierarchical phrase-based translation representations. In: Proceedings of EMNLP, pp. 1373–1383 (2011)
10. Kaplan, R.M., Kay, M.: Regular models of phonological rule systems. Comput. Linguist. **20**(3), 331–378 (1994)
11. Kirsten, D.: A Burnside approach to the termination of Mohri's algorithm for polynomially ambiguous min-plus-automata. ITA **42**(3), 553–581 (2008)
12. Kirsten, D.: Decidability, undecidability, and pspace-completeness of the twins property in the tropical semiring. Theor. Comput. Sci. **420**, 56–63 (2012)
13. Kirsten, D., Lombardy, S.: Deciding unambiguity and sequentiality of polynomially ambiguous min-plus automata. In: STACS, pp. 589–600 (2009)
14. Klimann, I., Lombardy, S., Mairesse, J., Prieur, C.: Deciding unambiguity and sequentiality from a finitely ambiguous max-plus automaton. Theor. Comput. Sci. **327**(3), 349–373 (2004)
15. Kuich, W., Salomaa, A.: Semirings, Automata, Languages. EATCS Monographs on Theoretical Computer Science, vol. 5. Springer, Germany (1986)
16. Mohri, M.: Finite-state transducers in language and speech processing. Comput. Linguist. **23**(2), 269–311 (1997)
17. Mohri, M.: On the disambiguation of finite automata and functional transducers. Int. J. Found. Comput. Sci. **24**(6), 847–862 (2013)
18. Mohri, M., Pereira, F.C.N., Riley, M.: Speech recognition with weighted finite-state transducers. In: Handbook on Speech Proc. and Speech Comm. Springer, Heidelberg (2008)
19. Mohri, M., Riley, M.: An efficient algorithm for the n-best-strings problem. In Interspeech (2002)

20. Mohri, M., Riley, M.D.: On the disambiguation of weighted automata. ArXiv 1405.0500, May 2014
21. Schalkwyk, J., Beeferman, D., Beaufays, F., Byrne, B., Chelba, C., Cohen, M., Kamvar, M., Strope, B.: Your word is my command: Google search by voice: A case study. In: Advances in Speech Recognition, pp. 61–90. Springer, Heidelberg (2010)
22. Schmidt, E.M.: Succinctness of description of context-free, regular and unambiguous languages. Ph.D. thesis, Dept. of Comp. Sci., University of Aarhus (1978)

Checking Whether an Automaton Is Monotonic Is NP-complete

Marek Szykuła[✉]

Institute of Computer Science, University of Wrocław,
Joliot-Curie 15, 50-383 Wrocław, Poland
msz@cs.uni.wroc.pl

Abstract. An automaton is monotonic if its states can be arranged in a linear order that is preserved by the action of every letter. We prove that the problem of deciding whether a given automaton is monotonic is NP-complete. The same result is obtained for oriented automata, whose states can be arranged in a cyclic order. Moreover, both problems remain hard under the restriction to binary input alphabets.

Keywords: Automaton · Monotonic · Oriented · Complexity · Np-complete · Linear order · Cyclic order · Partial order · Order-preserving · Transition semigroup

1 Introduction

We deal with complete deterministic finite (semi)automata $\mathcal{A} = \langle Q, \Sigma, \delta \rangle$, where Q is the set of states, Σ is the input alphabet, and $\delta \colon Q \times \Sigma \to Q$ is the transition function defining the action of Σ on Q. This action naturally extends to the action of $\delta(q, w)$ words for any $q \in Q$, $w \in \Sigma^*$.

Monotonic automata are those that admit a linear order of the states. The same qualification is applied to transformation semigroups. Formally, an automaton $\mathcal{A}$ is *monotonic* if there exists a linear order $\leq$ of Q such that if $p \leq q$ then $\delta(p, a) \leq \delta(q, a)$, for all $p, q \in Q$ and $a \in \Sigma$. We call such an order $\leq$ an *underlying linear order* of $\mathcal{A}$. It is clear that if the actions of all letters preserve the order, then also the actions of all words do so.

The class of monotonic automata is a subclass of aperiodic ones [21], which recognize precisely *star-free* languages, and form one of the fundamental classes in the theory of formal languages. An automaton is *aperiodic* if no transformation of any word has a nontrivial cycle. Checking whether an automaton is aperiodic is known to be PSPACE-complete [8]. On the other hand, checking whether an automaton is *nonpermutational*, where no transformation acts like a permutation of a nontrivial subset of Q, can be easily done in $\mathcal{O}(|\Sigma| \times |Q|^2)$ time [14]. Such results may be useful in improving algorithms recognizing star-free languages to work better in particular cases. The complexity problems for

M. Szykuła—Supported in part by Polish NCN grant DEC-2013/09/N/ST6/01194.

F. Drewes (Ed.): CIAA 2015, LNCS 9223, pp. 279–291, 2015.
DOI: 10.1007/978-3-319-22360-5_23

various subclasses of regular languages are widely studied (see [6] for regogniz-
ing convex, and [15] for locally testable languages, and [13] for a survey). The
languages of monotonic automata do not have bounded level in the *dot-depth
hierarchy* of star-free languages [4].

Monotonic semigroups were studied by Gomes and Howie [11] for their maximum size (they use the term *order-preserving*). These semigroups play an important role as building-blocks in the constructions of the largest aperiodic semigroups
known so far ([5, 7]).

Monotonic automata have been considered, in particular, in connection with
the problems of synchronizing automata. An automaton is said to be *synchro-
nizing* if there is a word w such that $|Qw| = 1$; such a word is called a *reset
word*. The Černý conjecture, which is considered one of the most longstanding
open problem in automata theory, states that every synchronizing automaton
has a reset word of length at most $(|Q| - 1)^2$. Ananichev and Volkov [1] have
proved that a synchronizing monotonic automaton has a reset word of length at
most $|Q| - 1$. They have also proved the same bound for a larger class of *gen-
eralized monotonic* automata [2]. Volkov have introduced a still larger class of
weakly monotonic automata [28], which contains all aperiodic ones, and proved
that strongly connected automata in this class possess a synchronizing word of
length $|Q|(|Q| + 1)/6$. Finally, Grech and Kisielewicz have generalized this to
the class of automata *respecting intervals of a directed graph*, and they have
proved that the Černý conjecture holds for each automaton in this class, provided it holds for smaller *quotient* automata. These results could be also useful
in computational verification of the conjecture for automata of limited size, provided we could efficiently recognize and skip from computations automata that
belong to a class for which the conjecture has been proven [17,18]. Therefore it
is important to consider computational complexity of the related problems.

The term *monotonic* was also used by Eppstein [9] for automata whose states
can be arranged in a *cyclic order* that is preserved by the actions of the letters.
Following [1] we call such automata *oriented* automata. They form a broader class,
containing monotonic automata, which has certain applications in robotics (*part-
orienters*, see Natarajan [22]). Eppstein has established the tight upper bound for
the length of the shortest reset words of an oriented automaton $(|Q| - 1)^2$, and
provided an algorithm working in $\mathcal{O}(|\Sigma| \times |Q|^2)$ time for finding such a word.
However, this algorithm requires the cyclic order to be given.

Note that the problem of finding the shortest reset word is hard in general
[23] (also for approximation [3,10] and some restricted classes [20]). But due to
possible practical applications, there are many exponential algorithms that can
deal with fairly large automata and polynomial heuristics (e.g. [16,19,24,25,27]).
Also, hardness does not exclude a possibility of using a polynomial algorithm
for some easily tractable classes (cf. slowly synchronizing [16]).

Here we prove that the problem of checking whether a given automaton is
monotonic is NP-complete, even under restriction to binary alphabets (Sect. 2).
We also obtain that checking whether an automaton is oriented is NP-complete
under the same conditions (Sect. 3). It follows that, unfortunately, they are
hardly recognizable, and it is hard to find a preserved linear (cyclic) order of

a monotonic (oriented) automaton. In particular, we cannot efficiently apply the polynomial Eppstein algorithm [9] to compute a shortest reset word in the cases oriented automata, without knowing a cyclic order. On the other hand, checking whether an automaton admits a nontrivial *partial order* is easy (Sect. 4).

2 Monotonic Automata

The problem MONOTONIC can be formulated as follows: given an automaton $\mathcal{A}$, decide if $\mathcal{A}$ is monotonic. This is the unrestricted version, where the alphabet can be arbitrary large. For a given $k \geq 1$, the restricted problem to k-letter alphabets of the input automaton we call MONOTONIC$_k$.

We show that MONOTONIC is NP-complete, as well as MONOTONIC$_k$ for any $k \geq 2$. The problem is easy if the alphabet is unary.

Proposition 1. *A unary automaton is monotonic if and only if the transformation of the single letter does not contain a cycle of length ≥ 2. MONOTONIC$_1$ can be solved in $\mathcal{O}(|Q|)$ time, and a monotonic order can be found in $\mathcal{O}(|Q|)$ time if it exists.*

Proof. We simply check if the transformation of the single letter of $\mathcal{A}$ contains a cycle of length ≥ 2, that is $\delta(q_1, a) = q_2, \delta(q_2, a) = q_3, \ldots, \delta(q_\ell, a) = q_1$ for some distinct states $q_1, \ldots, q_\ell$. If so, then from $q_1 < q_2$ (or dually $q_1 > q_2$) it follows that $q_2 < q_3, \ldots, q_\ell < q_1$—a contradiction with that $<$ is an order. Thus the automaton is not monotonic.

Otherwise we have an acyclic digraph of the transformation, and we can fix some order on the connected components (sometimes called *clusters*). Each such a component form a rooted tree. We can perform an inverse depth-first search (DFS) starting from the root. Then $p \leq q$ if p is in a component before that of q, or they are in the same component but p was visited later than q during the inverse DFS in this component. So if $p \leq q$ from the same component, then $\delta(p, a)$ was visited later than $\delta(q, a)$, or $\delta(p, a) = \delta(q, a)$. Thus the order is preserved. These operations can be done in $\mathcal{O}(|Q|)$ time. □

Clearly, MONOTONIC is in NP, as we can guess an underlying linear order and check if the action of each letter preserves it (this can be done in $\mathcal{O}(|\Sigma| \times |Q|)$ time).

Proposition 2. MONOTONIC *is in NP.*

2.1 MONOTONIC Is NP-complete

We reduce MONOTONE-NAE-3SAT to MONOTONIC.

NAE-3SAT (NOT-ALL-EQUAL) is a variant of 3SAT, where a clause is satisfied if it contains at least one true and one false literal. The variant MONOTONE-NAE-3SAT additionally restricts instances so that every literal is a positive occurrence of a variable (negations are not allowed). From Schaefer's Theorem [26], we have that NAE-3SAT is NP-complete as well as MONOTONE-NAE-3SAT.

As an instance I of MONOTONE-NAE-3SAT we get a set of n boolean variables $\mathcal{V} = \{v_1, \ldots, v_n\}$, and a set of m clauses $\mathcal{C} = \{C_1, \ldots, C_m\}$, each one with exactly 3 literals. A literal is a positive occurrence of a variable v_i. The problem is to decide if there exists a satisfying assignment $\sigma \colon \mathcal{V} \to \{0, 1\}$ for I, that is, for each clause $C_i \in \mathcal{C}$, C_i contains at least one true literal ($v_j \in C_i$ with $\sigma(v_j) = 1$) and at least one false literal ($v_j \in C_i$ with $\sigma(v_j) = 0$). We can assume that each variable occurs at least one time, and no variable appears more than once in a clause. Note that the complement of a satisfying assignment for I is also satisfying.

Definition of $\mathcal{A}_I$. We construct $\mathcal{A}_I = \langle Q, \Sigma, \delta \rangle$ as follows. For each variable $v_i \in \mathcal{V}$ we create a pair of states p_i, q_i. We also add a unique state s (sink).

For a j-th clause $C_j = (v_f, v_g, v_h)$ (we fix the order of variables in clauses), we create the *clause gadget* as follows. We add three states x_j, y_j, z_j and three letters a_j, b_j, c_j, which correspond to the three occurrences of the variables v_f, v_g, v_h, respectively. The action of these letters is defined as follows:

- $\delta(p_f, a_j) = x_j$ and $\delta(q_f, a_j) = y_j$;
- $\delta(p_g, b_j) = y_j$ and $\delta(q_g, b_j) = z_j$;
- $\delta(p_h, c_j) = z_j$ and $\delta(q_h, c_j) = x_j$;
- $\delta(p_i, a_j) = p_i$ and $\delta(q_i, a_j) = q_i$, for $i < f$;
- $\delta(p_i, b_j) = p_i$ and $\delta(q_i, b_j) = q_i$, for $i < g$;
- $\delta(p_i, c_j) = p_i$ and $\delta(q_i, c_j) = q_i$, for $i < h$;
- $\delta(u, e) = s$, for the other states u and each $e \in \{a_j, b_j, c_j\}$.

So the actions of letters a_j, b_j, c_j send every state from $Q \setminus \{p_i, q_i\}$ either to itself or to s. The clause gadget is presented in Fig. 1.

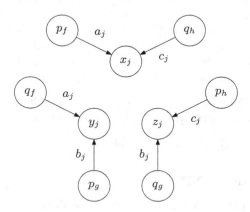

Fig. 1. The clause gadget for a j-th clause (v_f, v_g, v_h)

In Fig. 2 the construction of $\mathcal{A}_I$ is presented, with the action of a_1 as an example, in the case when variable v_2 is the first literal in clause C_1.

In summary, we have $|Q| = 2n + 3m + 1$ states and $|\Sigma| = 3m$ letters.

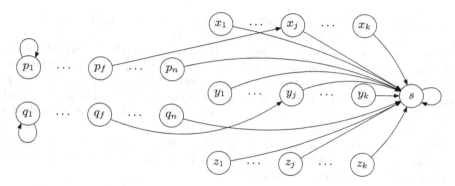

Fig. 2. The action of the letter a_j, where v_f is the first variable in C_j

Correctness of the Reduction

Theorem 1. $\mathcal{A}_I$ *is monotonic if and only if* I *has a satisfying assignment.*

Proof. Suppose that $\mathcal{A}_I$ is monotonic with the underlying linear order $\leq$. We define an assignment σ for I: $\sigma(v_i) = 0$ if $p_i < q_i$, and $\sigma(v_i) = 1$ otherwise. We show that σ is satisfying for I.

Assume for the contrary that there is a clause $C_j = (v_f, v_g, v_h)$, where all the three variables evaluate to 0. This means that $p_f < q_f$, $p_g < q_g$, and $p_h < q_h$. From that $\leq$ is preserved, we have:

- $\delta(p_f, a_j) = x_j < y_j = \delta(q_f, a_j)$;
- $\delta(p_g, b_j) = y_j < z_j = \delta(q_g, b_j)$;
- $\delta(p_h, c_j) = z_j < x_j = \delta(q_h, c_j)$.

Thus $x_j < y_j < z_j < x_j$, a contradiction with that $\leq$ is an order. The argument holds in the dual way in the case with all the three variables evaluated to 1. Hence, σ must be satisfying.

Now, suppose that there is a satisfying assignment σ. We define a linear order $\leq$ and show that it is preserved. To do so, we define $\tau\colon Q \to \mathbb{N}$, which for states $q \in Q$ assigns pairwise distinct natural numbers that will determine $\leq$.

First, for any $1 \leq i \leq n$ let:

- $\tau(p_i) = 2i - 1$ and $\tau(q_i) = 2i$ if $\sigma(v_i) = 0$;
- $\tau(p_i) = 2i$ and $\tau(q_i) = 2i - 1$ if $\sigma(v_i) = 1$.

For $u \in \{x_j, y_j, z_j\}$ we define $\tau(u) \in \{2n + 3j - 2, 2n + 3j - 1, 2n + 3j\}$, depending on the assignment of the variables in $C_j = (v_f, v_g, v_h)$. Assignment σ uniquely determines the relation between x_j, y_j, z_j in an underlying linear order. Each of the six satisfying combinations of $\sigma(v_f), \sigma(v_g), \sigma(v_h)$ defines an acyclic relation between x_j, y_j, z_j, which is enforced by the action of the letters a_j, b_j, c_j. For instance, if $\sigma(v_f) = 0$, then $p_f < q_f$, which implies $\delta(p_f, a_j) = x_j < y_j = \delta(q_f, a_j)$. If $\sigma(v_g) = 0$ then $y_j < z_j$. Then it must be $\sigma(v_h) = 1$ and $z_j > x_j$.

If $\sigma(v_g) = 1$ then $y_j > z_j$, and we have either $z_j < x_j < y_j$ if $\sigma(v_h) = 0$, or $x_j < z_j < y_j$ otherwise. This is dual for $\sigma(v_f) = 1$.

Finally we define $\tau(s) = 3n + 3m + 1$. Hence, in our order $\leq$, first there are states p_i, q_i sorted increasingly by i. The order between p_i and q_i depends on the assignment. Next, there are states from clause gadgets x_j, y_j, z_j sorted by j. The exact order on particular x_j, y_j, z_j depends on the assignment as described above. Finally s is the last state with $u \leq s$ for any $u \in Q$. The order is shown in Fig. 2 (from left to right).

Now we show that $\leq$ is indeed an underlying linear order. Consider a letter a_j for any $1 \leq j \leq k$, and let $C_j = (v_f, v_g, v_h)$. We show that for every pair of distinct states the order $\leq$ is preserved.

- For the pair p_f, q_f, if $p_f < q_f$ then also $\delta(p_f, a_j) = x_j < y_j = \delta(q_f, a_j)$, and if $p_f > q_f$ then $\delta(p_f, a_j) = x_j > y_j = \delta(q_i, a_j)$, since we have chosen the order of x_j, y_j, z_j to be consistent with σ, as described above.
- For p_f (or q_f) and $u \in Q \setminus \{p_f, q_f\}$, if $p_f < u$ then $\delta(p_f, a_j) = x_j < \delta(u, a_j) = s$. If $u < p_f$ then $\delta(u, a_j) = u < x_j = \delta(p_i, a_i^j)$. The same holds for q_f mapped to y_j.
- For distinct states $u, v \in Q \setminus \{p_f, q_f\}$ with $u < v$, if $\delta(u, a_j) = u$, then either $\delta(v, a_j) = v > u$ or $\delta(v, a_j) = s > u$. If $\delta(u, a_j) = s$ then also $\delta(v, a_j) = s$.

The same arguments work for letters b_j and c_j. It follows that any letter preserves $\leq$, so $\leq$ is an underlying linear order of $\mathcal{A}_I$. $\qquad\square$

We can state our main

Theorem 2. *The problem of checking whether a given automaton is monotonic is NP-complete.*

2.2   Reduction from MONOTONIC to MONOTONIC$_2$

Let $\mathcal{A} = \langle Q, \Sigma, \delta \rangle$ be an automaton with $Q = \{v_1, \ldots, v_n\}$ and $\Sigma = \{a_1, \ldots, a_k\}$ with $k \geq 3$. We construct a binary automaton $\mathcal{B}_\mathcal{A} = \langle Q_\mathcal{B}, \{a, b\}, \delta_\mathcal{B} \rangle$ such that $\mathcal{A}$ is monotonic if and only if $\mathcal{B}_\mathcal{A}$ is monotonic.

$Q_\mathcal{B}$ consists of kn states q_j^i for $1 \leq i \leq k, 1 \leq j \leq n$, and a unique state s (sink). Now we define the action of a. For each state q_j^i with $1 \leq i \leq k-1$ and $1 \leq j \leq n$, we define $\delta_\mathcal{B}(q_j^i, a) = \delta_\mathcal{B}(q_j^{i+1})$. For each q_j^k we define $\delta_\mathcal{B}(q_j^k, a) = s$. Finally $\delta(s, a) = s$. The action of b in each set $\{q_1^i, \ldots, q_n^i\}$ corresponds to the action of the i-th letter of Σ on Q: For $1 \leq i \leq k$ and $1 \leq j \leq n$, if $\delta(v_j, a_i) = v_g$ then we define $\delta_\mathcal{B}(q_j^i, b) = q_g^i$. Finally $\delta(s, b) = s$. The construction of $\mathcal{B}_\mathcal{A}$ is shown in Fig. 3.

Theorem 3. $\mathcal{B}_\mathcal{A}$ *is monotonic if and only if* $\mathcal{A}$ *is monotonic.*

Proof. Suppose that $\mathcal{A}$ is monotonic with the underlying linear order $\leq_\mathcal{A}$. We define the linear order $\leq_\mathcal{B}$ on the states of $\mathcal{B}_\mathcal{A}$. For $1 \leq i, f \leq k$ and $1 \leq j, g \leq n$, let $q_j^i \leq_\mathcal{B} q_g^f$ if and only if $i < f$, or $i = f$ and $v_j <_\mathcal{A} v_g$. Also, let $q_j^i <_\mathcal{B} s$

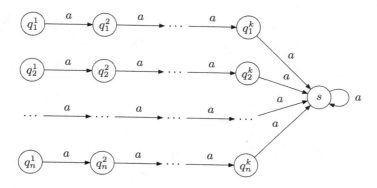

Fig. 3. The action of a in $\mathcal{B}_\mathcal{A}$

for each i, j. The order $\leq_\mathcal{B}$ is linear, since $\leq_\mathcal{A}$ is linear. We show that $\leq_\mathcal{B}$ is an underlying linear order of $\mathcal{B}_\mathcal{A}$.

Clearly, the actions of both letters preserve $\leq_\mathcal{B}$ on states q_j^i and s. Consider a pair q_j^i, q_g^f with $q_j^i \leq_\mathcal{B} q_g^f$. Then $i \leq f$ by definition. Consider the following cases:

- If $i < f$, then $\delta_\mathcal{B}(q_j^i, a) = \delta_\mathcal{B}(q_j^{i+1}, a) <_\mathcal{B} \delta_\mathcal{B}(q_g^f, a)$, since $\delta_\mathcal{B}(q_g^f, a)$ is either q_g^{f+1} or s. Also, for some x, y, $\delta_\mathcal{B}(q_j^i, b) = q_x^i <_\mathcal{B} q_y^f = \delta_\mathcal{B}(q_g^f)$, since $i < f$.
- If $i = f$, then $v_j <_\mathcal{A} v_g$ by definition. If $i = k$ then $\delta_\mathcal{B}(q_j^i, a) = \delta_\mathcal{B}(q_g^i, a) = s$; otherwise $\delta_\mathcal{B}(q_j^i, a) = q_j^{i+1} \leq_\mathcal{B} q_g^{i+1} = \delta_\mathcal{B}(q_g^i, a)$ from $v_j <_\mathcal{A} v_g$. Also, $v_j <_\mathcal{A} v_g$ implies $\delta(v_j, a_i) = v_x \leq_\mathcal{A} v_y = \delta(v_g, a_i)$ for some x, y. So $\delta_\mathcal{B}(q_j^i, b) = q_x^i \leq_\mathcal{B} q_y^i = \delta_\mathcal{B}(q_g^i, b)$.

Thus $\leq_\mathcal{B}$ is an underlying linear order of $\mathcal{B}_\mathcal{A}$.

Now, suppose that $\mathcal{B}_\mathcal{A}$ is monotonic with an underlying linear order $\leq_\mathcal{B}$. We define $\leq_\mathcal{A}$ on the states of $\mathcal{A}$: for $1 \leq j, g \leq n$, $v_j <_\mathcal{A} v_g$ if and only if $q_j^1 <_\mathcal{B} q_g^1$. Observe that for any $j \neq g$, $q_j^1 <_\mathcal{B} q_g^1$ implies $q_j^i <_\mathcal{B} q_g^i$ for each $2 \leq i \leq k$ due to the action of a. Consider two states v_j, v_g with $v_j < v_g$ and the i-th letter a_i. By definition $q_j^1 <_\mathcal{B} q_g^1$, and so $q_j^i <_\mathcal{B} q_g^i$. This implies $\delta_\mathcal{B}(q_j^i, b) = q_x^i \leq_\mathcal{B} q_y^i = \delta_\mathcal{B}(q_g^i, b)$ for some x, y, and it follows that $q_x^1 \leq_\mathcal{B} q_y^1$. Thus $\delta(v_j, a_i) = v_x \leq_\mathcal{A} v_y = \delta(v_g, a_i)$, and the order $\leq_\mathcal{A}$ is an underlying linear order of $\mathcal{A}$. □

As a corollary we obtain that MONOTONIC_2 is also NP-complete. We can reduce an instance of MONOTONE-NAE-3SAT with n variables and m clauses to a binary automaton with $3m(2n + 3m + 1) + 1$ states.

Corollary 1. *The problem of checking whether a given binary automaton $\mathcal{A}$ is monotonic is NP-complete.*

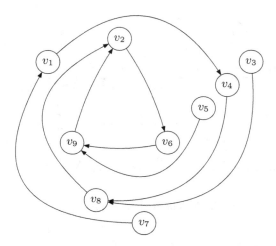

Fig. 4. The cyclic order $(v_1, v_2, \ldots, v_9)$ (clockwise) of a unary automaton

3   Oriented Automata

The following definition of oriented automata is due to Eppstein [9] (who used the term *monotonic*). An automaton is *oriented* if there is a cyclic order of the states preserved by the action of the letters. Formally, there is a cyclic order $q_1, \ldots, q_n$ such that for every $a \in \Sigma$, the sequence $\delta(q_1, a), \ldots, \delta(q_n, a)$, after removal of possibly adjacent duplicate states (the last is also adjacent with the first), is a subsequence of a cyclic permutation $q_i, \ldots, q_n, q_1, \ldots, q_{i-1}$ of the cyclic order, for some $1 \leq i \leq n$. Note that if $q_1, \ldots, q_n$ is a cyclic order then also $q_i, \ldots, q_n, q_1, \ldots, q_{i-1}$ is for every $1 \leq i \leq n$. Figure 4 presents a cyclic order of some unary oriented automaton. Every monotonic automaton is oriented, since if a linear order is preserved, then it is also preserved as a cyclic order. But the converse does not necessarily hold.

Let ORIENTED be the problem of deciding if a given automaton is oriented. As before, we consider ORIENTED$_k$ with the restriction to k-letter alphabets. Again, ORIENTED$_1$ can be easily solved in $\mathcal{O}(n)$ time due to the following

Proposition 3. *A unary automaton is oriented if and only if all cycles in the transformation of the single letter have the same length. There is an algorithm solving the problem* ORIENTED$_1$ *and finding a cyclic order if it exists, and working in $\mathcal{O}(|Q|)$ time.*

Proof. Let a be the single letter of the alphabet and $n = |Q|$. Suppose that $(c_1, \ldots, c_k)$ and $(d_1, \ldots, d_\ell)$ are two cycles in the transformation of a, with $1 \leq k < \ell$. Then, the transformation of a^k has the cycle $(d_1, d_k, \ldots, d_{(m-1)k \mod \ell})$ of length m, for some $2 \leq m \leq \ell$. On the other hand it has the fixed point c_1. Let $q_1, \ldots, q_{n-1}, q_n = c_1$ be a cyclic order of the states of the automaton. Since the transformation of a^k has a cycle of length ≥ 2 (which does not involve c_1), there are two states q_i, q_j with $i < j < n$, $\delta(q_i, a^k) = q_f$, and $\delta(q_j, a^k) = q_g$, such that $f > g$. It follows that $\delta(q_i, c_1) = q_f, \delta(q_j, c_1) = q_g, \delta(c_1, a^k) = c_1$ violates the cyclic order $q_1, \ldots, q_g, \ldots, q_f, \ldots, q_n = c_1$, since (q_f, q_g, c_1) is not a subsequence of any cyclic permutation of the cyclic order—a contradiction.

Assume now that we have m cycles of the same length k:

$$(c_1^1, \ldots, c_k^1), (c_1^2, \ldots, c_k^2), \cdots, (c_1^m, \ldots, c_k^m),$$

so c_i^j is the i-th state of the j-th cycle, and $\delta(c_i^j) = c_{i \bmod k+1}^j$. We can compute a cyclic order by breadth-first search (BFS) in the inverse digraph of the transformation of a. The constructed cyclic order will have the form

$$Q_1^1, c_1^1, \ldots, Q_1^m, c_1^m, \cdots, Q_k^1, c_k^1, \ldots, Q_k^m, c_k^m,$$

where Q_i^j are sequences of states that do not lie on a cycle. Let $\ell(q)$ (*level*) be the smallest integer i such that $\delta(q, a^i)$ is a state on a cycle. To simplify the notation, let $i \oplus j$ be $(i - 1 + j) \bmod k + 1$.

The algorithm starts from the list $(c_1^1, \ldots, c_1^m, \cdots, c_k^1, \ldots, c_k^m)$ of all cycle states; they are considered as visited in the 0-th step in this order. In the i-th step ($i \geq 1$), the algorithm processes the list of visited states from the $(i - 1)$-th step in the order in which they were visited. For each state p from the list, the algorithm computes all states q such that $\delta(q, a) = p$ and q is not a cycle state; so it visits precisely all the states q with $\ell(q) = i$. For every visited state q, it appends q to the end of the new list of visited states in the current step. For a visited q, we have the corresponding cycle state $c_g^f = \delta(q, a^{\ell(q)})$, from which q was reached (possibly indirectly). The algorithm appends q to the beginning of Q_j^f with $j = (nk + g - 1 - i) \oplus 1$; for example, if $g = 1$, then for $i = 1, 2, 3, \ldots, k - 1, k, k + 1, \ldots$ we have $j = k, k - 1, \ldots, 2, 1, k, \ldots$, respectively.

To illustrate the algorithm, consider the automaton from Fig. 4. We start from the list $(c_1, c_2, c_3) = (v_2, v_6, v_9)$ of the one cycle, and empty Q_1, Q_2, Q_3. In the first step, from state v_2 we reach v_8, from v_6 we do not reach any state, and from v_9 we reach v_5. Hence, $Q_3 = (v_8)$ as $v_2 = c_1$, and $Q_2 = (v_5)$ as $v_9 = c_3$. Then, in the second step, from v_8 we reach v_4 and v_3, and from v_5 we do not reach any state; hence, we append v_4 and v_3 to the beginning of Q_2, obtaining $Q_2 = (v_3, v_4, v_5)$. In the third step, from v_4 we reach v_1, so Q_1 becomes (v_1). Finally, in the last fourth step, from v_1 we reach v_7, obtaining $Q_3 = (v_7, v_8)$. The final order is so

$$Q_1, c_1, Q_2, c_2, Q_3, c_3 = v_1, v_2, v_3, v_4, v_5, v_6, v_7, v_8, v_9.$$

We can show that the resulted cyclic order is indeed preserved by the action of a. Observe that $\delta(c_i^j, a) = c_{i \oplus 1}^j$, and if $q \in Q_i^j$ then $\delta(q, a) \in Q_{i \oplus 1}^j$ or $\delta(q, a) = c_{i \oplus 1}^j$. Hence, the sequence Q_i^j, c_i^j is mapped into $Q_{i \oplus 1}^j, c_{i \oplus 1}^j$, and it remains to show that for each $Q_i^j = (p_1, \ldots, p_s)$, the sequence $\delta(p_1, a), \ldots, \delta(p_s, a), \delta(c_i^j, a)$ is a subsequence of $Q_{i \oplus 1}^j, c_{i \oplus 1}^j$. Consider $<$ as the order in these sequences, and let u, v be two states from $Q_i^j \cup \{c_i^j\}$ with $u < v$. If $v = c_i^j$ then we have

$\delta(u, a) \leq \delta(v, a) = c_{i \oplus 1}^j, a$. If $u = p_f$, $v = p_g$ then $u < v$ means that the algorithm appended u after v, so u was visited after v. They were directly reached from $\delta(u, a)$ and $\delta(v, a)$, respectively. If $\delta(v, a) = c_i^j$ then $\delta(u, a) \leq \delta(v, a)$ clearly holds, and if $\delta(u, a) = c_i^j$ then also $\delta(v, a) = c_i^j$. Otherwise, $\delta(u, a), \delta(v, a) \in Q_{i \oplus 1}^j$ and it follows that $\delta(u, a)$ was visited after $\delta(v, a)$ by the algorithm, so $\delta(u, a) > \delta(v, a)$. As usual breadth-first search, this procedure works in $\mathcal{O}(|Q|)$ time.   □

To show hardness, we reduce the NP-complete problems MONOTONIC and MONOTONIC$_k$ (with $k \geq 2$) to ORIENTED and ORIENTED$_k$, respectively.

Proposition 4. *Let* $\mathcal{A}_{+1}$ *be an automaton obtained from* $\mathcal{A} = \langle Q, \Sigma, \delta \rangle$ *by adding a unique state* s *with* $\delta(s, a) = s$ *for every* $a \in \Sigma$. *Then the following are equivalent:*

- $\mathcal{A}$ *is monotonic;*
- $\mathcal{A}_{+1}$ *is monotonic;*
- $\mathcal{A}_{+1}$ *is oriented.*

Proof. Clearly $\mathcal{A}_{+1}$ is monotonic if and only if $\mathcal{A}$ is monotonic, and if $\mathcal{A}_{+1}$ is monotonic then it is also oriented. It remains to show that if $\mathcal{A}_{+1}$ is oriented then $\mathcal{A}_{+1}$ is monotonic.

Assume that $\mathcal{A}_{+1}$ is not monotonic but is oriented, and let $q_1, \ldots, q_n, s$ be a preserved cyclic order of the states of $\mathcal{A}$. Since no state is mapped to s, except s, and s is mapped to itself under the action of every letter, $q_1, \ldots, q_n$ is a preserved cyclic order of the states of $\mathcal{A}$. Since $\mathcal{A}$ is not monotonic, $q_1, \ldots, q_n$ is not an underlying linear order of $\mathcal{A}$. So there are two states $q_i, q_j \in Q$ and $a \in \Sigma$, with $i < j$, $\delta(q_i, a) = q_f$, and $\delta(q_j, a) = q_g$, such that $f > g$. It follows that $\delta(q_i, a) = q_f, \delta(q_j, a) = q_g, \delta(s, a) = s$ violates the cyclic order $q_1, \ldots, q_g, \ldots, q_f, \ldots, q_n, s$ of the states of $\mathcal{A}$, since (q_f, q_g, s) is not a subsequence of any cyclic permutation of the cyclic order. Thus $\mathcal{A}_{+1}$ cannot be oriented and not monotonic.   □

Corollary 2. *The problem of checking whether a given automaton is oriented is NP-complete, even under the restriction to binary alphabets.*

4   Discussion

We have proved that checking whether an automaton is monotonic or oriented is NP-complete. However, several related problems remain open. The complexity of determining whether an automaton is *generalized monotonic* [2], and *weakly monotonic* [28] is not known. The class of generalized monotonic automata strictly contains the class of monotonic ones, and the class of weakly monotonic automata strictly contains the class of generalized monotonic ones. Also, it remains open what is the complexity of checking whether an automaton *respects intervals of a directed graph* [12]; this is the widest of the classes containing the classes of generalized and weakly monotonic automata.

It can be observed that if the alphabet is unary then the classes of generalized and weakly monotonic automata are precisely the class of monotonic automata. However, it is not difficult to check that automata $\mathcal{A}_I$ from the construction from Subsect. 2.1 are generalized, and so weakly monotonic, regardless of the instance I; thus our proof of NP-completeness of testing monotonicity does not work for these wider classes.

On the other hand, for the class of automata preserving a nontrivial *partial order*, the membership problem can be easily solved in polynomial time. An automaton preserves a partial order $\leq$, if $p \leq q$ implies $\delta(p,a) \leq \delta(q,a)$ for every $p,q \in Q$, $a \in \Sigma$. A partial order is nontrivial if at least one pair of states is comparable. In contrast to monotonic automata, not all pairs of states must be comparable, but at least one. This class contains monotonic, generalized monotonic, and weakly monotonic automata, but not oriented, and is a subclass of automata respecting intervals of a directed graph. From [12] it follows that if the Černý conjecture is true for all automata outside this class (admitting only trivial partial orders), then it is true for all automata.

Proposition 5. *Checking whether an automaton preserves a nontrivial partial order and finding it if exists can be done in $\mathcal{O}(|\Sigma| \times |Q|^6)$ time and $\mathcal{O}(|Q|^2)$ working space.*

Proof. For each pair of distinct states $p, q \in Q$, we try to construct a partial order $<$ with $p < q$. So at the beginning of constructing, all states are incomparable and we order $p < q$. When ordering a pair $x, y \in Q$ with $x < y$, we take all the consequences $\delta(x,a) < \delta(y,a)$ for every $a \in \Sigma$ with $\delta(x,a) \neq \delta(y,a)$. Of course, this also involves that $x' < y'$ for every $x' < \delta(x,a)$ and $y' > \delta(y,a)$. For each newly ordered pair we repeat the procedure of taking consequences. If a contradiction is found, that is, if we need to order $x < y$ but they have been already ordered so that $x > y$, the construction fails and we start from another pair p, q. If for some pair p, q all the consequences are taken without a contradiction, we have found a preserved partial order with $p < q$.

Clearly, if the algorithm finds a partial order, then $x \leq y$ implies $\delta(x,a) \leq \delta(y,a)$ as it has taken all the consequences, so the order is preserved. Conversely, if there exists a preserved nontrivial partial order $\leq$, then $p < q$ for some pair of states, and the consequences cannot lead to a contradiction. Hence, the algorithm will find the minimal partial order with $p < q$ that is preserved and is contained in $\leq$.

Concerning the complexity, we need to process $\mathcal{O}(|Q|^2)$ pairs. The constructed partial order can be simply stored as a directed acyclic graph. For every p, q, we start from the empty digraph with one edge (p, q). For each ordered pair $\{x, y\}$ we need to take or check $\mathcal{O}(|\Sigma|)$ consequences, and we order $\mathcal{O}(|Q|^2)$ pairs. Taking a consequence and updating the constructed partial order takes $\mathcal{O}(|Q|^2)$ time, due to the possibly quadratic size of $\{z \in Q \mid z < x\} \times \{z \in Q \mid z > y\}$. These together yield in $\mathcal{O}(|\Sigma| \times |Q|^6)$ time, and the need of storing digraphs yields in $\mathcal{O}(|Q|^2)$ space. $\qquad\square$

The algorithm from Proposition 5 may be modified for finding an underlying linear order of the given automaton. To do so, after finding a partial order that is not yet linear, we need to order another pair that is not yet comparable, say $\{x, y\}$. Here we must consider both possibilities $x < y$ and $x > y$ to check if one of them finally leads to a linear order. Hence, this results in super-exponential worst case running time. However, based on some of our experimental evidence, this algorithm is practically much more efficient than the naive checking of all linear orderings: in most cases of not monotonic automata we can find a contradiction quickly, without the need to enumerate directly all orderings.

References

1. Ananichev, D.S., Volkov, M.V.: Synchronizing monotonic automata. In: Ésik, Z., Fülöp, Z. (eds.) DLT 2003. LNCS, vol. 2710, pp. 111–121. Springer, Heidelberg (2003)
2. Ananichev, D.S., Volkov, M.V.: Synchronizing generalized monotonic automata. Theor. Comput. Sci. **330**(1), 3–13 (2005)
3. Berlinkov, M.V.: Approximating the minimum length of synchronizing words is hard. In: Ablayev, F., Mayr, E.W. (eds.) CSR 2010. LNCS, vol. 6072, pp. 37–47. Springer, Heidelberg (2010)
4. Brzozowski, J., Knast, R.: The dot-depth hierarchy of star-free languages is infinite. J. Comput. Sys. Sci. **16**(1), 37–55 (1978)
5. Brzozowski, J., Li, B., Liu, D.: Syntactic complexities of six classes of star-free languages. J. Automata Lang. Comb. **17**(2–4), 83–105 (2012)
6. Brzozowski, J., Shallit, J., Xu, Z.: Decision problems for convex languages. Inform. Comput. **209**(3), 353–367 (2011)
7. Brzozowski, J., Szykuła, M.: Large aperiodic semigroups. In: Holzer, M., Kutrib, M. (eds.) CIAA 2014. LNCS, vol. 8587, pp. 124–135. Springer, Heidelberg (2014)
8. Cho, S., Huynh, D.T.: Finite-automaton aperiodicity is PSPACE-complete. Theor. Comput. Sci. **88**(1), 99–116 (1991)
9. Eppstein, D.: Reset sequences for monotonic automata. SIAM J. Comput. **19**, 500–510 (1990)
10. Gerbush, M., Heeringa, B.: Approximating minimum reset sequences. In: Domaratzki, M., Salomaa, K. (eds.) CIAA 2010. LNCS, vol. 6482, pp. 154–162. Springer, Heidelberg (2011)
11. Gomes, G., Howie, J.: On the ranks of certain semigroups of order-preserving transformations. Semigroup Forum **45**, 272–282 (1992)
12. Grech, M., Kisielewicz, A.: The Černý conjecture for automata respecting intervals of a directed graph. Discrete Math. Theor. Comput. Sci. **15**(3), 61–72 (2013)
13. Holzer, M., Kutrib, M.: Descriptional and computational complexity of finite automata - a survey. Inf. Comput. **209**(3), 456–470 (2011)
14. Iván, S., Nagy-György, J.: On nonpermutational transformation semigroups with an application to syntactic complexity (2014). http://arxiv.org/abs/1402.7289
15. Kim, S.M., McNaughton, R., McCloskey, R.: A polynomial time algorithm for the local testability problem of deterministic finite automata. IEEE Trans. Comput. **40**(10), 1087–1093 (1991)
16. Kisielewicz, A., Kowalski, J., Szykuła, M.: Computing the shortest reset words of synchronizing automata. J. Comb. Optim. **29**(1), 88–124 (2015)

17. Kisielewicz, A., Szykuła, M.: Generating small automata and the Černý conjecture. In: Konstantinidis, S. (ed.) CIAA 2013. LNCS, vol. 7982, pp. 340–348. Springer, Heidelberg (2013)
18. Kisielewicz, A., Szykuła, M.: Synchronizing Automata with Large Reset Lengths (2014). http://arxiv.org/abs/1404.3311
19. Kudłacik, R., Roman, A., Wagner, H.: Effective synchronizing algorithms. Expert Sys. Appl. **39**(14), 11746–11757 (2012)
20. Martyugin, P.V.: Complexity of problems concerning reset words for some partial cases of automata. Acta Cybernetica **19**, 517–536 (2009)
21. McNaughton, R., Papert, S.A.: Counter-Free Automata, volume 65 of MIT Research Monographs. The MIT Press, Cambridge (1971)
22. Natarajan, B.K.: An algorithmic approach to the automated design of parts orienters. In: 27th Annual Symposium on Foundations of Computer Science, pp. 132–142 (1986)
23. Olschewski, J., Ummels, M.: The complexity of finding reset words in finite automata. In: Hliněný, P., Kučera, A. (eds.) MFCS 2010. LNCS, vol. 6281, pp. 568–579. Springer, Heidelberg (2010)
24. Rho, J.-K., Somenzi, F., Pixley, C.: Minimum length synchronizing sequences of finite state machine. In: Proceedings of the 30th ACM/IEEE Design Automation Conference (DAC1993), pp. 463–468 (1993)
25. Sandberg, S.: 1 Homing and synchronizing sequences. In: Broy, M., Jonsson, B., Katoen, J.-P., Leucker, M., Pretschner, A. (eds.) Advanced Lectures. LNCS, vol. 3472, pp. 5–33. Springer, Heidelberg (2005)
26. Schaefer, T.J.: The Complexity of Satisfiability Problems. In: Proceedings of the Tenth Annual ACM Symposium on Theory of Computing (STOC), pp. 216–226. ACM (1978)
27. Skvortsov, E., Tipikin, E.: Experimental study of the shortest reset word of random automata. In: Bouchou-Markhoff, B., Caron, P., Champarnaud, J.-M., Maurel, D. (eds.) CIAA 2011. LNCS, vol. 6807, pp. 290–298. Springer, Heidelberg (2011)
28. Volkov, M.V.: Synchronizing automata preserving a chain of partial orders. Theor. Comput. Sci. **410**(37), 3513–3519 (2009)

On the Semantics of Regular Expression Parsing in the Wild

Martin Berglund[1] and Brink van der Merwe[2]([⊠])

[1] Umeå University, Umeå, Sweden
mbe@cs.umu.se
[2] University of Stellenbosch, Stellenbosch, South Africa
abvdm@cs.sun.ac.za

Abstract. We introduce prioritized transducers to formalize capturing groups in regular expression matching in a way that permits straightforward modelling of and comparison with real-world regular expression matching library behaviors. The broader questions of parsing semantics and performance are discussed, and also the complexity of deciding equivalence of regular expressions with capturing groups.

1 Introduction

Few formal language research results have greater practical reach than regular expressions. As a result the practical implementations [5] have in many ways surged ahead of research, with new features which require underpinnings different from the original theory. Practical implementations perform matching as a form of parsing, using *capturing groups,* outputting what subexpression matched which substring. A popular implementation strategy, still very common [2], is a worst-case EXPTIME depth-first search for such parses. A more formal approach suggests using finite transducers, outputting annotations on the string to signify the nature of the match [9]. This is complicated by the matching semantics dictating a single output string for each input string, using rules to determine a "highest priority" match among the potentially exponentially many possible ones (for contrast e.g. [3] discusses non-deterministic capturing groups).

The pNFA (prioritized non-deterministic finite automaton) model of [2] (similar results were also published mere weeks later in [7]) provides the right level of abstraction to model the matching time behavior of regular expression matchers. However, for matchers based on an input directed depth first search, it does not provide an understanding of why practical regular expression matchers often (indirectly) use the pNFA model, and in particular, there is no notion of when one pNFA is equivalent to another. By adding output to pNFA to obtain pTr (prioritized transducers), we obtain a better understanding of the usefulness of the prioritized automata/transducer model, and we also have the notion of equivalence of pTr, which is defined in terms of equality of the underlying functions represented by the pTr. A regular expression to transducer construction is done in [9], but it is remarked that translating regular expression matching directly

© Springer International Publishing Switzerland 2015
F. Drewes (Ed.): CIAA 2015, LNCS 9223, pp. 292–304, 2015.
DOI: 10.1007/978-3-319-22360-5_24

into transducers is highly non-trivial. In Sect. 3, where we discuss conversion from regular expression to pTr, it will become clear that pTr are a perfect fit when converting regular expressions to transducers. We also discuss a linear-time matching algorithm for pTr (i.e. determining the image of input strings), generalizing e.g. [6] which operates directly on expressions, and mirroring work by Russ Cox [4] which is to a great extent not formally published.

The outline of the paper is as follows. In the next section, we define prioritized automata and transducers. After that, we show how to adapt the standard Thompson construction, from [10], for converting regular expressions to nondeterministic finite automata, to the more general setting of converting regular expressions to prioritized transducers. This is followed by a normal form for prioritized transducers, so called flattened prioritized transducers, that simplifies the following discussions on deciding equivalence of and parsing with pTr.

2 Definitions

Let $\text{dom}(f)$ and $\text{range}(f)$ denote the domain and range of a function f respectively. When unambiguous let a function f with $\text{dom}(f) = S$ generalize to S^* and $\mathcal{P}(S)$ element-wise. By $\mathcal{P}(S)$, we denote the power set of a set S. The cardinality of a (finite) set S is denoted by $|S|$. We denote by $\mathbb{N}$ the set of natural numbers, i.e. the set $\{1, 2, 3, \ldots\}$. The empty string is denoted ε. An alphabet Σ is a finite set of symbols with $\varepsilon \notin \Sigma$. We denote $\Sigma \cup \{\varepsilon\}$ by Σ^ε. For any string w let $\pi_S(w)$ be the maximal subsequence of w containing only symbols from S (e.g. $\pi_{\{a,b\}}(abcdab) = abab$). For sequences $s = (z_{1,1}, \ldots, z_{1,n}) \ldots (z_{m,1}, \ldots, z_{m,n}) \in (Z_1 \times \ldots \times Z_n)^*$, we denote by $\sigma_i(s)$ the subsequence of tuples obtained from s by deleting duplicates of tuples in s and only keeping the first occurrence of each tuple, where equality of tuples are based only on the value of the ith component of a tuple (e.g. $\sigma_1((1, a)(2, a)(1, b)(3, b)(2, c)) = (1, a)(2, a)(3, b)$). For each $k > 1$, we denote by B_k the alphabet of k types of brackets, which will be represented as $\{[_1,]_1, [_2,]_2, \ldots [_k,]_k\}$. The Dyck language D_k over the alphabet B_k is the set of strings representing well balanced sequences of brackets over B_k.

As usual, a regular expression over an alphabet Σ (where $\varepsilon, \emptyset \notin \Sigma$) is either an element of $\Sigma \cup \{\varepsilon, \emptyset\}$ or an expression of one of the forms $(E \mid E')$, $(E \cdot E')$, or (E^*), where E and E' are regular expressions. Some parentheses can be dropped with the rule that * (Kleene closure) takes precedence over $\cdot$ (concatenation), which takes precedence over $\mid$ (union). Further, outermost parentheses can be dropped, and $E \cdot E'$ can be written as EE'. The language of a regular expression E, denoted $\mathcal{L}(E)$, is obtained by evaluating E as usual, where $\emptyset$ stands for the empty language and $a \in \Sigma \cup \{\varepsilon\}$ for $\{a\}$. The size of E, denoted $|E|$, is the number of symbols appearing in E. A capturing group is any parenthesized subexpression, e.g. (E). The matching procedure will also produce information about which substring(s) are matched by each capturing group. Thus brackets in regular expressions are used both for precedence and capturing, and in Java[1] a

---

[1] Java is a registered trademark of Oracle and/or its affiliates. Other names may be trademarks of their respective owners.

non-capturing subexpression E is indicated by $(?\!:\!E)$. The precise matching and capturing semantics follow from Sect. 3. When we say that E matches a string w we mean that all of w is read by E, as opposed to $vwv' \in \mathcal{L}(E)$, for $v, v' \in \Sigma^*$, as some implementations do. This substring matching can be simulated in our model with the expression $.^{*?}(R).^*$, where $E^{*?}$ and E^* denotes respectively the lazy and greedy (both to be defined in Sect. 3) Kleene closure of E.

For later constructions we require a few different kinds of automata and transducers. First (non-)deterministic finite automata (and runs for them), followed by the prioritized finite automata from [2], which are used to model the regular expression matching behaviors exhibited by typical software implementations.

Definition 1. *A non-deterministic finite automaton (NFA) is a tuple* $A = (Q, \Sigma, q_0, \delta, F)$ *where: (i) Q is a finite set of* states; *(ii) Σ is the* input alphabet; *(iii) $q_0 \in Q$ is the* initial state; *(iv) $\delta : Q \times \Sigma^\varepsilon \to \mathcal{P}(Q)$ is the* transition function; *and (v) $F \subseteq Q$ is the set of* final states.

A is ε-free if $\delta(q, \varepsilon) = \emptyset$ for all q. A is deterministic if it is ε-free and $|\delta(q, \alpha)| \leq 1$ for all q and α. The state size of A is denoted by $|A|_Q$, and defined to be $|Q|$.

Definition 2. *For a NFA $A = (Q, \Sigma, q_0, \delta, F)$ and $w \in \Sigma^*$, a run for w is a string $r = s_0 \alpha_1 s_1 \cdots s_{n-1} \alpha_n s_n \in (Q \cup \Sigma)^*$, with $s_0 = q_0$, $s_i \in Q$ and $\alpha_i \in \Sigma^\varepsilon$ such that $s_{i+1} \in \delta(s_i, \alpha_{i+1})$ for $0 \leq i < n$, and $\pi_\Sigma(r) = w$. A run is accepting if $s_n \in F$. The language accepted by A, denoted by $\mathcal{L}(A)$, is the subset $\{\pi_\Sigma(r) \mid r$ is an accepting run in $A\}$ of Σ^*.*

Now for the prioritized NFA variant, as defined in [2].

Definition 3. *A prioritized non-deterministic finite automaton (pNFA) is a tuple $A = (Q_1, Q_2, \Sigma, q_0, \delta_1, \delta_2, F)$, where if $Q := Q_1 \cup Q_2$, we have: (i) Q_1 and Q_2 are disjoint finite sets of* states; *(ii) Σ is the* input alphabet; *(iii) $q_0 \in Q$ is the* initial state; *(iv) $\delta_1 : Q_1 \times \Sigma \to Q$ is the deterministic, but not necessarily* total, *transition function; (v) $\delta_2 : Q_2 \to Q^*$ is the* non-deterministic prioritized *transition function; and (vi) $F \subseteq Q_1$ are the* final states.

Remark 4. For a pNFA $A = (Q_1, Q_2, \Sigma, q_0, \delta_1, \delta_2, F)$ the *corresponding finite automaton* nfa(A) is given by nfa$(A) = (Q_1 \cup Q_2, \Sigma, q_0, \bar{\delta}, F)$, where $\bar{\delta}(q, \alpha) = \{\delta_1(q, \alpha)\}$ if $q \in Q_1$, and $\bar{\delta}(q, \varepsilon) = \{q_1, \ldots, q_n\}$ if $q \in Q_2$ with $\delta_2(q) = q_1 \ldots q_n$.

Next we define runs for pNFA. An accepting run for a string w in a pNFA A, is defined to be the highest priority accepting run of w in nfa(A), not repeating the same ε-transition in a subsequence of consecutive ε-transitions. Prioritized NFA are thus on a conceptual level closely related to unambiguous NFA, since in an pNFA there is at most one accepting run for an input string. The repeated ε-transition restriction is made to ensure that we consider only finitely many of the runs in nfa(A) for a given input string w, when determining the highest priority path (referred to as a run in A) for w, and also to ensure that regular expression matchers based on pNFA/pTr do not end up in an infinite loop during (attempted) matching.

Definition 5. *For a pNFA* $A = (Q_1, Q_2, \Sigma, q_0, \delta_1, \delta_2, F)$*, a path of* $w \in \Sigma^*$ *in* A*, is a run* $s_0 \alpha_1 s_1 \cdots s_{n-1} \alpha_n s_n$ *of* w *in* $nfa(A)$*, such that if* $\alpha_i = \alpha_{i+1} = \cdots = \alpha_{j-1} = \alpha_j = \varepsilon$*, with* $i \leq j$*, then* $(s_{k-1}, s_k) = (s_{l-1}, s_l)$*, with* $i \leq k, l \leq j$*, implies* $k = l$ – *i.e. a path is not allowed to repeat the same transition in a sequence of* ε-*transitions. For two paths* $p = s_0 \alpha_1 s_1 \cdots s_{n-1} \alpha_n s_n$ *and* $p' = s'_0 \alpha'_1 s'_1 \cdots s'_{m-1} \alpha'_m s'_m$ *we say that* p *is of higher priority than* p'*,* $p > p'$*, if* $p \neq p'$*,* $\pi_\Sigma(p) = \pi_\Sigma(p')$ *and either* p' *is a proper prefix of* p*, or if* j *is the first index such that* $s_j \neq s'_j$*, then* $\delta_2(s_{j-1}) = \cdots s_j \cdots s'_j \cdots$*. An accepting run for* A *on* w *is the highest-priority path* $p = s_0 \alpha_1 s_1 \cdots \alpha_n s_n$ *such that* $\pi_\Sigma(p) = w$ *and* $s_n \in F$*. The language accepted by* A*, denoted by* $\mathcal{L}(A)$*, is the subset of* Σ^* *defined by* $\{\pi_\Sigma(r) \mid r$ *is an accepting run in* $A\}$ *. Note that* $\mathcal{L}(A) = \mathcal{L}(nfa(A))$*.*

Our definition of pNFA, compared to the one in [2], is slightly less general, since we assume that $F \subseteq Q_1$, instead of $F \subseteq Q$. This restriction was introduced to simplify our definitions, and the more general pNFA can be converted to pNFA with $F \subseteq Q_1$, by introducing one new state $q_F \in Q_1$ and δ_2 transitions from the old accepting states $q \in Q_2$ to q_F, where we give the new δ_2 transitions for example the highest priority of all δ_2 transitions at q. In [7], an ordered NFA, very similar to our definition of pNFA, is defined, with a single set of states Q and a transition function $\delta : Q \times \Sigma \rightarrow Q^*$. We can simulate this with our pNFA, by decomposing $q \mapsto \delta(q, a)$ into $q \mapsto^{\delta_1(\_, a)} q_a \mapsto^{\delta_2(\_)} \delta(q, a)$, where $q \in Q_1$ and $q_a \in Q_2$. Note that we introduced pNFA (and runs in pNFA in Definition 5) mainly as an aid in defining runs in prioritized transducers in Definition 9 below.

Example 6. In Fig. 1(a), a Java based pNFA A for the regular expression $(a^*)^*$, constructed as described in [2], is given. The accepting run for the string a^n, in A, is $q_0 q_1 (q_2 a q_1)^n q_0 q_3$. Since there are for the input strings a^n, $n \geq 0$, exponentially many paths in A, a regular expression matcher using an input directed depth first search (without memoization as in Perl), such as the Java implementation, will take exponential time to attempt to match the strings $a^n x$, for $n \geq 0$.

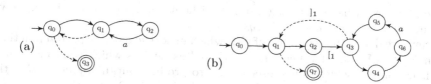

Fig. 1. (a) Java based pNFA for the regular expression $(a^*)^*$, i.e., the pNFA $A = (\{q_2, q_3\}, \{q_0, q_1\}, \{a\}, q_0, \{(q_2, a, q_1)\}, \{[q_0, (q_1, q_3)], [q_1, (q_2, q_0)]\}, q_3)$. (b) Java based pTr with $\Sigma_1 = \{a\}$ and $\Sigma_2 = \{[_1,]_1\}$, for $(a^*)^*$. Lower priority transitions are indicated by dashed edges.

Recall that a transducer T (see for example [11], Definition 3.1) is a tuple $(Q, \Sigma_1, \Sigma_2, q_0, \delta, F)$, where Q is a finite set of states, Σ_1 and Σ_2 the input and output alphabets respectively, $\delta \subseteq Q \times \Sigma_1^\varepsilon \times \Sigma_2^* \times Q$ the set of transitions, q_0 the initial state and F the set of final states. Accepting runs are defined as for NFA, but in a run when moving from state q to q' while reading input x and using the transition (q, x, y, q'), the string y is also produced as output. The *state size* of T, denote by $|T|_Q$, is the number of states in T, the *transition size*, $|T|_\delta$, is the sum of $(1 + |y|)$ over all transitions (q, x, y, q'), and the *size* of T, denoted by $|T|$, is $|T|_Q + |T|_\delta$. A transducer T defines a relation $\mathcal{R}(T) \subseteq \Sigma_1^* \times \Sigma_2^*$, containing all pairs (v, w) for which there is an accepting run reading input v and producing w as output while moving from the first to last state in the accepting run (i.e. v and w are the concatenation of the input symbols and output strings respectively, of all the transitions taken in the accepting run). As usual, we denote by $\text{dom}(T)$ the set $\{v \in \Sigma_1^* \mid (v, w) \in \mathcal{R}(T)\}$, and by $\text{range}(T)$ the set $\{w \in \Sigma_2^* \mid (v, w) \in \mathcal{R}(T)\}$. For functional transducers (see [8], Chapter 5), the relation $\mathcal{R}(T)$ is a function, and we write $T(v) = w$ if $(v, w) \in \mathcal{R}(T)$. Prioritized string transducers, defined next, also define relations, in this case contained in $\Sigma_1^* \times (\Sigma_1 \cup \Sigma_2)^*$, which are in fact functions, and the notation $\mathcal{R}(T)$, $\text{dom}(T)$, $\text{range}(T)$, and $T(v) = w$ if $(v, w) \in \mathcal{R}(T)$, will thus also be used.

Definition 7. *A prioritized non-deterministic finite transducer (pTr) is a tuple* $T = (Q_1, Q_2, \Sigma_1, \Sigma_2, q_0, \delta_1, \delta_2, F)$, *where if* $Q := Q_1 \cup Q_2$, *we have: (i)Q_1 and Q_2 are disjoint finite sets of states; (ii) Σ_1 is the input alphabet; (iii)Σ_2, disjoint from Σ_1, is the group identifier or output alphabet; (iv)$q_0 \in Q$ is the initial state; (v)$\delta_1 : Q_1 \times \Sigma_1 \to Q$ is the deterministic, but not necessarily total, transition function; (vi)$\delta_2 : Q_2 \to (\Sigma_2^* \times Q)^*$ is the non-deterministic prioritized transition and output function; and (vii)$F \subseteq Q_1$ are the final states.* The state size of T is $|T|_Q := |Q_1| + |Q_2|$, the δ_1 transitions size $|T|_{\delta_1} := \sum_{q \in Q_1, a \in \Sigma_1} |\delta_1(q, a)|$ where $|\delta_1(q, a)| = 1$ if $\delta_1(q, a)$ is defined and 0 otherwise, the δ_2 transitions size $|T|_{\delta_2} := \sum_{q \in Q_2} |\delta_2(q)|$ where $|\delta_2(q)|$ equals $\sum_i (1 + |w_i|)$ if $\delta_2(q) = (w_1, q_1) \dots (w_n, q_n)$ (and $|\delta_2(q)| = 0$ if $\delta_2(q) = \varepsilon$), the transitions size $|T|_\delta := |T|_{\delta_1} + |T|_{\delta_2}$, and finally, the size of T is $|T| := |T|_Q + |T|_\delta$.

Remark 8. It is only in Sect. 3, when we construct pTr from regular expressions, where the assumption $\Sigma_1 \cap \Sigma_2 = \emptyset$ is required.

Going forward, when discussing a pTr T without being specific on the tuple, we assume that $T = (Q_1, Q_2, \Sigma_1, \Sigma_2, q_0, \delta_1, \delta_2, F)$.

Next we define the semantics of pTr, which make them define partial functions from Σ_1^* to $(\Sigma_1 \cup \Sigma_2)^*$. The pTr are viewed as devices which consume strings in their domain, which is a subset of Σ_1^*, to produce output by *decorating* the input string with symbols from Σ_2. For a pTr $T = (Q_1, Q_2, \Sigma_1, \Sigma_2, q_0, \delta_1, \delta_2, F)$, pnfa$(T)$ is the pNFA $(Q_1, Q_2, \Sigma_1, q_0, \delta_1, \delta_2', F)$ obtained from T with $\delta_2'(q) = q_1 \dots q_n$ if $\delta_2(q) = (w_1, q_1) \dots (w_n, q_n)$ for some $w_i \in \Sigma_2^*$. For a pTr T, the runs in pnfa(T) determine the decorated output string, in $(\Sigma_1 \cup \Sigma_2)^*$, produced from a given input string in Σ_1^*. When applying the function δ_1 on (q, α), T produces α as output, where when using δ_2 on q with $\delta_2(q) = (w_1, q_1) \dots (w_n, q_n)$, one of the w_i's is produced as output.

Definition 9. *Let* $T = (Q_1, Q_2, \Sigma_1, \Sigma_2, q_0, \delta_1, \delta_2, F)$ *be a pTr, and let Q denote* $Q_1 \cup Q_2$ *and Σ the set $\Sigma_1 \cup \Sigma_2$. An accepting run for $v \in \Sigma_1^*$ in T is a string*

$r = s_0\alpha_1 s_1 \cdots s_{n-1}\alpha_n s_n \in (Q \cup \Sigma^\varepsilon)^*$, $s_i \in Q$ and $\alpha_i \in (\Sigma_1 \cup \Sigma_2)^*$, such that $\pi_{Q \cup \Sigma_1}(r)$ is an accepting run of v in $pnfa(T)$, and if $s_i \in Q_2$, with $i < n$, then the sequence of tuples defined by $\delta_2(s_i)$ contains (α_{i+1}, s_{i+1}) (and not necessarily at position $(i+1)$ of $\delta_2(s_i)$). The pTr T defines a partial function from Σ_1^* to Σ^* by $T(v) = w$, if there is an accepting run r for v in T with $\pi_\Sigma(r) = w$.

Remark 10. Note that for pTr we may assume that $\delta_2(q) = (w_1, q_1) \ldots (w_n, q_n)$ implies that the states $q_1, \ldots, q_n$ are pairwise distinct, since if $q_i = q_j$ with $i < j$, then if the remainder of the input is not accepted from q_i, it will also not be accepted from q_j, and thus (w_j, q_j) may be removed from $\delta_2(q)$.

Example 11. In Fig. 1(b), a pTr T for the regular expressions $(a^*)^*$, constructed by the procedure described in the next section, is given. In this case, only the substrings matched by the subexpression a^*, are captured, and the captured substrings are enclosed by the pair of brackets in $B_1 = \{[_1,]_1\}$. Also, $\text{dom}(T)$ is a^*, and $T(a^n) = [_1 a^n]_1$ for $n \geq 0$. It should be pointed out that the Java regular expression matcher in fact only prohibits duplicates of the ε transitions $q_1 \to q_2$ and $q_3 \to q_4$ (in Fig. 1(b)) in a sequence of ε transitions, and thus in the Java case we have $T(a^n) = [_1 a^n]_1[_1]_1$ for $n \geq 1$. In general, the Java matcher only prohibits duplicates of the ε transitions $f_1 \to q_1$ in the lazy and greedy Kleene closure in Figs. 2(c) and (d) in the next section. For the regular expressions $R = (a)(a^*)$ and $R' = (a^*)(a)$, we have $\mathcal{L}(R) = \mathcal{L}(R')$, but the corresponding pTr T and T' are not equivalent, since $T(a^n) = [_1 a]_1[_2 a^{n-1}]_2 \neq [_1 a^{n-1}]_1[_2 a]_2 = T'(a^n)$, for $n = 1$, or $n \geq 3$. Note that the same subexpression in a regular expression may capture more than one substring, for example, if T'' is a pTr for $(a^* | b)^*$, then $T''(a^p b^q a^r) = [_1 a^p]_1[_1 b]_1 \cdots [_1 b]_1[_1 a^r]_1$, for $p, q, r > 0$.

3 Converting Regular Expressions into pTr

Next we give a Java based construction to turn a regular expression E into an equivalent pTr $\bar{J}^p(E)$ (refer to Fig. 2 for reference). If for a pTr T, we denote by $u(T)$ the string transducer obtained by ignoring the priorities in T, then for $w \in \mathcal{L}(E)$, $u(\bar{J}^p(E))(w)$ gives all possible ways in which E can match w with capturing information indicated, while $\bar{J}^p(E)(w)$ selects the highest priority

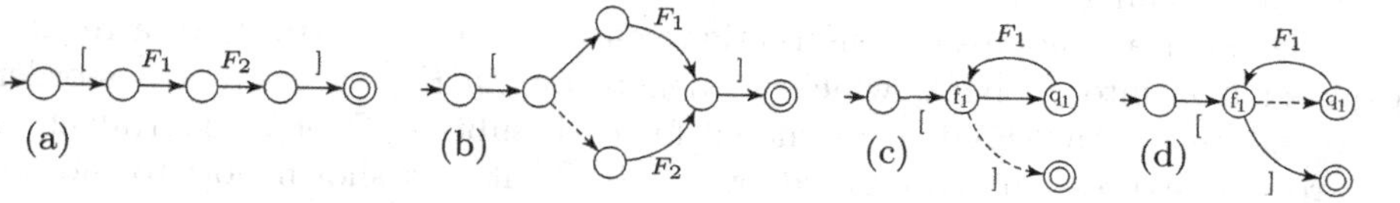

Fig. 2. Java based regular expression to pTr constructions for (a) $(F_1 F_2)$ (b) $(F_1 | F_2)$ (c) (F_1^*) and (d) $(F_1^{*?})$. Lower priority transitions are indicated by dashed edges. The pair of brackets $[,]$ are used to indicate the substring or substrings captured by each of $F_1 F_2$, $F_1 | F_2$, F_1^*, and $F_1^{*?}$ respectively.

match from $u(\bar{J}^p(E)(w))$. Due to space limitations, it is not possible to describe the matching semantics of regular expression with capturing groups in terms of a non-deterministic parser or other means, and show that the constructed pTr produces equivalent output, but we hope that it will at least be intuitively clear that this can be done. See also [2] for a thorough argument for the pNFA case, which may be extended to pTr with some effort. As indicated in Example 11, we opt to deviate from the Java matching semantics (and follow RE2 matching semantics [4]) in cases where the Java matcher follows a non-empty capture of a subexpression F (with F being part of a larger subexpression F^*), by an empty capture with F. Our construction is similar to the Java based regular expression to pNFA constructions given in [2] (and the classical Thompson construction [10]), with the additional detail of adding a group opening symbol on the transition leaving the initial state, and a group closing symbol on the transition incoming to the final state, for the pTr constructed for each subexpression of E. Where required, a new initial state and/or final state is added to the constructions from [2], so that there is only a single δ_2 transition from the initial state, and similarly, only a single incoming δ_2 transition to the final state of a pTr.

We denote the set of subexpressions of E by $\mathrm{SUB}(E)$. Assume $F_1, \ldots, F_k$ are the subexpressions in $\mathrm{SUB}(E)$, with the order obtained from a preorder traversal of the parse tree of E, or equivalently, ordered from left to right, based on the starting position of each subexpression in the overall regular expression. Note if the same subexpression appears more than once in E, we regard these occurrences as distinct elements in $\mathrm{SUB}(E)$. Also, let $t : \mathrm{SUB}(E) \to \mathbb{N}$ be defined by $t(F_i) = i$. To simplify our exposition of the regular expression to pTr construction procedure, we assume that matches by all subexpressions are captured, and that the pair of brackets $[_i,]_i \in B_k$ indicates matches by the i-th subexpression. The more general case of placing brackets only around the substrings matched by subexpressions that is marked as capturing subexpressions, is obtained by replacing some of the pairs of brackets by ε (in our pTr constructions) and renumbering the remaining brackets appropriately.

For a regular expression E we define a prioritized transducer $T := \bar{J}^p(E)$ such that $\mathrm{dom}(T) = \mathcal{L}(E)$, and $\mathrm{range}(T)$ is contained in the shuffle of $\mathrm{dom}(T)$ and the Dyck language D_k. In fact, taking some F which is a subexpression of E, and $v \in \mathrm{dom}(T)$, if $T(v)$ contains the substring $[_{t(F)}w]_{t(F)}$, where $[_{t(F)}$ and $]_{t(F)}$ are matching brackets, then $\pi_{\Sigma_1}(w) \in \mathcal{L}(F)$ (recall that $\pi_{\Sigma_1}(w)$ is obtained from w by deleting all brackets). Also, all output symbols from Σ_1 in $T(v)$, are between matching brackets.

The classical Thompson construction converts the parse tree T of a regular expression E into an NFA, which we denote by $Th(E)$, by doing a postorder traversal on T. An NFA is constructed for each subtree T' of T, equivalent to the regular expression represented by T'. In [2] it was shown how to modify this construction to obtain a Java based pNFA denoted by $J^p(E)$, instead of the NFA $Th(E)$, from E. Here we take it one step further, and modify the construction of $J^p(E)$ to return a pTr, denoted by $\bar{J}^p(E)$, from E. Just as in the case of the constructions for $Th(E)$ and $J^p(E)$, we define $\bar{J}^p(E)$ recursively on the parse tree for E. For each subexpression F of E, $\bar{J}^p(F)$ has a single initial state with no incoming transitions and a single outgoing δ_2 transition, and a single final state with a single incoming δ_2 transition and no outgoing transitions. The constructions of $\bar{J}^p(\emptyset)$, $\bar{J}^p(\varepsilon)$, $\bar{J}^p(a)$, and $\bar{J}^p(F_1 \cdot F_2)$, given that $\bar{J}^p(F_1)$ and $\bar{J}^p(F_2)$ are already constructed, are defined as for $Th(E)$, splitting the state set into Q_1 and Q_2 in the obvious way. We also place the symbol $[_{t(F)} \in \Sigma_2$ on the δ_2 transition leaving the initial state of $\bar{J}^p(F)$ and $]_{t(F)}$ on the transition incoming to the final state of $\bar{J}^p(F)$ (adding a new initial and/or final state if required).

When we construct $\bar{J}^p(F_1|F_2)$ from $\bar{J}^p(F_1)$ and $\bar{J}^p(F_2)$, and $\bar{J}^p(F_1^*)$ from $\bar{J}^p(F_1)$, the priorities of newly introduced δ_2-transitions require attention. We also consider the lazy Kleene closure $F_1^{*?}$. In the constructions (i) and (ii) below, we assume $\bar{J}^p(F_i)$ $(i \in \{1,2\})$ has initial state q_i and the final state f_i. Furthermore, δ_2 denotes the prioritized transition function in the newly constructed pTr $\bar{J}^p(F)$. All non-final states in $\bar{J}^p(F)$ that are in $\bar{J}^p(F_i)$ inherit their outgoing transitions from $\bar{J}^p(F_i)$. (i) If $F = F_1|F_2$ then $\bar{J}^p(F)$ is constructed by introducing new initial and final states $q_0 \in Q_2$ and $f_0 \in Q_1$, an additional new state $q' \in Q_2$, merging the states $f_1, f_2 \in Q_1$ into a state denoted by $f \in Q_2$, and defining $\delta_2(q_0) = ([_{t(F)}, q')$, $\delta_2(q') = (\varepsilon, q_1)(\varepsilon, q_2)$ and $\delta_2(f) = (]_{t(F)}, f_0)$. (ii) If $F = F_1^*$ then we add new initial and final states $q_0 \in Q_2$ and f_0 to Q_1, and change f_1 from being a state in Q_1, to be in Q_2. We define $\delta_2(q_0) = ([_{t(F)}, f_1)$ and $\delta_2(f_1) = (\varepsilon, q_1)(]_{t(F)}, f_0)$. The case $F = F_1^{*?}$ is the same, except that $\delta_2(f_1) = (]_{t(F)}, f_0)(\varepsilon, q_1)$. Thus $\bar{J}^p(F^*)$ tries F as often as possible whereas $\bar{J}^p(F^{*?})$ does the opposite.

Example 12. In Fig. 3(a), a pTr T for the regular expression $(\varepsilon \,|\, b)^*(b^*)$ is given. This regular expression has a subexpression F^*, such that F matches ε. This is the so called problematic case in regular expressions matching, briefly discussed in [9]. In this example, the subexpression $(\varepsilon \,|\, b)^*$ will first match only ε, and will attempt to match more of the input string only if an overall match can not be achieved. Thus for the given pTr T, we have that $T(b) = [_1]_1[_2b]_2$. Regular expression matchers, such as RE2 [4], uses different matching semantics in the problematic case. The problematic case is also present in regular expressions with no explicit ε symbols, such as $(a^* \,|\, b)^*(b^*)$. In Fig. 3(c) a pTr is given again for $(\varepsilon \,|\, b)^*(b^*)$, but this time obtained by using the modified greedy Kleene closure construction in Fig. 3(b). Note that $T'(b) = [_1b]_1[_2]_2$, corresponding to how RE2 only matches non-empty words with F in a subexpression F^*.

4   A Normal Form for Prioritized Transducers

To simplify later constructions, we introduce flattening for pTr in this section. The main simplification obtained by flattening is that δ_2 loops such as $q_1 \to q_2 \to q_3 \to q_1$ in Example 11 are removed, making it unnecessary to require that there are no repetition of the same δ_2 transition in a subsequence of transitions without δ_1 transitions, as in Definition 5. These δ_2 loops are found in pTr obtained from problematic regular expressions, as discussed in Examples 11 and 12.

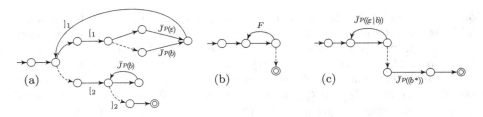

Fig. 3. (a) Java based pTr for $(\varepsilon|b)^*(b^*)$, (b) alternative F^* construction, and (c) pTr for $(\varepsilon|b)^*(b^*)$ using alternative F^* construction. Lower priority transitions are indicated by dashed edges.

Definition 13. *A pTr* $T = (Q_1, Q_2, \Sigma_1, \Sigma_2, q_0, \delta_1, \delta_2, F)$ *is flattened if* $\delta_2(q) \in (\Sigma_2^* \times Q_1)^*$ *for all* $q \in Q_2$.

We denote by $r_T(Q_2)$ the subset of Q_2 defined by $Q_2 \cap (\{q_0\} \cup \{\delta(q, \alpha) \mid q \in Q_1, \alpha \in \Sigma_1\})$, i.e. all Q_2 states reachable from a Q_1 state in one transition, and also the state q_0 if it is in Q_2. We denote the flattened pTr constructed in the proof of the next theorem, and equivalent to T, by flat(T).

Theorem 14. flat(T) *can be constructed in time* $\mathcal{O}(|Q_1||\Sigma_1| + |r_T(Q_2)||T|_{\delta_2})$.

Proof. We start with some preliminaries required to define flat(T). For a pTr T, a sequence $p_1 \cdots p_n$ is a δ_2-*path* if $\delta_2(p_i) = \cdots (w_{i+1}, p_{i+1}) \cdots$ for all $1 \leq i < n$ and $(p_i, p_{i+1}) = (p_j, p_{j+1})$ only if $i = j$. The string $w_2 \cdots w_n$, obtained from the definition of a δ_2-path, is denoted by $o_T(p_1 \cdots p_n)$. Let P_T be the set of δ_2-paths. For $p_1 \cdots p_n, p_1' \cdots p_m' \in P_T$ having $p_1 = p_1'$, we define $p_1 \cdots p_n > p_1' \cdots p_m'$ if and only if either (i) $p_1' \cdots p_m'$ is a proper prefix of $p_1 \cdots p_n$, or (ii) the least i such that $p_i \neq p_i'$ is such that $\delta_2(p_{i-1}) = \cdots p_i \cdots p_i' \cdots$. Note this is similar to the definition of priorities of paths in Definition 5, but in this case restricted to δ_2-paths, and allowing any starting state in Q_2. Let $P_{q,q'} = \max\{p_1 \cdots p_n \in P_T \mid p_1 = q, p_n = q'\}$, that is, the highest priority δ_2-path from q to q' (if it exists).

We let flat(T) be $(Q_1, r_T(Q_2), \Sigma_1, \Sigma_2, q_0, \delta_1, \delta_2', F)$, where δ_2' is defined as follows. For $q \in Q_2'$, let $P_{q,q_1} < \cdots < P_{q,q_n}$ be all highest-priority δ_2-paths which end in a state $q_i \in Q_1$, ordered according to priority. We define $\delta_2'(q) := (o_T(P_{q,q_1}), q_1) \cdots (o_T(P_{q,q_n}), q_n)$. To compute δ_2', with duplicate tuples removed as in Remark 10, in time $O(|r_T(Q_2)||T|_{\delta_2})$, repeat the following procedure for each $q \in r_T(Q_2)$: Determine the highest priority δ_2-path starting at q and ending in a state in Q_1. If the ending state is $q_1 \in Q_1$ (determining P_{q,q_1}), remove q_1 and all transitions going to or coming from q_1 from T, to obtain the pTr T_{q_1}. Repeat the procedure in T_{q_1}, successively finding all $P_{q,q'}$ with q fixed, in order. Note that computing $r_T(Q_2)$ takes $\mathcal{O}(|Q_1||\Sigma_1|)$ time. $\square$

Example 15. For the pTr T corresponding to the regular expression $(a^*)^*$ and discussed in Example 11, the pTr flat(T) is given in Fig. 4. Note that the flattening procedure removed the δ_2 loop $q_1 \rightarrow q_2 \rightarrow q_3 \rightarrow q_1$ from T. As noted in

Example 11, Java matchers do not keep track of all δ_2 transitions in order to avoid repeated δ_2 transitions. When using this Java way of determining which paths are legal and which not, the flattened procedure in the proof of Theorem 14 can be modified, and when applied to T, we obtain an almost identical flattened pTr, but with output $]_1 [_1]_1$ on the transition from q_5 to q_7.

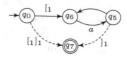

Fig. 4. flat(T) for T in Fig. 1(b). The dashed edges are lower priority.

Remark 16. Note that $|\text{flat}(T)|_Q \leq |T|_Q$, $|\text{flat}(T)|_{\delta_1} = |T|_{\delta_1}$ and $|\text{flat}(T)|_{\delta_2} \leq |T|_{\delta_2}$, i.e. flat($T$) is of the same size or smaller than T.

Remark 17. All Q_2 states, with the exception of q_0 when $q_0 \in Q_2$, can be removed from a flattened pTr. To see this, redefine δ_2 to be the identity on Q_1 and let $\delta = \delta_2 \circ \delta_1$. Thus we can redefine a pTr to have a single transition function $\delta : Q_1 \times \Sigma_1 \rightarrow (Q_1 \times \Sigma_2^*)^*$ (except for the transitions from q_0 if $q_0 \in Q_2$), if we are willing to allow prioritized non-determinism on input from Σ_1.

5   Equivalence and Parsing with Prioritized Transducers

Regular expressions R and R' are equivalent if the pTr $\bar{J}^p(R)$ and $\bar{J}^p(R')$ are equivalent. In general, deciding equivalence of string transducers is undecidable, but in [8] it is shown that equivalence of functional transducers is decidable, but PSPACE-complete. In [9], the equivalence of regular expressions through transducers, is approached by first formulating the semantics of regular expression matching as a non-deterministic parser, then transforming the parser into first a transducer with regular lookahead, and then into a functional transducer without lookahead. For non-problematic regular expressions R, a functional transducer of size $2^{\mathcal{O}(|R|)}$ is obtained. Thus to decide equivalence of regular expressions with capturing groups, equivalence is decided on the corresponding functional transducers. We obtain a similar result for a larger class of regular expressions and regular expression matching semantics, through equivalence of pTr.

Theorem 18. *A pTr T can be converted into an equivalent functional transducer T_F with $|T_F|_Q = |T|_Q 2^{|T|_Q}$, and $|T_F|$ in $\mathcal{O}(|T| 2^{|T|_Q})$.*

Proof (Sketch). Let $T = (Q_1, Q_2, \Sigma_1, \Sigma_2, q_0, \delta_1, \delta_2, F)$ and A the NFA obtained from T by ignoring output and priorities of T. Let Q_A be the states of A, which is of course just $(Q_1 \cup Q_2)$. The set of states of T_F is $(Q_1 \cup Q_2) \times 2^{Q_A}$. On the $(Q_1 \cup Q_2)$ part of the states of T_F, T_F behaves like T with priorities ignored, and the subsets of Q_A, which form the 2nd component of the states of T_F, is used to take priorities of δ_2 transitions in T into account. Each time we are at a state $q \in Q_2$ in T with $\delta_2(q) = (w_1, q_1) \cdots (w_k, q_k)$, T_F chooses non-deterministically a transition $q \rightarrow q_i$ (with output w_i) in the first component of states of T_F, and keep track with subsets from Q_A, that the input would not have been accepted if we took $q \rightarrow q_j$ for $j < i$. Note as we reach the next state on a path taken in T, we keep on tracking, in the 2nd component of the states of T_F, all states of T that could be reached, on the given input, from higher priority transitions we did not take at previous Q_2 states encountered. The accept states of T_F are the states (q, X) with $q \in F$ and $X \cap F = \emptyset$. Note T_F is a functional transducer, since the relation defined by T is a function. $\square$

Corollary 19. *(a) For prioritized transducers T_1 and T_2, equivalence can be decided in time $\mathcal{O}((|T_1|2^{|T_1|_Q} + |T_2|2^{|T_2|_Q})^2)$. (b) Equivalence of regular expressions R_1 and R_2 with capturing groups, can be decided in time $2^{\mathcal{O}(|R_1|+|R_2|)}$.*

Proof. (a) For pTr T_1 and T_2, first check that $\text{dom}(T_1) = \text{dom}(T_2)$, which can obviously be done in the stated time complexity bound. Now convert T_1 and T_2 into functional transducers T_{F_1} and T_{F_2}, and use Theorem 1.1 in [8] that states that the complexity of deciding if the transducer $T_1 \cup T_2$ is functional (and thus that T_1 and T_2 are equivalent, since $\text{dom}(T_1) = \text{dom}(T_2)$), is quadratic in the number of transitions in $T_1 \cup T_2$. For (b) use (a) and the fact that for a regular expression R, $|\bar{J}^p(R)| \leq c|R|$, for some constant c. $\square$

Remark 20. Note that deciding equivalence of pTr and regular expressions with capturing groups is at least PSPACE-hard, since it is PSPACE-complete already to check if the domains of pTr are equal.

Remark 21. The transducer construction in the proof of Theorem 18 must be close to ideal, as the worst-case state complexity of a transducer T_F equivalent to a pTr T is bounded from below by $2^{|T|_Q}$. This is so since one can for any NFA A, construct a pTr T, with $\Sigma_2 = \{\beta, \beta'\}$, having $T(w) = \beta w$ for $w \in \mathcal{L}(A)$ and $T(w) = \beta' w$ for all $w \notin \mathcal{L}(A)$ (a similar example is obtained by constructing a pTr for the regular expression $(R)|(\Sigma_1^*)$, with R corresponding to A). Simply let $\delta_2(q_0) = (\beta, q_A)(\beta', q_{\Sigma_1^*})$, with q_0 the initial state of T, q_A the initial state of A, and $q_{\Sigma_1^*}$ a sink accept state. Now consider the class of NFA A, for which the complement of $\mathcal{L}(A)$ can only be recognized by NFA with at least $2^{|A|_Q}$ states. Then if T_F' is obtained from T_F by removing transitions having β as output, $\text{dom}(T_F')$ will be equal to the complement of $\mathcal{L}(A)$. Thus T_F' and also T_F, will require at least $2^{|A|_Q}$ states.

Some specifics of real-world matchers can be generalized away, such as the Σ_2 subsequences a real-world pTr outputs always forming a Dyck language (as in Sect. 3). One which we need to consider however, as including it saves memory in the parsing algorithm, is that the strings in range(T) are not output in practice, but rather matchers will walk through a string w in dom(T), and will *once the string has been accepted*, output for each symbol in $\alpha \in \Sigma_2$ in $T(w)$, the *index* of the *last* occurrence of α in $T(w)$. This limits the possible memory usage, notably it means that the amount of data output by a matching an input string w with T is bounded by $|\Sigma_2| \log(|w|)$.

Definition 22. *For a pTr T with $w \in dom(T)$, let $T(w) = v_0\alpha_1 \cdots v_{n-1}\alpha_n v_n$, where $v_i \in \Sigma_2^*$ and $\alpha_i \in \Sigma_1$ for each i. Then the* slim parse output *of T on w is a function $s^T : \Sigma_2 \to \{\bot, 0, \ldots, n\}$ such that for each $\beta \in \Sigma_2$ we have $\beta \in v_{f(\beta)}$, but $\beta \notin v_{f(\beta)+j}$ for any $j \in \mathbb{N}$. If $\beta \notin v_1 \cdots v_n$, $s^T(\beta) = \bot$.*

For a pTr T and a string $w = \alpha_1 \cdots \alpha_n \in dom(T)$, we next describe a linear (in the length of the input string) algorithm to compute the slim parse output of T, where T is flattened. Let $f, f', f_\bot : \Sigma_2 \to \mathbb{N} \cup \{\bot\}$. For $v \in \Sigma_2^*$ and $k \in \mathbb{N}$, define $f' := U(f, v, k)$ by letting $f'(\beta) = k$ for $\beta \in v$, and $f'(\beta) = f(\beta)$ otherwise. Also, $f_\bot(\beta) = \bot$ for all $\beta \in \Sigma_2$. Define $\Delta(q, f, i) = (q_1, U(f, \beta_1, i)) \cdots (q_n, U(f, \beta_n, i))$ when $q \in Q_2$ and $\delta_2(q) = (\beta_1, q_1) \cdots (\beta_n, q_n)$, and $\Delta(q, f, i) = (q, f)$ for $q \in Q_1$. Now for the steps in the algorithm. (i) Let $S_0 = \Delta(q_0, f_\bot, 0)$. (ii) Given $S_i = (q_1, f_1) \cdots (q_m, f_m)$, where $i < n$, then $S_{i+1} = \sigma_1(\Delta(q'_1, f_1, i) \cdots \Delta(q'_m, f_m, i))$, where $q'_j = \delta_1(q_j, \alpha_{i+1})$ for each j. (iii) If $S_n = (q_1, f_1) \ldots (q_m, f_m)$ and i is the smallest index such that $q_i \in F$, then $s^T(w) = f_i$. If no state q_i is in F, the string w is rejected.

Theorem 23. *The slim parsing algorithm runs in linear time in the length of the input string, and is correct, i.e. with input a pTr T and $w \in dom(T)$, it returns the slim parse for T on w, and if $w \notin dom(T)$, it rejects the input w.*

Proof (Sketch). Since highest priority paths in a flattened pTr may be determined by DFS, a pTr can be translated into a deterministic stack machine with output. Each S_i can be associated with the stack content at a particular stage of the DFS, with the left-most tuple being the top element of the stack. It follows from the argument in Remark 10 that duplicates of tuples with the same state can be removed from the stack. In contrast to a stack machine, the given algorithm simply processes all stack elements in parallel. Clearly, from the description of the slim parsing algorithm, it runs in linear time in the length of the input string. □

6   Conclusions and Future Work

In this paper we brought together several different angles on regular expressions into one formal framework. This enables us to talk both about the matching behaviors of less than ideal real-world matchers as in [2], while allowing a modelling of the special features of those matchers without being tied to their algorithmic choices. Still, there is ample room for continued work. For example, there are a *lot* of additional operators in regex libraries that should be analyzed. A special example is pruning operators, such as atomic subgroups, and the cut operator of [1], which interact deeply with the matching procedure. From a theoretical perspective, the next step should be to determine the precise complexity class for equivalence in Corollary 19.

References

1. Berglund, M., Björklund, H., Drewes, F., van der Merwe, B., Watson, B.: Cuts in regular expressions. In: Béal, M.-P., Carton, O. (eds.) DLT 2013. LNCS, vol. 7907, pp. 70–81. Springer, Heidelberg (2013)
2. Berglund, M., Drewes, F., van der Merwe, B.: Analyzing catastrophic backtracking behavior in practical regular expression matching. In: Ésik, Z., Fülöp, Z., (eds.) Proceedings of the 14th International Conference on Automata and Formal Languages, pp. 109–123 (2014)
3. Câmpeanu, C., Salomaa, K., Yu, S.: A formal study of practical regular expressions. Int. J. Found. Comput. Sci. 14(6), 1007–1018 (2003)
4. Cox, R.: Implementing regular expressions (2007). http://swtch.com/rsc/regexp/. (Accessed 3 March 2015)
5. Friedl, J.: Mastering Regular Expressions, 3rd edn. O'Reilly Media Inc., Sebastopol (2006)
6. Frisch, A., Cardelli, L.: Greedy regular expression matching. In: Díaz, J., Karhumäki, J., Lepistö, A., Sannella, D. (eds.) ICALP 2004. LNCS, vol. 3142, pp. 618–629. Springer, Heidelberg (2004)
7. Rathnayake, A., Thielecke, H.: Static analysis for regular expression exponential runtime via substructural logics. CoRR, abs/1405.7058 (2014)
8. Sakarovitch, J.: Elements of Automata Theory. Cambridge University Press, New York (2009)
9. Sakuma, Y., Minamide, Y., Voronkov, A.: Translating regular expression matching into transducers. J. Applied Logic 10(1), 32–51 (2012)
10. Thompson, K.: Regular expression search algorithm. Commun. ACM 11(6), 419–422 (1968)
11. Wang, J.: Handbook of Finite State Based Models and Applications. 1st edn. Chapman & Hall/CRC, Boca Raton (2012)

Tool Demonstration Papers

Introducing Code Adviser: A DFA-driven Electronic Programming Tutor

Abejide Ade-Ibijola[1]([envelope]), Sigrid Ewert[1], and Ian Sanders[2]

[1] School of Computer Science, University of the Witwatersrand,
Johannesburg, South Africa
researcher@abejide.com, sigrid.ewert@wits.ac.za
[2] School of Computing, University of South Africa, Florida, South Africa
sandeid@unisa.ac.za

Abstract. In this paper, we demonstrate a software system called Code Adviser that attempts to *understand* and find semantic bugs in student programs written in C++ programming language. To do this, Code Adviser has to take a model solution from a lecturer (or expert), generate many variations of the model solution, and compare student programs to the most similar model solution. The student's program to be checked for correctness is normalized, granulated and abstracted to a *string* of semantic tokens — we call this the *abstraction stage*. Similarly, the model solutions are taken through the abstraction stage and the program strings representing all model solutions are abstracted to deterministic finite automaton (DFA) for the programming problem. Code Adviser then uses some algorithms to make inference on student's program correctness. If the student's program string is accepted by the problem's DFA, it is reported as correct. Else, we make inferences on what the bug could be. Code Adviser is a promising proof of concept, and more work is currently being done to improve its inference and make it available to student programmers.

Keywords: Bug detection · Semantic bugs · Program strings

1 Introduction

Computer-aided teaching of programming at first year still remains a difficult task for one simple reason — *dealing with the large variability in programs.* Many tools have attempted to perform a variety of compilation-related tasks on computer programs using techniques such as lexical analysis, program slicing, control and data-flow analysis, and input-output (IO) analysis. These techniques are useful for validating, assessing or marking student programs. Some tools built on these techniques are: Cloud Coder [8,12], Auto Grader [7], Lackwit [11], and JRipples [4].

There have been other attempts to support the student's comprehension process by abstracting student programs to a friendly visual or textual representation using tools such as: NOPRON [1] for narrating program steps; tools

© Springer International Publishing Switzerland 2015
F. Drewes (Ed.): CIAA 2015, LNCS 9223, pp. 307–312, 2015.
DOI: 10.1007/978-3-319-22360-5_25

based on the ATS[1] technology [5] for summarizing large programs; and Code Crawler [10], DynaRIA [3] and SHriMP Views [13] for visualisation.

However, the comprehension task is an Artificial Intelligence task that involves *understanding* the problem, knowing all possible solutions, comparing the student's solution to the repository of solutions and pointing out the semantic bugs in the student's program — *just like the human expert would do*. For this, the popular approach is the knowledge base approach — often aided by other previously stated techniques. This involves making inferences based on program beacons [6], keeping static bug clichès [2], and keeping repositories of possible programs with all possible bugs [9]. Methods used for this task often struggle to cope with the variability in student programs for the comprehension task. Code Adviser attempts to achieve this task using a different approach.

2   How Code Adviser Works

In this section we demonstrate how Code Adviser works with the aid of a simple diagram shown in Fig. 1.

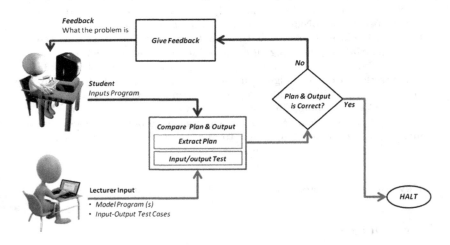

Fig. 1. Automatic tutoring model

In order to understand student programs, detect semantic bugs and suggest possible repairs, Code Adviser:

1. takes a model program for a programming problem written in C++ with a number of test cases, cleans up and granulates the model program,
2. generates all possible variations of the model program,
3. constructs a DFA from the solution space, taking each solution as a string,

---

[1] Automated Text Summarisation.

4. attempts to compare a buggy student program to the finite list of program strings accepted by the problem's DFA, and
5. depending on the student's plan and output correctness, it reports on discovered bugs and suggest repairs or declares the student's program as correct.

3 Acting Intelligent

To seemingly act intelligent, Code Adviser had to manipulate both the student's and expert's program strings to give specific feedbacks that is not only helpful to the student, but also demonstrates the *wealth* of knowledge it has about the student's program. Code Adviser achieved this by:

1. reporting the percentage of similarity of the student's and expert's programs by calculating the Levenshtein distance between the program strings of both programs,
2. discussing with the student in *first person* with statements such as *Ask me* ..., *I think you should* ..., and *Your program is* ...,
3. explaining bugs in the student's terms, i.e. with the variables in the student's programs and not the ones in the model solution, and
4. pointing to the buggy lines.

4 One Step Bug Repair

Code Adviser does a *top-down one-step repair*[2]. It starts with the first mismatched program statement in the novice's program and prompts the student to correct it. If entire program bugs has been corrected, it informs the novice and terminates. Else, it points the student to the next bug.

5 Discussing Bug Repair

When Code Adviser discovers a bug, it *explains* how to repair the bug to the novice programmer using bug discussion modules. Hence, if the bug repair is to add a new line that displays the value of the variable called sum, Code Adviser says:

I think you should insert an additional line after Line <line number here>.
The new line should display the value of sum.
Try this and seek my advise again.

More examples of how the Code Adviser handles the bug discussion are presented in Sect. 6.

---

[2] Repairing bugs, one bug at a time, starting from the first line in a program (the initial symbol) to the last line.

6   Code Adviser in Action

In this section, we demonstrate how `Code Adviser` report bugs and discusses bug repairs. In Fig. 2, `Code Adviser` uses a grid view control to show each program line with a feedback and a general note on program's correctness. In this case the program is correct and the student gets a corresponding message.

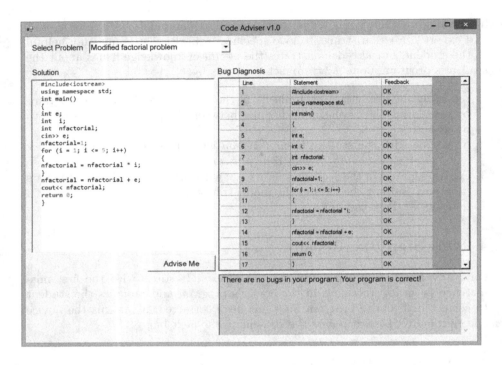

Fig. 2. Code Adviser finding no bugs in program

In Fig. 3 we show how `Code Adviser` reports a semantically buggy program, pointing to the line and using a bug discussion module to explain what needs to be done to fix the bug.

7   Reflections on Automatic Tutoring

This paper has presented a brief demonstration of `Code Adviser` a tool that supports the teaching of programming at first year undergraduate level. `Code Adviser` is *not a robust tool*, it is a proof of concept that we have used to demonstrate how a tool can be used to tutor student programmers based on DFAs of alternative solutions and bug detection algorithms.

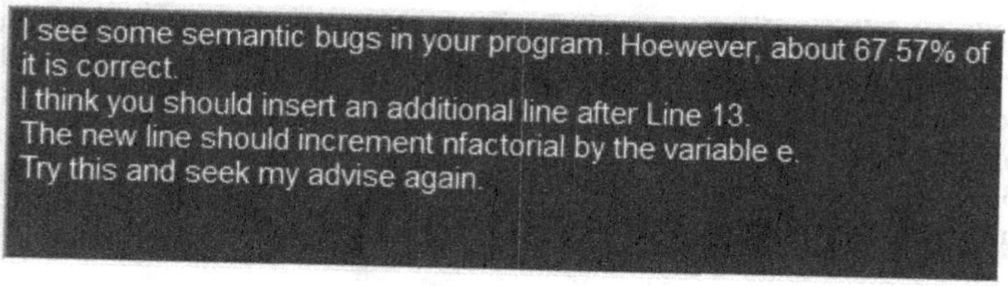

I see some semantic bugs in your program. Hoewever, about 67.57% of it is correct.
I think you should insert an additional line after Line 13.
The new line should increment nfactorial by the variable e.
Try this and seek my advise again.

Fig. 3. Code Adviser suggesting a repair

Acknowledgment. This work is based on research supported by the National Research Foundation (NRF) of South Africa. Any opinion, findings and conclusions or recommendations expressed in this material are those of the authors and therefore the NRF does not accept liability in regard thereto.

References

1. Ade-Ibijola, A., Ewert, S., Sanders, I.: Abstracting and narrating novice programs using regular expressions. In: Proceedings of the Annual Conference of the South African Institute for Computer Scientists and Information Technologists. pp. 19–28. ACM (2014)
2. Al-Omari, H.M.A.: Conceiver: A program understanding system. Ph.D. thesis, University Kebangsaan Malaysia (1999)
3. Amalfitano, D., Fasolino, A.R., Polcaro, A., Tramontana, P.: Dynaria: a tool for ajax web application comprehension. In: IEEE 18th International Conference on Program Comprehension, pp. 46–47. IEEE (2010)
4. Buckner, J., Buchta, J., Petrenko, M., Rajlich, V.: JRipples: a tool for program comprehension during incremental change. In: International Workshop on Program Comprehension. vol. 5, pp. 149–152 (2005)
5. Haiduc, S., Aponte, J., Marcus, A.: Supporting program comprehension with source code summarization. In: ACM/IEEE 32nd International Conference on Software Engineering, vol. 2, pp. 223–226 (2010)
6. Harris, N., Cilliers, C.: A program beacon recognition tool. In: 7th International Conference on Information Technology Based Higher Education and Training. pp. 216–225. IEEE (2006)
7. Helmick, M.T.: Interface-based programming assignments and automatic grading of Java programs. ACM SIGCSE Bull. **39**(3), 63–67 (2007)
8. Hovemeyer, D., Hertz, M., Denny, P., Spacco, J., Papancea, A., Stamper, J., Rivers, K.: Cloudcoder: building a community for creating, assigning, evaluating and sharing programming exercises. In: Proceeding of the 44th ACM Technical Symposium on Computer Science Education. pp. 742–742. ACM (2013)
9. Johnson, W., Soloway, E.: PROUST: knowledge-based program understanding. IEEE Trans. Software Eng. **3**, 267–275 (1985)
10. Lanza, M., Ducasse, S., Gall, H., Pinzger, M.: Codecrawler-an information visualization tool for program comprehension. In: Proceedings of the 27th International Conference on Software Engineering, pp. 672–673. IEEE (2005)
11. O'Callahan, R., Jackson, D.: Lackwit: a program understanding tool based on type inference. In: International Conference on Software Engineering, Vol. 97, pp. 338–348 (1997)

12. Papancea, A., Spacco, J., Hovemeyer, D.: An open platform for managing short programming exercises. In: Proceedings of the 9th Annual International ACM Conference on International Computing Education Research, pp. 47–52. ACM (2013)
13. Storey, M., Best, C., Michand, J.: SHriMP views: An interactive environment for exploring java programs. In: Proceedings of the 9th International Workshop on Program Comprehension, pp. 111–112. IEEE (2001)

BSP: A Parsing Tool for Ambiguous Regular Expressions

Angelo Borsotti[1], Luca Breveglieri[1],
Stefano Crespi Reghizzi[2]([✉]), and Angelo Morzenti[1]

[1] Dipartimento di Elettronica, Informazione E Bioingegneria (DEIB), Politecnico di
Milano, Piazza Leonardo Da Vinci N. 32, 20133 Milano, Italy
angelo.borsotti@mail.polimi.it,
{luca.breveglieri,stefano.crespireghizzi,angelo.morzenti}@polimi.it
[2] Dipartimento di Elettronica, Informazione E Bioingegneria (DEIB), CNR - IEIIT,
Politecnico di Milano, Piazza Leonardo Da Vinci N. 32, 20133 Milano, Italy

Abstract. *BSP* (Berry-Sethi Parser) is a new *SW* tool for parsing
ambiguous regular expressions (r.e.). Given a r.e., the *BSP* tool generates a *DFA*. Then the *DFA* processes the given input string, recognizes
it and outputs, depending on user choice, all the syntax trees or just the
one selected by either the *Greedy* or the *POSIX* criterion. The *BSP* tool
is a *HTML* page including *JavaScript* code, and can be executed by any
browser. It is self-documented and is intended for educational purposes.
See http://github.com/breveglieri/ebs (see also [1] for details).

1 Tool Description

The *BSP* (Berry-Sethi Parser) *SW* tool is a web-based *HTML* page, available
on the *GitHub* system at http://github.com/breveglieri/ebs. This tool allows a
user to interactively play with regular expressions (r.e.): matching and parsing
strings, displaying *DFA*'s and parse trees, watching animations of the matching
and parsing algorithm, and comparing the parse trees obtained by different disambiguation criteria. As such, the intended use of this tool is educational, with
no particular attention paid to the optimization or efficiency of the parser *DFA*
construction or of the string parsing process.

The tool provides a graphic representation (as well as a textual one) of the -
Abstract Syntax Trees *(AST)* of the r.e.'s , the parse trees of the strings, and the
DFA's that perform as parsers. The graphical representations obtained for such
objects tend to be somehow large, exceeding the size of the browser windows
when the r.e.'s are complex or the strings are long. Anyway, this is not too
annoying, since the intent is to allow the users to reason about r.e.'s and to
figure out the outcome of the algorithms, which is best done with small inputs.

The tool implements the parsing algorithm published in [1], which is a generalization of the classical Berry-Sethi algorithm to build a pure recognizer. Our
tool adds the capability of parsing the string and building its syntax tree(s), as

Work partially supported by *PRIN* "Automi e Linguaggi Formali", Italy.

F. Drewes (Ed.): CIAA 2015, LNCS 9223, pp. 313–316, 2015.
DOI: 10.1007/978-3-319-22360-5_26

well as to select a specific tree when the parsed string is ambiguous. To internally verify its results, it implements also the Okui and Suzuki algorithm [4].

Tool input. The input field "*RE*" holds the r.e., which has this syntax:

$$\text{empty string, } a, \ r\,r, \ r \mid r, \ (r), \ [r], \ (r)^+, \ (r)^*, \ a^+, \ a^* \quad a \in \Sigma$$

Specifications:

- empty string means no character at all (corresponds to ε)
- a is a character that is not a meta-symbol: $\mid$, $($ $)$, $[$ $]$, $*$ and $+$
- $r\,r$ is the concatenation of two or more regular expressions
- $r \mid r$ is the alternative (union) of two or more regular expressions
- (r) is a group that represents the enclosed regular expression
- $(r)^+$ is the enclosed regular expression repeated one or more times
- $(r)^*$ is the enclosed regular expression repeated zero or more times
- a^+ is an abbreviated form of $(a)^+$ (similarly for a^*)
- spaces (blanks) can be used everywhere

The field "input" holds the string to be recognized (matched) and/or parsed. The use of terminal characters whose glyphs have a direct visual representation (such as letters or digits) is preferable to the ones that do not (e.g., combining Unicode characters), which might not be represented clearly by certain web browsers.

Disambiguation policy. The radio buttons "Posix" and "Greedy" allow the user to choose which disambiguation criterion has to be used to select the *prior parse tree*: the *Greedy* one is defined in [2] and the *POSIX* one is defined in [3]. Notice that disambiguation is done only when the marking of the regular expression is of the type "fully marked" (see below).

RE marking. To construct the parser *DFA*, the tool applies a new algorithm that, as the classic Berry-Sethi one, starts from a *marked* r.e. Three kinds of r.e. marking can be chosen by using the radio buttons "fully marked", "partly marked" or "numerically marked":

fully marked all the elements of the r.e. are marked with a multilevel index, which identifies the corresponding inner node or leaf in the *AST* of the r.e.
partly marked all the elements of the r.e. are marked with a multilevel index, except concatenations and alternatives (which can be indirectly identified by their left and right immediate neighbouring symbols)
numerically marked as partly marked, but the marks are simple numbers

Notice that only the "fully marked" option allows the tool to select the prior tree in accordance with the chosen disambiguation criterion. The other two marking options are a simplification, still sufficient to properly visualize the syntax tree(s).

Tree drawing. The checkbox "show all trees" makes the tool display all the parse trees of the input string at the end of the page. The user can then select them two by two and compare them (see below).

Matching and / or parsing. The "go" button makes the tool match the input string and parse it, i.e., build its syntax tree(s). Also the r.e., annotated with the chosen marking, and its *AST*, are displayed. The "*DFA*" button allows the user to see the parser *DFA* (see below). Under the *AST* and *DFA* drawings, the linearized forms of the syntax tree(s) of the matched string are shown, followed by the prior tree (only when the option "fully marked" is checked), or by all the syntax trees (only when the option "show all trees" is checked), in graphic form.

Parser *DFA*. Upon pressing the "*DFA*" button, a textual representation of the parser *DFA* is shown, including (see [1] for an explanation of the terms):

- the *Initial* set
- the *Follow* sets
- the list of the *states*, each state with:
  - the list of its *items*, each with its finished string and item identifiers
  - the list of *transitions*, each with its label and target state
- the *final* states are flagged "final"

All this syntactic information in the node is used by the parser to build the syntax tree(s), and to select the prior one if requested (details in [1]).

A graphical representation of the *DFA* follows. It is an interactive one, which allows the user to perform these operations:

- enlarge it on the left by clicking the left border and dragging it
- enlarge it at the bottom by clicking the bottom border and dragging it
- move the states by clicking and dragging them (arcs can be moved similarly)
- reshape the arcs by hovering on them and then moving one of the three knots that are displayed for a few seconds

Parsing Animation. When the *DFA* is displayed, two more buttons appear: "play" and "stop". The first starts an animation that shows how the *DFA* is traversed during recognition. A road runner walks the arcs that lead one to travel from the initial state to a final state. The traversed states blink. The runner stops in a final state. On traversing a state, a box is displayed below the *DFA* and shows the items in the state. The current state being traversed has a blinking box. This phase is highlighted on the right of the "stop" button by the word "matching", and likewise the next phases are indicated. Then the next phases are shown:

- the items that belong to some path starting from the initial state and ending in a final item of a final state, are depicted in acqua and made blinking as they are visited (this is the "marking" phase)
- the items that belong to paths that are not part of the prior tree, are then colored in beige (this is the "pruning" phase)

The "stop" button allows the user to pause the matching process at any time.

Tree Comparison. When all the trees are displayed and there are at least two of them (ambiguous input string), it is possible to compare pairs of trees:

- by hovering a tree with the mouse, a dashed box appears around it, and by clicking on it, the border changes to solid, indicating that the tree has been selected
- hovering on a second tree and clicking on it, makes it selected and compared with the other one
- the winner, i.e., the one with higher priority, is flagged with a king crown
- in both trees, the path that starts from the root and ends in the node that represents the first distinguishing node (i.e., the first one, visiting the tree in preorder, that is present in one tree and not in the other) has yellow nodes showing the position index (of the corresponding node in the *AST*) and the length of its yield

Selecting another tree automatically deselects the last one selected and immediately compares the new tree.

Suggested Browsers. The tool is tested to work with Chrome and Firefox.

2   Future Developments

An efficient implementation in *Java* of the same algorithm is under way. We plan to use it to compare the performances (e.g., execution time) of our parsing algorithm with respect to the performances of the existing ones.

References

1. Borsotti, A., Breveglieri, L., Crespi Reghizzi, S., Morzenti, A.: From ambiguous regular expressions to deterministic parsing automata. In: Drewes, F. (ed.) CIAA. LNCS, vol. 9223, pp. 35–48. Springer, Heidelberg (2015)
2. Frisch, A., Cardelli, L.: Greedy regular expression matching. In: Díaz, J., Karhumäki, J., Lepistö, A., Sannella, D. (eds.) ICALP 2004. LNCS, vol. 3142, pp. 618–629. Springer, Heidelberg (2004)
3. IEEE: std. 1003.2, POSIX, regular expression notation, section 2.8 (1992)
4. Okui, S., Suzuki, T.: Disambiguation in regular expression matching via position automata with augmented transitions. In: Domaratzki, M., Salomaa, K. (eds.) CIAA 2010. LNCS, vol. 6482, pp. 231–240. Springer, Heidelberg (2011)

Author Index

Printed in the United States
By Bookmasters